CA-Clipper®

Programming and
Utilities Guide
Version 5.2

Manual ©1992
Computer Associates International, Inc.

All Rights Reserved

Computer Associates International, Inc.,
Publisher

All trade names referenced herein are either trademarks or registered trademarks of their respective companies.

Contents

Chapter 1: Introduction

Part 1: Basic Concepts 1-1
Part 2: Programming 1-1
Part 3: Utilities 1-2
Glossary 1-2

Chapter 2: Basic Concepts

In This Chapter 2-1
The Structure of a CA-Clipper Program 2-2
    A Typical CA-Clipper Program 2-2
    The Main Procedure Definition 2-3
    Function and Procedure Calls 2-4
    Variable Declaration 2-5
    Function and Procedure Definitions 2-5
    Preprocessor Directives 2-6
    Comments 2-6
    Continuation 2-7
    Reserved Words 2-8
Functions and Procedures 2-8
    Defining Functions and Procedures 2-9
    Calling Functions and Procedures 2-10
    Passing Parameters 2-11
    Passing by Value 2-12
    Passing by Reference 2-13
    Passing Arrays and Objects as Arguments 2-13
    Argument Checking 2-15
    Passing Arguments from the DOS Command line 2-15

Returning Values From User-Defined Functions	2-16
Recursion	2-17
Control Structures	2-17
Looping Structures	2-17
Decision-making Structures	2-19
Error Handling Structures	2-20
Variables	2-20
Lexically Scoped Versus Dynamically Scoped Variables	2-20
Lifetime and Visibility of Variables	2-21
Declaring Variables	2-22
The Scope of a Declaration	2-22
Referring to Variables	2-23
Ambiguous Variable References	2-23
Creation and Initialization	2-24
Local Variables	2-25
Static Variables	2-26
Private Variables	2-27
Public Variables	2-28
Field Variables	2-29
Expressions	2-29
Data Types	2-30
Character	2-30
Memo	2-33
Date	2-34
Numeric	2-37
Logical	2-39
NIL	2-40
Operators	2-41
Definitions	2-42
Error Handling	2-43
String Operators	2-43
Date Operators	2-43
Mathematical Operators	2-44
Relational Operators	2-44
Logical Operators	2-45
Assignment Operators	2-46

Increment and Decrement Operators	2-49
Special Operators	2-50
Operator Precedence	2-51
The Macro Operator (&)	2-55
Text Substitution	2-55
Compile and Run	2-56
Relationship To Commands	2-56
Macros and Arrays	2-58
Arrays	2-61
Creating Arrays	2-61
Addressing Array Elements	2-62
Assigning Values to Array Elements	2-63
Multidimensional Arrays	2-64
Literal Arrays	2-65
Arrays as Function Arguments and Return Values	2-65
Traversing an Array	2-66
Empty Arrays	2-67
Determining the Size of an Array	2-68
Comparing Arrays	2-68
Changing the Size of an Array	2-69
Inserting and Deleting Array Elements	2-69
Copying Elements and Duplicating Arrays	2-70
Sorting an Array	2-70
Searching an Array	2-71
Code Blocks	2-72
Defining a Code Block	2-72
Operations	2-73
Executing a Code Block	2-73
Scope of Variables Within a Code Block	2-73
Using Code Blocks	2-75
Evaluation of Macros in Code Blocks	2-76
Storing and Compiling Code Blocks at Runtime	2-77
Objects and Messages	2-78
Classes	2-78
Instances	2-78
Instance Variables	2-79

Programming and Utilities Guide   v

Sending Messages	2-79
Accessing Exported Instance Variables	2-80
The Database System	2-81
Work Areas	2-82
Database Files	2-84
Index Files	2-87
The Input/Output System	2-90
Console Operations	2-90
Full-screen Operations	2-91
Controlling Screen Color	2-91
Controlling Output Destination	2-92
The Keyboard System	2-94
Changing the Keyboard Buffer Size	2-94
Putting Characters in the Keyboard Buffer	2-94
Reading Characters from the Keyboard Buffer	2-95
Controlling Predefined Keys	2-95
Reassigning Key Definitions	2-96
Clearing the Keyboard Buffer	2-96
The Low-level File System	2-97
Opening a File	2-97
Reading from a File	2-97
Writing to a File	2-98
Manipulating the File Pointer	2-98
Closing a File	2-99
Error Detection	2-99

Chapter 3: The Runtime Environment

In This Chapter	3-1
Setting the Workstation Environment	3-1
Files and Buffers	3-2
Temporary Files	3-4
Specifying the Location of COMMAND.COM	3-5
Configuring the Serial Port	3-5
Setting the Application Program Environment	3-6
Application Program Command line	3-7

The CLIPPER Environment Variable.. 3-8
Saving/Restoring EMM Page Frame—BADCACHE 3-8
Preventing Extended Cursor Use—CGACURS 3-8
Specifying Number of Dynamic Overlay File Handles—DYNF 3-8
Configuring Expanded Memory—E 3-9
Specifying the Number of Files—F 3-9
Displaying Memory Configuration Details at Startup—INFO 3-10
Preventing Detection of Idle Time—NOIDLE 3-10
Specifying Maximum Swap File Size—SWAPK 3-11
Specifying Swap File Location—SWAPPATH 3-11
Specifying Temporary File Location—TEMPPATH 3-11
Excluding Available Memory—X 3-11
Setting the Network Environment 3-12
Network Hardware Requirements 3-12
Assigning Rights 3-12
Setting Up Network Devices 3-13
The Application Batch File 3-14
Summary 3-14

Chapter 4: Network Programming

In This Chapter 4-1
Definition of a Local Area Network 4-2
Local Area Network Requirements for CA-Clipper 4-2
CA-Clipper Network Features 4-3
Network Commands 4-3
Network Functions 4-4
Programming in a Network Environment 4-5
Attempting a Lock 4-5
How to Open Files for Sharing 4-7
Other Commands that Open Files 4-8
Resolving a Failed Lock 4-10
The Mechanics of Locking 4-11
When to Lock Records and Files 4-11
When to Obtain Exclusive Use 4-13
Overlays on a Network 4-13

Update Visibility 4-14
    The Observers 4-14
    Update Rules 4-15
    When Does a Commit Write to Disk? 4-16
    Additional Visibility Issues 4-16
Network Printing 4-17
    Setting Up the Network Printer 4-18
    Program Design Considerations 4-18
    Printing to a File 4-19
Summary 4-20

Chapter 5: Introduction to TBrowse

In This Chapter 5-1
TBrowse 5-1
Basic Browse Operations 5-2
    Creating TBrowse Objects 5-2
    Main Loop 5-3
Optimization of the Browse 5-6
    Quicker Response Time 5-6
    Multi-user Usage 5-7
    Repositioning the Record Pointer 5-8
    Using TBrowse:cargo and TBColumn:cargo 5-10
Browsing with Get 5-11
    Determining Whether the Record Has Moved 5-13
Adding Color 5-14
    TBrowse:colorSpec 5-14
    TBColumn:defColor 5-15
    TBColumn:colorBlock 5-16
    Controlling the Highlight 5-18
    TBrowse:colorRect() 5-19
Controlling the Scope 5-20
    Viewing a Specific Key Value 5-22
    Browsing Search Results 5-25
Summary 5-27

Chapter 6: The Get System

In This Chapter..6-1
Objects and the Get System...6-2
Using Objects..6-2
Get Class...6-2
    Get:block..6-4
    Setting and Retrieving Values...6-5
    Get:cargo..6-5
    Get Class Protocol 6-6
Get System..6-7
The Read Layer 6-8
Extending the Read Layer 6-10
    Nested Reads 6-10
    Using GetActive()...6-10
    Using Code Blocks for WHEN/VALID 6-13
    Processing the Entire GetList 6-15
Creating a New Read Layer 6-15
    Basic Guidelines 6-15
    Important Implementation Rules..6-16
    Implementation Steps 6-17
    Example: READ VALID 6-17
    Example: Gets With Messages 6-20
The Reader Layer...6-23
Creating New Reader Layers 6-24
    Basic Guidelines 6-24
    Important Implementation Rules..6-25
    Example: Incremental Date Get 6-26
The Get Function Layer 6-28
Creating New Get Function Layers 6-30
Summary...6-31
    Newly Documented Methods and Instance Variables.............................6-32
    Get System Functions...6-32
GETACTIVE() function 6-33
GETAPPLYKEY() function..6-34
GETDOSETKEY() function..6-35

GETPOSTVALIDATE() function 6-36
GETPREVALIDATE() function 6-37
GETREADER() function 6-38
READFORMAT() function 6-39
READKILL() function 6-40
READUPDATED() function 6-41

Chapter 7: Error Handling Strategies

In This Chapter 7-2
Overview of Exception Handling Concepts 7-2
    Error Scoping 7-3
    Raising Errors 7-5
    Handling Errors 7-6
CA-Clipper Exception Mechanisms 7-8
    The SEQUENCE Construct 7-8
    The Posted Error Block 7-13
    SEQUENCE vs. Error Block 7-15
Error Objects 7-16
    A Model Example 7-18
    The "System Error" Line-up: Error.ch 7-19
    Carrying Error-Processing Protocols 7-19
A Baseline Strategy 7-21
    A Comprehensive Example 7-23
    What Belongs in RECOVER? 7-25
Network Processing 7-29
    Transact.prg 7-30

Chapter 8: CA-Clipper Compiler—CLIPPER.EXE

In This Chapter 8-1
Invoking the CA-Clipper Compiler 8-2
Specifying Options with CLIPPERCMD 8-2
    Examples 8-3
The Compiler Script File 8-3
The Compiler Return Code 8-4

How the CA-Clipper Preprocessor Works 8-4
How CA-Clipper Compiles 8-5
The Compile and Link Batch File 8-6
Header Files 8-6
Output Files 8-7
    Object Files 8-7
    Temporary Files 8-7
    Preprocessed Output Listing 8-8
Changing the Size of the Environment 8-8
Compiler Options 8-9
    Examples 8-14
Summary 8-14

Chapter 9: CA-Clipper Linker—RTLink.EXE

In This Chapter 9-1
Overview of .RTLink 9-2
The .RTLink Files 9-3
Invoking .RTLink 9-4
    The FREEFORMAT Interface 9-4
    The POSITIONAL Interface 9-7
Configuring the .RTLink Defaults 9-11
    The RTLINKCMD Environment Variable 9-11
    RTLINK.CFG 9-12
    Configuration Options 9-12
The .RTLink Return Code 9-13
The Compile and Link Batch File 9-13
Output Files 9-14
    Prelinked Library Files (.PLL) 9-14
    Prelinked Transfer Files (.PLT) 9-15
    Executable Files (.EXE) 9-15
    Map Files (.MAP) 9-16
    Overlay Files (.OVL) 9-16
    Temporary Files 9-16
    Information Files (.INF) 9-17
How .RTLink Searches for Files 9-17

Library Files (.LIB) 9-18
　　Object Files (.OBJ) 9-18
　　Script Files (.LNK) 9-19
　　Prelinked Transfer Files (.PLT) 9-19
Linker Options 9-19
　　Precedence of Options 9-20
　　Abbreviations 9-20
　　Entering Numbers 9-21
　　The NIL Keyword 9-21
　　Batch Mode Options 9-21
　　Case Sensitivity Options 9-22
　　Configuration Options 9-22
　　Dynamic Overlaying Options 9-23
　　Help and Debugging Options 9-24
　　Incremental Linking Options 9-26
　　Miscellaneous Options 9-27
　　Prelink Options 9-28
　　Static Overlaying Options 9-30
　　Dynamic Overlaying of CA-Clipper Code 9-32
　　Dynamic Modules 9-33
　　Resident Modules 9-33
Prelinking 9-34
　　Deciding to Use a Prelinked Library 9-34
　　Building a Prelinked Library 9-35
　　Building a Dependent Executable 9-38
Incremental Linking 9-38
Static Overlaying 9-39
　　Deciding to Use Static Overlays 9-40
　　Designing Static Overlays 9-40
　　External Static Overlays 9-41
　　Examples 9-42
Summary 9-42

Chapter 10: CA-Clipper Debugger—CLD.EXE

In This Chapter 10-2
Starting the Debugger 10-2
    Preparing Your Programs for Debugging 10-3
    Invoking the Debugger 10-5
    How the Debugger Searches for Files 10-8
    Using a Script File 10-8
    Getting Help 10-10
    Leaving the Debugger 10-10
The Debugger Display 10-11
    The Debugger Windows 10-11
    Dialog Boxes 10-26
    The Debugger Menus 10-27
    The Function Keys 10-32
Debugging a Program 10-32
    Executing Program Code 10-33
    Inspecting Data and Expressions 10-37
    Inspecting Program Code 10-40
    Accessing DOS 10-45
Menu Command Reference 10-46
? | ?? 10-47
Animate 10-49
BP 10-50
Callstack 10-52
Delete 10-53
DOS 10-55
File DOS 10-56
File Exit 10-57
File Open 10-58
File Resume 10-60
Find 10-61
Go 10-62
Goto 10-63
Help 10-64
Input 10-66

List	10-67
Locate Case	10-68
Locate Find	10-69
Locate Goto	10-71
Locate Next	10-72
Locate Previous	10-73
Monitor All	10-74
Monitor Local	10-75
Monitor Private	10-76
Monitor Public	10-77
Monitor Sort	10-78
Monitor Static	10-79
Next	10-80
Num	10-81
Options Codeblock	10-82
Options Color	10-83
Options Exchange	10-84
Options Line	10-85
Options Menu	10-86
Options Mono	10-87
Options Path	10-88
Options Preprocessed	10-89
Options Restore	10-90
Options Save	10-91
Options Swap	10-92
Options Tab	10-93
Output	10-94
Point Breakpoint	10-95
Point Delete	10-96
Point Tracepoint	10-97
Point Watchpoint	10-98
Prev	10-99
Quit	10-100
Restart	10-101
Resume	10-102
Run Animate	10-103

Run Go 10-104
Run Next 10-105
Run Restart 10-106
Run Speed 10-107
Run Step 10-108
Run To 10-109
Run Trace 10-110
Speed 10-111
Step 10-112
TP 10-113
View 10-114
View App 10-115
View Callstack 10-116
View Sets 10-117
View Workareas 10-118
Window Iconize 10-119
Window Move 10-120
Window Next 10-121
Window Prev 10-122
Window Size 10-123
Window Tile 10-124
Window Zoom 10-125
WP 10-126

Chapter 11: Program Mainentance—RMAKE.EXE

In This Chapter 11-1
Invoking RMAKE 11-2
RMAKE Options 11-2
The RMAKE Environment Variable 11-4
    Example 11-4
The RMAKE Return Code 11-5
How RMAKE Works 11-5
How RMAKE Searches for Files 11-6
    Make Files 11-6
    Target and Dependency Files 11-7

The Make File 11-8
    Using Quotation Marks 11-8
    Line Continuation 11-9
    Comments 11-9
    Dependency Rules 11-9
    Inference Rules 11-11
    Directives 11-13
    Macros 11-16
Examples 11-19
Notes 11-19
Summary 11-21

Chapter 12: CA-Clipper Program Editor—PE.EXE

In This Chapter 12-1
Invoking the Program Editor 12-1
Navigation and Editing 12-2
Leaving the Program Editor 12-3
The PE System Architecture 12-3
Summary 12-4

Chapter 13: Database Utility—DBU.EXE

In This Chapter 13-1
Invoking the Database Utility 13-1
The Main DBU Screen 13-2
    The Menu Bar 13-3
    The Message and Prompt Area 13-4
    Dialog Boxes 13-4
    Windows 13-6
    Work Areas 13-7
Leaving DBU 13-10
The DBU Menus 13-11
    F1 Help 13-11
    F2 Open 13-12
    F3 Create 13-14

> F4 Save 13-19
> F5 Browse 13-20
> F6 Utility 13-22
> F7 Move 13-27
> F8 Set 13-29

Summary 13-33

Chapter 14: Report and Label Utility—RL.EXE

In This Chapter 14-1
Loading the Report and Label Utility 14-2
Creating and Modifying Reports 14-2
> Creating or Modifying a Report 14-2
> Defining Report Columns 14-3
> Deleting a Column 14-5
> Inserting a New Column 14-5
> Locating a Column 14-6
> Defining Report Options 14-6
> Defining Groups 14-7
> Saving the Report Definition 14-9
> Printing a Report 14-9
> Reporting from Related Work Areas 14-10

Creating and Modifying Labels 14-10
> The Label Editor Screen 14-11
> Defining the Label Dimensions and Formatting 14-12
> Standard Label Formats 14-13
> Defining the Label Contents 14-13
> Saving the Label Design 14-15
> Printing the Labels 14-15

Leaving RL 14-15
The RL System Architecture 14-16
Notes 14-16
Summary 14-17

Programming and Utilities Guide   xvii

Chapter 15: The Guide to CA-Clipper—NG.EXE

In This Chapter 15-1
Loading the Instant Access Engine 15-1
    Using Memory-Resident Mode 15-2
    Using Pass Through Mode 15-2
    Accessing the Instant Access Engine 15-3
How the Instant Access Engine Searches for Files 15-3
Using the Access Window 15-4
    The Menu Bar 15-4
    Selecting Menus and Menu Items 15-5
    Sizing and Moving the Access Window 15-5
    Getting Help 15-6
Viewing a Documentation Database 15-6
    The Short Entry List 15-7
    Searching a Short Entry List 15-7
    Expanding an Entry—Moving Down a Level 15-8
    See Also References 15-8
    Moving Up a Level 15-9
    Selecting a New Documentation Database 15-10
Instant Access Engine Navigation Keys 15-11
Configuring the Instant Access Engine 15-12
    Toggling Color 15-12
    Toggling Auto Lookup 15-12
    Changing the Hot key 15-13
    Saving the New Configuration 15-14
Leaving the Instant Access Engine 15-14
    Exiting the Instant Access Engine 15-14
    Uninstalling the Instant Access Engine 15-15
Summary 15-15

Glossary

Index

Chapter 1
Introduction

This is the *Programming and Utilities* guide for *CA-Clipper*. It contains conceptual information on programming and creating applications using the *CA-Clipper* development system, as well as usage information for each utility in the package.

Further information on many of the topics covered in this guide can be found in the *Reference* guide. In addition, you will find much of the utilities portion of the *Programming and Utilities* guide packaged in other forms to make access more convenient.

For online information accessible while operating your program editor or any other development utility, use *The Guide To CA-Clipper* Utilities database (see *The Guide To CA-Clipper* chapter of this guide for documentation). For a brief summary of the utilities, use the *Quick Reference* guide.

Part 1: Basic Concepts

This section consists of a single chapter that builds on the *Language Reference* chapter of the *Reference* guide by focusing on various aspects of the programming language and explaining how some of the language components fit together. Although not as exhaustive as the *Language Reference*, this chapter gives you a different view of the language, based on systems and other logical groupings, that may help you better understand how to use *CA-Clipper* to accomplish you programming tasks.

Part 2: Programming

This section is divided into several chapters, each of which gives you specific details on a particular aspect of the *CA-Clipper* system that may require further programming. It contains conceptual information on programming and creating applications using the following subsystems:

TBrowse, The Get System, and Error Handling, as well as chapters on The Runtime Environment and Network Programming. Further reference information can also be found in the *Language Reference* chapter of the *Reference* guide.

Part 3: Utilities

This section documents each utility supplied with the *CA-Clipper* package in a separate chapter. The complete command line syntax as well as all options are completely specified for each utility. In addition, usage information, including navigation and selection, is provided for the menu-driven utilities. The following utilities are covered:

- *CA-Clipper* Compiler—CLIPPER.EXE
- *CA-Clipper* Linker—RTLINK.EXE
- *CA-Clipper* Debugger—CLD.EXE
- Program Maintenance—RMAKE.EXE
- Program Editor—PE.EXE
- Database Utility—DBU.EXE
- Report and Label Utility—RL.EXE
- The Guide To *CA-Clipper*—NG.EXE

The symbols and conventions used to document the command line syntax and options for these utilities are the same as those used in the *Reference* guide. For instance, square brackets ([ ]) delimit optional items, angle brackets (< >) delimit user-supplied items, and required keywords are shown in uppercase letters. If you are unfamiliar with the conventions, refer to the introduction of the *Reference* guide for further details.

Glossary

The Glossary is a comprehensive dictionary of terms used throughout the *CA-Clipper* documentation. Each glossary entry consists of the item name, the identity of one or more categories to which the item belongs, and a short definition.

Chapter 2
Basic Concepts

The *Language Reference* chapter of the *Reference* guide introduced you to programming in *CA-Clipper* by defining all of the components (e.g., commands, functions, and operators) that go together to make up the programming language. This chapter builds on the *Language Reference* material by focusing on various aspects of the programming language and explaining how some of the language components fit together. Although not as exhaustive as the *Language Reference*, this chapter gives you a different view of the language, based on systems and other logical groupings, that may help you better understand how to use *CA-Clipper* to accomplish your programming tasks.

In This Chapter

This chapter provides an overview of the *CA-Clipper* language constructs and its major systems. The following topics are covered:

- Structure of a *CA-Clipper* Program
- Functions and Procedures
- Control Structures
- Variables
- Expressions
- Data Types
- Operators
- The Macro Operator
- Arrays
- Code Blocks
- Objects and Messages
- Database, Input/Output, Keyboard, and Low-level File Systems

The Structure of a CA-Clipper Program

A *CA-Clipper* program is a collection of statements that adhere to the rules defined by the *CA-Clipper* language. A program statement can take many different forms including:

- Command invocation
- Function call (e.g., library, pseudo, or user-defined)
- Procedure call
- Preprocessor directive
- Control structure or declaration statement
- Assignment statement
- Comment

A program is stored in a text file with a (.prg) extension. In general, white space (e.g., blanks, tabs) is ignored by the compiler, allowing you to format your programs for readability. However, the compiler interprets the carriage return/linefeed pair as the end of a statement (see the Continuation section for exceptions).

A Typical CA-Clipper Program

The following example program, Menu.prg, is used throughout the discussion in this section to illustrate the various components of a *CA-Clipper* program and to give you an idea of what a typical program looks like:

```
// Menu.prg -- Display a menu and process a menu choice

// Manifest constant definitions
#define ONE     1
#define TWO     2
#define THREE   3

// Filewide variable declarations go here

PROCEDURE Main

   // Local variable declarations
   LOCAL nChoice

   // Continuous loop redisplays main menu after each
   // function return
   DO WHILE .T.

      // Function call to display main menu
      nChoice := MainMenu()
```

```
        // Case structure to make function call
        // based on return value of MainMenu()
        DO CASE

        // Functions called in the CASE structure are
        // defined in Database.prg, which should be
        // compiled and linked with this application
        CASE nChoice = ONE
            AddRecs()
        CASE nChoice = TWO
            EditRecs()
        CASE nChoice = THREE
            Reports()
        OTHERWISE
            EXIT              // Break out of continuous loop
        ENDCASE
    ENDDO
    RETURN

FUNCTION MainMenu(menuChoice)
    CLEAR
    SET WRAP ON
    SET MESSAGE TO 23 CENTER
    @  6, 10 PROMPT "Add"    MESSAGE "New Account"
    @  7, 10 PROMPT "Edit"   MESSAGE "Change Account"
    @  8, 10 PROMPT "Reports" MESSAGE "Print Account Reports"
    @ 10, 10 PROMPT "Quit"   MESSAGE "Return to DOS"
    MENU TO menuChoice
    CLEAR
    RETURN (menuChoice)
```

The Main Procedure Definition

The main program file in any *CA-Clipper* application usually has a main procedure (or function) definition located at the top of the program (.prg) file. In our example, the main procedure definition begins with the following statement:

```
PROCEDURE Main
```

This statement marks the beginning of the procedure and gives it a name. After the application is compiled and linked, it is this procedure that serves as a startup routine when the application is executed. In fact, if your application is designed with a main procedure, to prevent the compiler from generating a startup procedure of its own, you must compile with the /N option.

Like other procedure and function definitions, the main procedure consists of statements, commands, and function calls that perform a particular task. Typically, variable declarations are placed at the top of the procedure.

Following the variable declarations is the main procedure body which usually consists of one or more function calls. Depending on the nature of the application, you may place the function calls in a control structure as in Menu.prg:

```
DO WHILE .T.
   nChoice := MainMenu()
   DO CASE
   CASE nChoice = ONE
      AddRecs()
   CASE nChoice = TWO
      EditRecs()
   CASE nChoice = THREE
      Reports()
   OTHERWISE
      EXIT
   ENDCASE
ENDDO
```

Finally, the end of the main procedure may be marked with a RETURN statement. The RETURN statement is not a required part of any procedure definition but is included for readability. If no RETURN statement is present, the next PROCEDURE or FUNCTION statement (or the end of file) is used to indicate the end of the previous definition.

Function and Procedure Calls

Within a routine, calls to other routines are made using the routine name. For functions, the name must be followed by its arguments enclosed in parentheses. A function call is an expression whose data type is determined by its return value. Therefore, a function call may be part of another expression. Function calls are the same regardless of how the function is defined. For example, calls to the standard CA-Clipper library functions are identical to your own user-defined function calls.

The following line of code from Menu.prg illustrates a function call as part of another statement:

```
nChoice := MainMenu()
```

Procedures, on the other hand, always return NIL. Thus, a call to a procedure is normally made as a separate program statement. A procedure can, however, accept parameters. Parameters are passed to a procedure using the same syntax as a function (e.g., a parameter list enclosed in parentheses).

Thus, in Menu.prg, AddRecs, EditRecs, and Reports could be defined as functions or procedures without affecting the calling convention.

Variable Declaration

In *CA-Clipper*, dynamic variables can be created and initialized without formal declaration statements. The default scope of an undeclared dynamic variable is private to the routine in which the variable is first initialized. In addition, *CA-Clipper* is a weak-typed language which means that the data type of a variable is never declared, only the *scope*, or visibility.

Lexically scoped variables must be declared. In Menu.prg, there is one declaration statement that sets up a local variable to receive the value of a menu choice:

```
LOCAL nChoice
```

This statement declares the variable *nChoice* as a local variable. Except in one case (i.e., the PUBLIC statement), variable declarations assign NIL to the variable as its initial value.

Where you place a variable declaration in a program file helps determine its scope. For instance, STATIC declarations, that are made before the first procedure or function definition, have a filewide scope (i.e., are visible to all routines in the file), whereas those made within a routine definition are local to that routine. See the Variables section in this chapter for more information on variable declaration and scoping.

Function and Procedure Definitions

Function and procedure definitions are also part of a typical program. In Menu.prg, there is only one function definition which is shown below:

```
FUNCTION MainMenu( nChoice )
   CLEAR
   SET WRAP ON
   SET MESSAGE TO 23 CENTER
   @  6, 10 PROMPT "Add"     MESSAGE "New Account"
   @  7, 10 PROMPT "Edit"    MESSAGE "Change Account"
   @  8, 10 PROMPT "Reports" MESSAGE "Print Account Reports"
   @ 10, 10 PROMPT "Quit"    MESSAGE "Return to DOS"
   MENU TO nChoice
   CLEAR
   RETURN (nChoice)
```

This definition is typical, consisting of the declared function name with a parameter list in parentheses, followed by the function body, and ending with a return statement that passes a value to the calling routine. All function definitions must return a value in *CA-Clipper*.

The definition of a procedure is like that of a function except for these differences: the keyword PROCEDURE precedes the procedure name and parameters, and a return value is unnecessary since procedures automatically return NIL.

Normally, routines defined within a program file are only used by other routines in that file. More generic routines are usually stored in a separate file, often called a procedure or function file, so they can be shared by many programs. In Menu.prg, the functions AddRecs(), EditRecs(), and Reports() are called but not defined in the program. They are, instead, defined in Database.prg which should be compiled and linked with the Menu.prg program.

Preprocessor Directives

In addition to statements, preprocessor directives can be part of a *CA-Clipper* program. These directives are instructions to the compiler rather than statements that are compiled. Preprocessor directives always begin with a hash symbol (#). In Menu.prg, there are several manifest constant definitions at the beginning of the file. These directives assign constant values to identifier names that can be used in statements throughout the program. Whenever the compiler encounters the identifier name it substitutes the associated value:

```
#define ONE     1
#define TWO     2
#define THREE   3
```

Comments

You can formulate program comments in several ways, depending on the effect that you want. For long comments that span several lines, or comment blocks, use the following:

```
/* The slash-asterisk introduces a comment block.  All
text following this special comment indicator is ignored by the
compiler, including carriage returns, until an asterisk-slash is
encountered to indicate the end of the comment block.*/
```

For single-line comments, use the double-slash (//) to introduce the comment. All text following this comment symbol until the next carriage return/linefeed is treated as a comment to be ignored by the compiler. You can also use the asterisk (*) symbol for single-line comments, as in dBASE programs. In our example, there are several comments formulated in this way throughout the code, one of which is shown below:

```
// Continuous loop redisplays main menu after each
// function return
```

Comments need not be in a block or on a line by themselves. The double-slash can also be used to place a comment at the end of another statement line as in the following example:

```
EXIT            // Break out of continuous loop
```

You may also use the double-ampersand (&&) for inline comments. With both inline comment symbols, the compiler ignores all text that follows the symbol until the next carriage return/linefeed.

Continuation

Since the *CA-Clipper* compiler interprets the carriage return/linefeed pair as marking the end of a program statement, it might appear that each statement must be on a single-line and, furthermore, that no more than one statement can be on a line. However, you can use the semicolon (;) character to specify multiline statements and multistatement lines.

To continue a statement on more than one line (i.e., form a multiline statement), place a semicolon at the end of the first line before the carriage return/linefeed, and put the remainder of the statement on a new line. When the compiler encounters a semicolon followed by a carriage return/linefeed, it continues reading the statement on the next line. There is no example of statement continuation in Menu.prg, but the prompt commands could have been formatted as follows:

```
@  6, 10 PROMPT "Add";
   MESSAGE "New Account"
@  7, 10 PROMPT "Edit";
   MESSAGE "Change Account"
@  8, 10 PROMPT "Reports";
   MESSAGE "Print Account Reports"
@ 10, 10 PROMPT "Quit";
   MESSAGE "Return to DOS"
```

You can repeat this method of line continuation several times to break a single statement into several lines. Using this feature, you can format very long expressions to make them more readable.

To form a multistatement line, simply place a semicolon, without a carriage return/linefeed, between the statements. When the compiler encounters such a semicolon, it knows to read the next statement from the same line. For example, the case structure in Menu.prg could have been coded as follows:

Functions and Procedures

```
DO CASE
CASE nChoice = ONE   ; AddRecs()
CASE nChoice = TWO   ; EditRecs()
CASE nChoice = THREE ; Reports()
OTHERWISE ; EXIT
ENDCASE
```

Reserved Words

In *CA-Clipper* there are several words that are reserved for internal use by *CA-Clipper*, itself. These are listed in the *Reserved Word* appendix of the *Error Messages and Appendices* guide. Reserved words cannot be used as identifier names in program statements. In addition to these reserved words, it is illegal for an identifier to start with an underscore.

Functions and Procedures

User-defined functions and procedures are the basic building blocks of *CA-Clipper* programs. Every program is made of one or more procedures and user-defined functions. These building blocks consist of a group of statements that perform a single task or action. They are similar to functions in C, Pascal, or other major languages.

The visibility of function and procedure names falls into two classes. Functions and procedures visible anywhere in a program are referred to as *public* and are declared with FUNCTION or PROCEDURE statements. Functions and procedures that are visible only within the current program (.prg) file are referred to as *static* and are declared with STATIC FUNCTION or STATIC PROCEDURE statements. Static functions and procedures are said to have *filewide* scope.

Static functions and procedures are quite useful for a number of reasons. First, they limit visibility of a function or procedure thereby restricting access to the function or procedure. Because of this, subsystems defined within a single program (.prg) file can provide an access protocol with a series of public function or procedures, and conceal the implementation details of the subsystem within static functions and procedures. Second, since the static function or procedure references are resolved at compile time, they preempt to public functions and procedures which are resolved at link time. This assures that within a program file, a reference to a static function or procedure executes that routine if there is a name conflict with a public function.

Defining Functions and Procedures

Functions and procedure definitions are quite similar with slightly different requirements for return values.

User-defined functions can be defined anywhere in a program (.prg) file, but definitions cannot be nested. A function definition has the following basic form:

```
[STATIC] FUNCTION <identifier> [(<local parameter list>)]
   [<variable declarations>]
   .
   . <executable statements>
   .
   RETURN <return exp>
```

A function definition consists of a function declaration statement with an optional list of declared parameters. Following the function declaration are a series of optional variable declaration statements such as LOCAL, STATIC, FIELD, or MEMVAR. As mentioned before, a function must return a value, therefore a RETURN statement with a return value must be specified somewhere in the body of the function. A function definition begins with the FUNCTION declaration statement and ends with the next FUNCTION or PROCEDURE statement or end of file.

If the user-defined function declaration begins with the STATIC keyword, the function is visible only to procedures and user-defined functions declared within the same program (.prg) file.

The following is a typical example of a user-defined function definition:

```
FUNCTION AmPm( cTime )
   IF VAL(cTime) < 12
      cTime += " am"
   ELSEIF VAL(cTime) = 12
      cTime += " pm"
   ELSE
      cTime := STR(VAL(cTime) - 12, 2) + ;
         SUBSTR(cTime, 3) + " pm"
   ENDIF
   RETURN cTime
```

Defining Procedures

Procedures are identical to user-defined functions with the exception that no return value is required and therefore no RETURN statement is required. Like user-defined functions, procedures can be defined anywhere in a program (.prg) file, but definitions cannot be nested. The following is the general form of a procedure definition:

Functions and Procedures

```
[STATIC] PROCEDURE <identifier> [(<local parameter list>)]
   [<variable declarations>]
   .
   . <executable statements>
   .
   [RETURN]
```

A procedure definition begins with a PROCEDURE declaration statement and ends with the next FUNCTION or PROCEDURE statement, or end of file.

If the procedure declaration begins with the STATIC keyword, the procedure is visible only to procedures and user-defined functions declared within the same program (.prg) file.

Calling Functions and Procedures

Although functions and procedures are quite similar they can be called in different ways.

Calling a Function

User-defined functions you create and specify are not different from the supplied standard functions and therefore can be called in the same way. You can specify functions either in expressions or as statements. For example, the following are all legitimate function calls:

```
// Expression
? "This is the " + Ordinal(DATE()) + "day"
Report()                              // Statement
result := Report("Quarterly")         // Expression
```

When a function is called, it must always be specified including the open and close parentheses. If arguments are to be passed to called functions, they are specified between the parentheses and separated by commas. For example:

```
? MajorFunc("One", "Two", "Three")
```

Calling a Procedure

A procedure can be called using function-calling syntax by specifying the procedure call as a statement. Here, you call the procedure as you would a *CA-Clipper* user-defined function specified as a statement, like this:

```
<procedure>([<argument list>])
```

The second way is using the DO...WITH command. This method is not recommended since it passes arguments by reference as a default.

Passing Parameters

When you invoke a procedure or user-defined function, you can pass values and references to it. This facility allows you to create black box routines that can operate on data without any direct knowledge of the calling routine. The following discussion defines the various aspects of passing parameters in *CA-Clipper*.

Arguments and Parameters

Passing data from a calling routine to an invoked routine involves two perspectives, one for the calling side and one for the receiving side. On the calling side, the values and references passed are referred to as *arguments* or actual parameters. For example, the following function call passes two arguments, a constant and a variable:

```
LOCAL nLineLength := 80
? Center("This is a string", nLineLength)
```

On the receiving side, the specified variables are referred to as *parameters* or formal parameters. For example, here the variables are specified to receive the string to center as well as the line length:

```
FUNCTION Center( cString, nLen )
   RETURN PADC( cString, nLen, " " )
```

The specified receiving variables are placeholders for values and references obtained from the calling routine. When a procedure or user-defined function is called, the values and references specified as arguments of the routine's invocation are assigned to the corresponding receiving parameter in the invoked routine.

In *CA-Clipper*, there are two ways to specify parameters depending on the storage class of the parameter you want to use. Parameters specified as a part of the procedure or user-defined function declaration are declared *parameters* and are the same as local variables. They are visible only within the called routine and have the same lifetime as the routine. Parameters specified as arguments of the PARAMETERS statement are created as private variables and hide any private or public variables or arrays of the same name inherited from higher-level procedures and user-defined functions.

Note: In *CA-Clipper*, you cannot mix declared parameters and a PARAMETERS statement within the same procedure or user-defined function definition. Attempting this generates a fatal compiler error.

In *CA-Clipper*, parameters are received in the order passed and the number of arguments does not have to match the number of parameters specified. In fact, arguments may be skipped within the list of

Functions and Procedures

arguments when a routine is invoked. Received parameters without corresponding arguments are initialized to NIL. For example, the following function call skips two arguments from within the list of arguments, thereby initializing them to NIL values:

```
? TestProc("Hello",,,"There")

FUNCTION TestProc( param1, param2, param3,;
         param4, param5 )
   ? param1, param2, param3, param4
// Result: Hello NIL NIL There
   RETURN NIL
```

When a routine is called, the PCOUNT() function is updated with the position of the last argument specified within the list of arguments. This includes skipped arguments, but not ones left off the end of the list. In the example above, PCOUNT() returns 4.

Passing by Value

Passing an argument by value means the argument is evaluated and its value is copied to the receiving parameter. Changes to a received parameter are local to the called routine and lost when the routine terminates. In *CA-Clipper*, all variables, expressions, and array elements are passed by value as a default if the function-calling syntax is used. This includes variables containing references to arrays, objects, and code blocks.

As an example, the following code passes a database field and a local variable to a procedure by value:

```
LOCAL nNumber := 10
USE Customer NEW
SayIt( Customer->Name, nNumber )
? nNumber                                 // Result: 10
RETURN

PROCEDURE SayIt( fieldValue, nValue )
   ? fieldValue, ++nValue                 // Result: Smith 11
   RETURN
```

The importance of pass-by-value is that the called routine cannot change the caller's data by changing the value of the received parameter. This increases modularity by relegating the responsibility to the calling routine to determine whether it allows its data to be changed by a called routine.

Passing by Reference

Passing an argument by reference means that a reference to the value of the argument is passed instead of a copy of the value. The receiving parameter then refers to the same location in memory as the argument. If the called routine changes the value of the receiving parameter, it also changes the argument passed from the calling routine.

Variables other than field variables and array elements can be passed by reference if prefaced by the pass-by-reference operator (@) and the function or procedure is called using the function-calling syntax. For example:

```
nX := nY := 1
Increment(@nX, nY)
? nX                                    // Result: 2

PROCEDURE Increment( nNumber, nIncrement )
   nNumber := nNumber + nIncrement
   RETURN
```

In this example, the change made to the *nNumber* parameter in the called routine is reflected in the *nX* argument after the Increment() procedure is called.

As you can see, passing arguments by reference can be quite dangerous if parameters are inadvertently assigned new values in the called routine. Because of this, passing by reference should only be used in special cases such as returning a modified value from a procedure or user-defined function. A good example of this is the FREAD() function which takes a buffer variable passed by reference, fills the variable with characters read from a binary file, and returns the number of bytes read.

Note that arguments passed to routines called with the DO...WITH statement are passed by reference as a default. Note also that other dialects commonly pass arguments by reference as a default. In *CA-Clipper*, this practice is strongly discouraged and therefore use of the DO statement is not recommended. All DO invocations should be replaced with function-calling syntax and variables passed by reference should be prefaced with the pass-by-reference operator (@).

Passing Arrays and Objects as Arguments

When using function-calling syntax, variables containing references to arrays and objects are passed by value as are variables containing values of any other data type. This may seem confusing since it reveals that passing references involves a level of indirection. But when a

Functions and Procedures

variable containing a reference is passed to a routine, a copy of the reference is passed instead of the actual reference.

If, as an example, an array is passed to a routine by value and the called routine changes the value of one of the array elements, the change is reflected in the array after the return, since the change was made to the actual array via the copied reference to it. If, however, a reference to a new array is assigned to a variable passed by value, the new reference is discarded upon return to the caller and the array reference contained in the original variable is unaffected.

Conversely, if the variable containing a reference to an array or object is passed by reference by preceding the argument variable with the pass-by-reference operator (@), any changes to the reference are reflected in the calling routine upon return. For example:

```
// Change the value of an array element
a := {1, 2, 3}
changeElement(a)
? a[1]                  // Result: 10

// Change an array; does not work
a := {1, 2, 3}
changeArray(a)
? a[1]                  // Result: 1

// Change an array by passing by reference
changeArray(@a)
? a[1]                  // Result: 4

FUNCTION changeElement( aArray )
   aArray[1] := 10
   RETURN NIL

FUNCTION changeArray( aArray )
   aArray := {4, 5, 6}
   RETURN NIL
```

Although, this behavior appears different from previous versions, it has always been the case. The confusion lies in the different way arrays are handled in *CA-Clipper*. In previous versions of *CA-Clipper*, you could not assign, return, or otherwise transport a reference to an array except by passing it to a procedure or user-defined function. Therefore, there was never any way to know how arrays were actually passed to subroutines. Because of their behavior, we always assumed they were passed by reference. In *CA-Clipper*, array references are treated the same as other data types, thereby revealing the actual and true rules of parameter passing.

Argument Checking

In *CA-Clipper*, there is no argument checking, and therefore the number of parameters does not have to match the number of arguments passed. Arguments can be skipped or left off the end of the argument list. A parameter not receiving a value or reference is initialized to NIL. If arguments are specified, PCOUNT() returns the position of the last argument passed.

Since arguments can be skipped, PCOUNT() is not always an accurate gauge of what arguments have been specified. To ascertain this information, compare the parameter in question to NIL. If it is equal, you can either supply a default value or generate an argument error. The following example demonstrates this concept:

```
FUNCTION TestParams( param1, param2, param3 )
   IF param2 = NIL
      ? "Parameter was not passed"
      param2 := "default value"
   ENDIF
   .
   . <statements>
   .
   RETURN NIL
```

In addition to the NIL test, VALTYPE() can be used to test for a parameter receiving a value as well as the proper data type, like this:

```
FUNCTION TestParams( cParam1, nParam2 )
   IF VALTYPE(nParam2) != "N"
      ? "Parameter was not passed or invalid type"
   ENDIF
   .
   . <statements>
   .
   RETURN NIL
```

Note: Remember to use VALTYPE() instead of TYPE() to test the data type of declared parameters. A TYPE() applied to a declared parameter will always return "U."

Passing Arguments from the DOS Command line

Arguments can be passed directly from the DOS command line to a program's main procedure. Parameters are specified either in a PARAMETERS statement or declared as a part on the main procedure's declaration. Remember, if you declare the main procedure, you must compile the main program (.prg) file with the /N option.

Arguments passed are all received as character strings. Specify multiple arguments with each argument separated by a space. If an argument

Functions and Procedures

contains an embedded space, it must be enclosed in quote marks in order to be passed as one string. For example, the following DOS command line passes two arguments when invoking PROG.EXE:

```
C>PROG "CA-CLIPPER COMPILER" 5
```

The main procedure receiving parameters from the DOS command line should test the number of passed parameters with PCOUNT() to assure that critical parameters have been specified.

Returning Values From User-Defined Functions

As defined, a user-defined function must return a value. To do this, it must contain a RETURN statement with an argument. A RETURN statement performs two actions: first, it terminates processing of the current routine transferring control to the calling routine; and second, if the current routine is a user-defined function, it returns any value specified as the argument of the RETURN statement to the calling routine. The return value can be a constant or an expression. For example:

```
RETURN ("Today is " + DTOC(DATE()))
```

In *CA-Clipper*, a user-defined function can return a value of any data type including arrays, objects, code blocks, and NIL. For example, the following user-defined function returns an array to a calling routine:

```
FUNCTION NumArrayNew
   LOCAL aNumArray := AFILL(ARRAY(10), 1, 1)
   RETURN aNumArray
```

RETURN statements can occur anywhere in a user-defined function definition, allowing processing to be terminated before the function definition ends. For example, the following code fragment terminates processing if a condition is true (.T.), and continues if the condition is false (.F.):

```
FUNCTION OpenFile( cDbf )
   USE (cDbf)
   IF NETERR()
      RETURN (.F.)
   ELSE
   .
   . <statements>
   .
   RETURN (.T.)
```

Note that RETURN is limited by the fact that it can return only one value. More than one value can be returned if arguments are passed by reference, although this is not a preferred solution. An aggregate data structure can be defined as an array containing other arrays, and arrays,

2-16 CA-Clipper

as mentioned above, can be passed throughout a program as a single value.

If the user-defined function's return value is not used, it is discarded. This is the reason you can specify a user-defined function as a statement.

Recursion

A procedure or user-defined function is recursive if it contains an invocation of itself. This can be either direct or indirect when a function calls another function that again calls the original function.

For example, the following example is a user-defined function that uses recursion to calculate the factorial of a specified number. A factorial is the product of all positive integers from one to a specified number. The factorial of 3 therefore is 1 * 2 * 3 which is 6.

```
FUNCTION Factorial( nFactorial )
   IF nFactorial = 0
      RETURN (1)
   ELSE
      RETURN (nFactorial * Factorial(nFactorial - 1))
   END
```

Control Structures

CA-Clipper supports several control structures that let you change the sequential flow of program execution. These structures allow you to execute code based on logical conditions and to repeatedly execute code any number of times.

In *CA-Clipper*, all control structures may be nested inside of all other control structures as long as they are nested properly. Control structures have a beginning and ending statement, and a nested structure must fall between the beginning and ending statements of the structure in which it is nested. This section summarizes all of the control structures in the *CA-Clipper* language, giving examples and suggested uses for each.

Looping Structures

Looping structures are designed for executing a section of code more than once. For example, you may want to print a report four times. Of

Control Structures

course, you could simply code four REPORT FORM commands in sequence, but this would be messy and would not solve the problem if the number of reports was variable.

FOR...NEXT

The FOR...NEXT control structure, or FOR loop, repeats a section of code a particular number of times. To solve the printing problem, you could do this:

```
FOR i := 1 TO 4
    REPORT FORM Accounts TO PRINTER
NEXT
```

FOR...NEXT is typical of all *CA-Clipper* control structures. It consists of a statement that defines the conditions of the structure and marks its beginning, followed by one or more executable statements that represent the body of the structure and, finally, a statement to end the structure.

FOR...NEXT sets up a loop by initializing a counter variable to a numeric value. The counter is incremented (or decremented) each time through the loop after the body statements are performed until it reaches a specified value. By default, as in the previous example, the increment value is one, but the increment can also be specified as part of the FOR statement. Furthermore, all of the FOR parameters may be numeric expressions allowing for a variable number of loop iterations. FOR...NEXT is commonly used to process arrays on an element by element basis.

DO WHILE...ENDDO

Another type of loop is one that is based on a condition rather than on a particular number of repetitions. For example, suppose you wanted to perform a complex process using records in a database file with a particular customer number. Without a conditional looping structure, it would be very difficult, if not impossible, to solve this problem.

The DO WHILE...ENDDO control structure (also called a DO WHILE loop) processes a section of code as long as a specified condition is true (.T.). To solve the database problem, you could do the following:

```
USE Accounts INDEX Accounts
SEEK 3456
DO WHILE Accounts->AcctNum = 3456
    .
    . <processing statements>
    .
    SKIP
ENDDO
```

2-18   CA-Clipper

Control Structures

This example illustrates a typical use of DO WHILE to process database file records. Note that SKIP is used as part of the loop body to advance the pointer to the next record. In a DO WHILE loop, the loop body must contain some statement that alters the loop condition; otherwise, the loop will execute forever.

EXIT and LOOP

EXIT and LOOP are special statements that can only be used inside a DO WHILE or FOR loop. EXIT transfers control out of the loop, and LOOP transfers control to the beginning of the loop. These statements are used as part of a conditional control structure to control looping behavior under unusual circumstances.

Decision-making Structures

Decision-making structures allow you to execute one or more program statements based on a condition. For example, you may want to execute a different function based on a menu choice. *CA-Clipper* has two such structures, IF...ENDIF and DO CASE...ENDCASE, but they are identical in functionality.

IF...ENDIF

The IF...ENDIF control structure executes a section of code if a specified condition is true (.T.). The structure can also specify alternative code to execute if the condition is false (.F.).

The following example illustrates IF...ENDIF in a user-defined function that tests for an empty array:

```
FUNCTION ZeroArray( aEntity )
   IF VALTYPE(aEntity) = "A" .AND. EMPTY(aEntity)
      RETURN (.T.)
   ENDIF
   RETURN (.F.)
```

The next example uses IF...ENDIF to process a menu choice:

```
IF nChoice = 1
   Func1()
ELSEIF nChoice = 2
   Func2()
ELSEIF nChoice = 3
   Func3()
ELSE
   QUIT
ENDIF
```

ELSEIF specifies an alternative condition to test if the previous condition was not met, allowing multiple levels of control. It replaces nested IF...ENDIF structures. ELSE is an alternative path to execute if none of the other conditions in the structure is met.

DO CASE...ENDCASE

DO CASE and IF are equivalent structures with slightly different syntax representations. For example, the following DO CASE code segment is functionally equivalent to the previous IF example:

```
DO CASE
CASE nChoice = 1
    Func1()
CASE nChoice = 2
    Func2()
CASE nChoice = 3
    Func3()
OTHERWISE
    QUIT
ENDCASE
```

Neither decision-making structure has advantages over the other. Which one you use is purely a matter of personal preference.

Error Handling Structures

BEGIN SEQUENCE...END is a specialized control structure often used for runtime error and exception handling.

Variables

This section describes the basic nature of variables in *CA-Clipper*. Variables are placeholders for values and references that have a defined lifetime and visibility, and a given name. When a variable name is referenced, the value it contains is returned.

In *CA-Clipper*, there are several types of variables. Variables are organized into *storage classes* that determine how the variable is stored, how long it lives, and where in a program its name can be used. Each storage class has a statement that either declares the variable names or creates the variable at runtime.

Lexically Scoped Versus Dynamically Scoped Variables

If you are familiar with *CA-Clipper*, you know that previous versions supported public, private, and field variables. These variables have what is referred to as *dynamic scope*. Dynamically scoped variables, including names, are created and maintained completely at runtime. Nothing is resolved at compile time and there is automatic inheritance of variables by called procedures and user-defined functions. These classes of variables are characteristic of interpreted systems and exist in *CA-Clipper* in order to be code compatible with interpreted dialects.

In *CA-Clipper*, two new classes of variables have been introduced that have what is called *lexical scope*. These classes of variables are resolved completely at compile time and have rigidly defined scoping rules, as described below.

Lexically scoped variables have been introduced for several reasons:

- Dynamically scoped variables violate the principle of modularity since to understand the operation of a routine, you must understand the operation of all the routines that call it. Each routine is a modularly constructed program that should be understood in the context of its own definition, and the program (.prg) file it is located in.

- Lexically scoped variables are more efficient and much faster than dynamically scoped variables since they are resolved completely at compile time. Dynamically scoped variables, by contrast, must be resolved each time they are referenced in a program.

We highly recommend the use of lexically scoped variables and their substitution for all uses of dynamically scoped variables.

Lifetime and Visibility of Variables

During execution, a variable, even though declared at compile time or referred to in an executable statement, doesn't actually exist until some portion of the computer's internal memory is allocated to contain its value. This is known as *creating* (or *instantiating*) the variable. Some variables are created automatically, while others must be explicitly created.

Once created, a variable continues to exist (and possesses a value, if one has been assigned) until it's memory is released. Some variables are released automatically, while others must be explicitly released. Some variables are never released. The duration of a variable's life is referred to as its *lifetime*.

Variables

Visibility refers to the conditions under which a variable is accessible to the program during execution. Some variables, even though they have been created and assigned a value, may not be visible under certain conditions.

During execution, a single variable name may be associated simultaneously with several different variables, some or all of which may be visible. Declarations can be used to ensure that occurrences of a particular name in the source code refer to the desired variable.

Declaring Variables

Variable declarations are not executable statements. Instead, they *declare*, at compile time, the names of program variables and inform the compiler of assumptions that can be made about them. Although the declarations are not themselves executable, the assumptions they produce affect the object code generated by the compiler for references to the variables declared.

You can declare variables by naming them in STATIC, LOCAL, MEMVAR, and FIELD declaration statements. They can also be declared by naming them as declared parameters (i.e., listing them in parentheses) in PROCEDURE or FUNCTION statements, or in code block definitions.

Declarations are optional for some kinds of variables (private, public and field variables) and mandatory for others (static and local variables, and declared parameters).

In conjunction with the /W compiler option, declarations allow compile-time checking for undeclared variables.

All declaration statements must occur before executable statements.

The Scope of a Declaration

A declaration's *scope* determines the parts of the program to which the declaration applies. Declaration scope is determined lexically. That is, declarations apply only to the compilation of certain sections of source code. (Note that the scope of a declaration is not necessarily the same as either the *visibility* or the *lifetime* of the variable it declares.)

For the purpose of determining declaration scope, source code is viewed as a series of discrete lexical *units*: files, procedures and code blocks. A *file* is all of the source code within a single disk file submitted for compilation. Note that code included via the #include preprocessor directive is considered part of the file containing the #include directive. A procedure is all source code from a PROCEDURE or FUNCTION statement until the next PROCEDURE or FUNCTION statement, or until the end of the file. A block is all source code within the block delimiters that define the block. Some units can be nested inside of others; procedures occur within files, for example, and code blocks occur within procedures or other code blocks.

A declaration's scope is the lexical unit in which it occurs and any nested units. For example, a declaration that occurs within a procedure definition applies to that procedure and any blocks within it. A declaration that occurs within a code block (a block parameter), applies to that code block and any code blocks nested within it.

STATIC, FIELD and MEMVAR declarations may also appear outside of any procedure or code block by specifying them before any PROCEDURE or FUNCTION statements in the source file in which case the declaration applies to the entire source file. Variables declared this way are said to have *filewide scope*. Note that to declare variables with filewide scope, the program (.prg) file must be compiled with the /N option.

A declaration in an inner unit may declare a variable with the same name as one declared in an outer unit. In this case, the inner declaration supersedes the outer, but only within the inner unit. For instance, a declaration which applies to an entire file can be superseded by a declaration in one of the procedures within that file. The superseding declaration affects only the procedure in which it occurred.

Referring to Variables

In the program source code, variables are referred to by name. When the compiler sees an occurrence of a particular name, it generates object code to assign or retrieve the value of the variable with that name.

Ambiguous Variable References

During execution, it is possible for a single name to be associated simultaneously with several different variables, some or all of which may be visible. In the absence of a declaration or an explicit qualifying alias, it may be impossible for the compiler to tell which variable a particular name will actually refer to when the program is executed. The occurrence of such a name is known as an *ambiguous variable reference*.

Variables

When the compiler encounters an ambiguous variable reference, it generates special object code (*lookup code*) which, when executed, checks to see if a variable exists with the specified name. If the name exists, it checks what kind of variable it is and whether it is visible. The order in which the various possibilities are checked depends on how the variable was referred to.

Avoiding Ambiguous Variable References

Avoiding ambiguous variable references at compile time allows the compiler to generate more efficient object code. It also increases the dependability of the program, since ambiguous references behave differently at runtime depending on a variety of circumstances.

Declarations can be used to ensure that occurrences of a particular name in the source code refer to the desired variable. Alternatively, references to variables may be explicitly *qualified* by preceding them with the alias operator.

The /V compiler switch causes ambiguous variable references to be treated as if they were preceded by the special qualifier MEMVAR->. The /W compiler switch is used to generate warnings of ambiguous variable references.

Qualifying Variable References

The alias operator (->) can be used to explicitly qualify a variable reference. The following combinations are legal:

```
<alias>-><name>
```

Refers to a field in the work area designated by *<alias>*. A runtime error occurs if *<name>* is not a field in the designated work area.

```
FIELD-><name>
```

Refers to a field in the currently selected work area. A runtime error occurs if *<name>* is not a field in the currently selected work area.

```
MEMVAR-><name>
```

Refers to either a PRIVATE variable (if one exists) or a PUBLIC variable. If no such variable exists and the reference was an attempt to assign a value, a PRIVATE variable is created at the current activation level. Otherwise a runtime error occurs.

Creation and Initialization

You may create variables and supply an initial value in the same statement. The following statements allow initializers:

- STATIC
- LOCAL
- PRIVATE
- PUBLIC

Initializers for static variables must be compile-time constant expressions. That is, the expressions must consist entirely of constants and simple operators (no variables or function calls).

Initializers for static variables are computed at compile time and are assigned to the variables before the beginning of execution.

Initializers for other variables can be any valid *CA-Clipper* expression.

Initializers for other variables are computed just prior to the creation of the variables, and assigned to the variables immediately afterward. Note that this allows a private variable to be initialized to the value held by an existing private or public variable with the same name.

If you specify no initializer, a variable is given an initial value of NIL, except for a public variable, which is given an initial value of false (.F.). For example:

```
STATIC lFirstTime := .T.                       // .T.
LOCAL i := 0, j := 1, nMax := MAXROW()         // 0, 1, 24
//  This is a string
PRIVATE cString := "This is a string"

PUBLIC lFinishedYet                            // .F.
```

The FIELD and MEMVAR statements do not allow initializers since they do not specify the creation of variables.

Local Variables

Local variables must be declared explicitly at compile time with the LOCAL statement. Local variables can be declared like this:

```
FUNCTION SomeFunc
   LOCAL nVar := 10, aArray[10][10]
   .
   . <executable statements>
   .
```

Variables

```
NextFunc()
RETURN (.T.)
```

In the above example, the variable *nVar* has been declared as LOCAL and defined as having a value of 10. When the function NextFunc() is executed, *nVar* still exists but cannot be accessed. As soon as execution of SomeFunc() is completed, *nVar* is destroyed. Any variable with the same name in the calling program is unaffected.

Local variables are created automatically each time the procedure in which they were declared is activated. They continue to exist and retain their values until the end of the activation (i.e., until the procedure returns control to the code which invoked it). If a procedure is invoked recursively (e.g., calls itself), each recursive activation creates a new set of local variables.

The visibility of a local variable is identical to the scope of its declaration. That is, the variable is visible anywhere in the source code to which its declaration applies. If a procedure is invoked recursively, only the locals created for the most recent activation are visible.

Static Variables

Static variables work much like local variables, but retain their values throughout execution and can be declared with initializers. Static variables must be declared explicitly at compile time with the STATIC statement.

Their scope depends on where they were declared. If they are declared within the body of a function, their scope is limited to that function. If they are declared outside of any function, their scope is the entire compile. Compile with the /N option to suppress the implicit procedure declaration. For example:

```
FUNCTION SomeFunc
   STATIC myVar := 10
   .
   . <statements>
   .
   NextFunc()
   RETURN (.T.)
```

In this example, the variable *myVar* has been declared as STATIC and initialized to a value of 10. When the function NextFunc() is executed, *myVar* still exists but cannot be accessed. Unlike variables declared as LOCAL or PRIVATE, *myVar* will still exist when execution of SomeFunc() completes; however, it cannot be accessed except by subsequent executions of SomeFunc().

Static variables are created automatically before the start of execution. They continue to exist and retain their values throughout the execution of the program. They cannot be released.

An initial value can be supplied in the STATIC statement. If no initial value is supplied, the static variable is initialized to NIL.

Private Variables

Declarations are optional for private variables. If desired, they may be declared explicitly at compile time with the MEMVAR statement.

Private variables are created dynamically at runtime with the PRIVATE and PARAMETERS statements. Additionally, assigning a value to a previously uncreated variable will automatically create a private variable. Once created, a private variable continues to exist and retain its value until termination of the procedure activation in which it was created (i.e., until the procedure which created it returns to the invoking procedure). At that time, it is automatically released. Private variables can also be explicitly released using the RELEASE statement.

It is possible to create a new private variable with the same name as an existing private variable. However, the new (duplicate) variable can only be created at a lower activation level than any existing private variable with the same name. The new private *hides* any existing privates (or publics, see below) with the same name.

Once created, a private variable is visible to the entire program until it is released (either explicitly or automatically, see above) or hidden by the creation of another private variable with the same name. In the latter case, the new (duplicate) private hides any existing private with the same name and the existing private becomes inaccessible until the new private is released.

In simpler terms, a private variable is visible within the creating procedure and any procedures called by the creating procedure, unless such a called procedure creates its own private variable with the same name. For example:

```
FUNCTION SomeFunc
   PRIVATE myVar := 10
   .
   . <statements>
   .
   NextFunc()
   RETURN (.T.)
```

In this example, the variable *myVar* is created as a private variable and initialized with a value of 10. When the function NextFunc() is executed, *myVar* still exists and, unlike a local variable, can be accessed by NextFunc(). When SomeFunc() terminates, *myVar* is released and any previous definition becomes accessible again.

When a private variable is created and not initialized to a value, it is given an initial value of NIL.

Public Variables

Declarations are optional for public variables. If desired, they may be declared explicitly at compile time with the MEMVAR statement.

Create public variables dynamically at runtime with the PUBLIC statement. They continue to exist and retain their values until the end of execution unless they are explicitly released with the RELEASE statement.

It is possible to create a private variable with the same name as an existing public variable. You may not, however, create a public variable with the same name as an existing private variable.

Once created, a public variable is visible to the entire program until it is explicitly released or hidden by the creation of a private variable with the same name. Then the new private variable *hides* the existing public variable of the same name and the public variable becomes inaccessible until the new private is released. For example:

```
FUNCTION SomeFunc
   PUBLIC myVar := 10
   .
   . <statements>
   .
   NextFunc()
   RETURN (.T.)
```

In this example, *myVar* is created as a public variable and initialized with a value of 10. When the function NextFunc() executed, *myVar* still exists and can be accessed. Unlike LOCAL and PRIVATE variables, *myVar* will still exist when execution of SomeFunc() is completed.

Declaring a variable PUBLIC when the variable does not already exist creates a new logical variable with an initial value of false (.F.).

Field Variables

Declarations are optional for field variables. You may declare them explicitly at compile time with the FIELD statement.

Field variables are really just synonyms for database fields of the same name. Thus, they typically exist and possess values even before the program begins execution, and they continue to exist after the program terminates. A field variable's value depends, at any given time, on which record of the associated database is the *current* record.

Once a database file is opened, the corresponding field variables are visible to the entire program. They remain visible until the database file is closed.

Field variables are initialized with empty values when records are appended to the database file.

When a variable is declared as a FIELD, all subsequent unqualified references to that variable will cause the program to look for a database field in the current work area with that name.

Expressions

In *CA-Clipper*, all data items are identified by type for the purpose of forming expressions. The most basic data items (i.e., variables, constants, and functions) are assigned data types depending on how the item is created. For example, the data type of a field variable is determined by its database file structure, the data type of a constant value depends on how it is formed, and the data type of a function is defined by its return value.

These basic data items are the simplest expressions in the *CA-Clipper* language. More complicated expressions are formed by combining the basic items with operators. This section defines all of the valid *CA-Clipper* data types and operators that you will use to form expressions, as well as any special rules you need to know.

Data Types

The data types supported in the *CA-Clipper* language are:

- Array
- Character
- Code Block
- Numeric
- Date
- Logical
- Memo
- NIL

In *CA-Clipper*, an array structure is as a separate data type, and there are distinct operations for arrays that are invalid for other data types. Code blocks are also a data type, but the operations that you can perform using them is limited. Separate sections in this chapter discuss Arrays and Code Blocks.

This section discusses each of the simple data types, providing you with the following information: when to use the data type; how to form its constants, or literals; what limitations apply; and what operations are available.

Character

The character data type identifies data items you want to manipulate as fixed length strings. The *CA-Clipper* character set is defined as the entire printable ASCII character set (i.e., CHR(32) through CHR(126)), the extended ASCII graphics character set (i.e., CHR(128) through CHR(255)), and the null character, CHR(0).

Valid character strings consist of zero or more characters in the defined character set with a maximum of 65,535 characters per string (i.e., 64K minus one character used as a null terminator).

Character string literals, or constants, are formed by enclosing a valid string of characters within a designated pair of delimiters. In the *CA-Clipper* language, the following delimiter pairs are designated:

- two single quotes (e.g., 'one to open and one to close')

- two double quotes (e.g., "one to open and one to close")

- left and right square brackets (e.g., [left to open and right to close])

Note: To express a null string, use only the delimiter pair with no intervening characters—including spaces. For example, "" and [] both represent a null string.

Since all of the designated delimiter characters are part of the valid character set, the delimiter characters themselves may be part of a character string. If you want to include any of these characters in a character string literal, you must use an alternate character for the delimiter. For example, the following string:

I don't want to go.

could be expressed as:

"I don't want to go."

Similarly, the string:

She said, "I have no idea."

can be represented with:

'She said, "I have no idea."'

This constraint should not be too limiting since you have three pairs of delimiters from which to choose.

The following table describes all operations in the *CA-Clipper* language that are valid for the character data type. These operations act on one or more character expressions to produce a result. The result is not necessarily a character data type:

Data Types

Character Operations

Operation	Description
+	Concatenate
-	Concatenate without intervening spaces
=	Compare for equality
==	Compare for exact equality
!=,<>, #	Compare for inequality
<	Compare for sorts before
<=	Compare for sorts before or the same as
>	Compare for sorts after
>=	Compare for sorts after or the same as
= or STORE	Assign
:=	Inline assign
+=	Inline concatenate (+) and assign
-=	Inline concatenate (-) and assign
$	Test for substring existence
REPLACE	Replace field value
ALLTRIM()	Remove leading and trailing spaces
ASC()	Convert to numeric ASCII code equivalent
AT()	Locate substring position
CTOD()	Convert to date
DESCEND()	Convert to complemented form
EMPTY()	Test for null or blank
ISALPHA()	Test for initial letter
ISDIGIT()	Test for initial digit
ISLOWER()	Test for initial lowercase letter
ISUPPER()	Test for initial upper case letter
LEFT()	Extract substring from the left
LEN()	Compute length
LOWER()	Convert letters to lowercase
LTRIM()	Remove leading spaces
PADC()	Pad with leading and trailing spaces
PADC()	Pad with leading and trailing spaces
PADL()	Pad with leading space
RAT()	Locate substring position starting from the right
REPLICATE()	Duplicate
RIGHT()	Extract substring from the right

Data Types

Character Operations (cont.)

Operation	Description
RTRIM()	Remove trailing spaces
SOUNDEX()	Convert to soundex equivalent
SPACE()	Create a blank string
STRTRAN()	Search and replace substring
STUFF()	Replace substring
SUBSTR()	Extract substring
TRANSFORM()	Convert to formatted string
TYPE()	Evaluate data type using macro substitution
UPPER()	Convert letters to upper case
VAL()	Convert to numeric
VALTYPE()	Evaluate data type directly

Note: The memo data type discussed later in this section represents variable length character data. It is a true variable length data type that can only exist in the form of a database field. You may also use the character operations listed in this section with memo fields.

Character strings are compared according to the ASCII collating sequence. Memo fields can also be compared to character strings. When EXACT is OFF two character strings are compared according to the following rules; assume two character strings A and B where the expression to test is (A = B):

- If B is null, return true (.T.).
- If LEN(B) is greater than LEN(A), return false (.F.).
- Compare all characters in B with A. If all characters in B equal A, return true (.T.); otherwise return false (.F.).

With EXACT ON, two strings must match exactly, except for trailing blanks.

Memo

The memo data type is used to represent variable length character data. It is a true variable length data type that can only exist in the form of a database field. A memo field uses ten characters in the database record and is used as a pointer to the actual data which is stored in a separate (.dbt) file.

The contents of a memo (.dbt) file are handled in blocks of 512 bytes. Each memo field in the database (.dbf) file contains the block number in

ASCII which identifies the memo field data location. If the memo field contains no data there is no number in the database (.dbf) file. When the user writes to a memo field, the next available block is used and its number is stored in the block number field.

In *CA-Clipper*, when you change a memo field with less than 512 bytes, the existing block is used until it is filled. Once full, the block is discarded and copied to a new location.

Besides the fact that the length may vary in a memo field from one record to another, memo fields are identical to character fields. The character set is identical, the 65,535 maximum character limitation stands, and comparisons are performed in the same way.

Since it can apply only to a field, there is no literal representation for the memo data type. However, you can assign character string literals to memo fields with REPLACE.

The following table lists all operations in the *CA-Clipper* language that are valid for the memo data type. The operations listed here are those designed to work specifically with memo fields, but you may also use them with long character strings:

Memo Operations

Operation	Description
HARDCR()	Replace soft with hard carriage returns
MEMOEDIT()	Edit contents
MEMOLINE()	Extract a line of text
MEMOREAD()	Read from a disk file
MEMOTRAN()	Replace soft and hard carriage returns
MEMOWRIT()	Write to a disk file
MLCOUNT()	Count lines
MLPOS()	Compute position

In addition to these specialized memo operations, you may use all of the character operations listed previously with memo fields.

Date

The date data type is used to identify data items that represent calendar dates. In *CA-Clipper*, you can manipulate dates in several ways, such as finding the number of days between two dates and determining what the date will be ten days from now. The date character set is defined as the digits from zero to nine and a separator character specified by SET DATE.

CA-Clipper adds the SET EPOCH command to the language to change the default century assumption. See the entry for this command in the *Language Reference* chapter of the *Reference* guide for more information.

Form dates by stringing together digits and separators in a particular order that depends on the current SET DATE value. By default, SET DATE is AMERICAN and dates are of the form *mm/dd/[cc]yy*: *mm* represents the month and must be a value between 1 and 12; *dd* represents the day which must fall between 1 and 31; *cc*, if specified, represents the century and must be between 0 and 29; *yy* represents the year which must fall between 0 and 99; and / is the separator.

CA-Clipper supports all valid dates in the range 01/01/0100 to 12/31/2999 as well as a null, or blank, date. You may specify the century as part of a literal date regardless of the status of SET CENTURY; however, you will not be able to enter non-twentieth century dates in @...GET variables, nor will you be able to display the century unless SET CENTURY is ON.

Dates do not have a straightforward literal representation like character strings and numbers. Instead, you must use the CTOD() function to convert a character string constant expressed in date form into an actual date value. To do this, you form the date according to the rules described above and enclose it in one of the character string delimiter pairs. Then, you use the character date as the CTOD() argument as illustrated in the examples below:

```
CTOD("1/15/89")
CTOD('03/5/63')
CTOD([09/28/1693])
CTOD("01/01/0100")
```

CTOD() checks its argument to be sure it is a valid date. For example, "11/31/89" and "02/29/90" would not be valid dates because November has only 30 days, and 1990 is not a leap year. CTOD() converts an invalid date to a null date.

Note: To express a blank, or null date, use a null string as the argument for CTOD() (e.g., CTOD("")).

The following table lists all operations in the *CA-Clipper* language that are valid for the date data type. These operations act on one or more date expressions to produce a result. The result is not necessarily a date data type:

Data Types

Date Operations

Operation	Description
+	Add a number to a date
-	Subtract from a date
++	Increment
--	Decrement
= or ==	Compare for equality
!=, <>, or #	Compare for inequality
<	Compare for earlier
<=	Compare for earlier or the same as
>	Compare for later
>=	Compare for later or the same as
= or STORE	Assign
:=	Inline assign
+=	Inline add and assign
-=	Inline subtract and assign
REPLACE	Replace field value
CDOW()	Compute day of week name
CMONTH()	Compute month name
DAY()	Extract day number
DESCEND()	Convert to complemented form
DOW()	Compute day of week number
DTOC()	Convert to character in SET DATE format
DTOS()	Convert to character in sorting format
EMPTY()	Test for null
MONTH()	Extract month number
TRANSFORM()	Convert to formatted string
TYPE()	Evaluate data type using macro substitution
VALTYPE()	Evaluate data type directly
YEAR()	Extract entire year number, including century

Dates are compared according to their chronological order. Any nonblank date will always compare as greater than a blank date. Less than and/or equality comparisons of a nonblank date to a blank date will always return false (.F.).

Numeric

The numeric data type is used to identify data items that you want to manipulate mathematically (e.g., perform addition, multiplication, and other mathematical functions). The *CA-Clipper* numeric character set is defined as the digits from zero to nine, the period to represent a decimal point, and the plus and minus to indicate the sign of the number.

With the exception of fields, *CA-Clipper* stores numeric values using the IEEE standard double precision floating point format, making the range of numbers from 10^{-308} to 10^{308}. Numeric precision is guaranteed up to 16 significant digits, and formatting a numeric value for display is guaranteed up to a length of 32 (i.e., 30 digits, a sign, and a decimal point). This means that numbers longer than 32 bytes may be displayed as asterisks, and digits other than the 16 most significant ones are displayed as zeroes.

When a value is stored in a numeric field of a database (.dbf) file, it is converted from IEEE format to a displayable representation. When a numeric field is retrieved from a file, it is converted back to IEEE format before any operations are performed on it. Since the displayable format of a numeric value is guaranteed up to a length of 32 bytes, this is the largest recommended numeric field length.

Form numeric literals by stringing together one or more of the following: a single leading plus or minus sign, one or more digits to represent the whole portion of the number, a single decimal point, and one or more digits to represent the fractional part of the number. Some valid numeric constants are shown below:

- 1234
- 1234.5678
- -1234
- +1234.5678
- -.5678

Note: Unlike character strings, literal numeric values are not delimited. If you enclose a number—or any other string of characters—in delimiters, it becomes a character string.

The following table lists all operations in the *CA-Clipper* language that are valid for the numeric data type. These operations act on one or more numeric expressions to produce a result. The result is not necessarily a numeric data type:

Data Types

Numeric Operations

Operation	Description
+	Add or Unary Positive
-	Subtract or Unary Negative
*	Multiply
/	Divide
%	Modulus
++	Increment
--	Decrement
= or ==	Compare for equality
!=, <>, or #	Compare for inequality
<	Compare for less than
<=	Compare for less than or equal
>	Compare for greater than
>=	Compare for greater than or equal
= or STORE	Assign
:=	Inline assign
+=	Inline add and assign
-=	Inline subtract and assign
*=	Inline multiply and assign
/=	Inline divide and assign
^	Inline exponentiate and assign
%=	Inline modulus and assign
REPLACE	Replace field value
ABS()	Compute absolute value
CHR()	Convert to ASCII character equivalent
DESCEND()	Convert to complemented form
EMPTY()	Test for zero
EXP()	Exponentiate with *e* as the base
INT()	Convert to integer
LOG()	Compute natural logarithm
MAX()	Compute maximum
MIN()	Compute minimum
MOD()*	Compute dBASE III PLUS modulus
ROUND()	Round up or down

Data Types

Numeric Operations (cont.)

Operation	Description
SQRT()	Compute square root
STR()	Convert to character
TRANSFORM()	Convert to formatted string
TYPE()	Evaluate data type using macro substitution
VALTYPE()	Evaluate data type directly

Numbers are compared according to their actual numeric value. Note that the = operator and the == operator behave identically when comparing numbers, they both check for equality to the maximum precision allowed by the IEEE 8-byte floating point format.

Numeric operations, including comparisons, are completely unaffected by any SET command. If numeric operations or comparisons appear to produce unexpected results, it is probably because of the automatic display formatting of floating point values. The display formatting is affected by the SET FIXED and SET DECIMALS commands.

Logical

The logical data type identifies data items that are Boolean in nature. Typical logical data items are those with values of true or false, yes or no, or on or off. In *CA-Clipper*, the logical character set consists of the letters y, Y, t, T, n, N, f, and F.

Though only two logical values are possible, in *CA-Clipper* there are several ways to represent these two values. To form a literal logical value, enclose one of the characters in the defined logical character set between two periods. The periods are delimiters for logical values just as quote marks are for character strings.

The literal values .y., .Y., .t., and .T. represent true. The literal values .n., .N., .f., and .F. represent false. The preferred literal representations are .T. and .F.

The following table lists all operations in the *CA-Clipper* language that are valid for the logical data type. These operations act on one or more logical expressions to produce a result. The result is not necessarily a logical data type:

Data Types

Logical Operations

Operation	Description
.AND.	And
.OR.	Or
.NOT. or !	Negate
= or ==	Compare for equality
!=, <>, or #	Compare for inequality

For the purpose of comparing logical values, false (.F.) is always less than true (.T.).

NIL

NIL is a new data type that allows you to manipulate uninitialized variables without generating a runtime error. It is a data type with only one possible value, the NIL value.

The literal representation for NIL is the string of characters "NIL" without delimiters. When referenced by a display command or function, this is how NIL values display. Note also that NIL is a reserved word in *CA-Clipper*.

The following table lists all valid operations in the *CA-Clipper* language for the NIL data type. Except for these operations, attempting to operate on a NIL results in a runtime error. For example, you cannot REPLACE a field with NIL, nor can you concatenate NIL to a character string:

NIL Operations

Operation	Description
= or ==	Compare for equality
!=, <>, or #	Compare for inequality
<	Compare for less than
<=	Compare for less than or equal
>	Compare for greater than
>=	Compare for greater than or equal
= or STORE	Assign to a nonfield variable
:=	Inline assign to a nonfield variable
EMPTY()	Test for NIL
TYPE()	Evaluate data type using macro substitution
VALTYPE()	Evaluate data type directly

For the purpose of comparison, NIL is the only value that is equal to NIL. All other values are greater than NIL.

In addition to the NIL operations, there are several declaration statements that automatically assign a NIL value to variables and array elements. These statements are listed in the following table and discussed below:

NIL Declaration Assignments

Operation	Description
DECLARE	Assign NIL to array elements
LOCAL	Assign NIL to local variables and array elements
PARAMETERS	Assign NIL to missing parameters
PRIVATE	Assign NIL to private variables and array elements
PUBLIC	Assign NIL to public array elements
STATIC	Assign NIL to static variables and array elements

All variables, with the exception of PUBLIC and FIELD variables, are initialized to NIL when declared or created without an initializer. PUBLIC variables are initialized to false (.F.) when created, and FIELD variables cannot contain NIL values at all. When you create an array with a declaration statement, including PUBLIC, all elements are initialized to NIL.

When you invoke a function or procedure, if you omit an argument either by leaving an empty spot next to a comma or by leaving an argument off the end, a NIL value is passed for that argument. Note, however, that the function or procedure cannot distinguish between a NIL value that is passed explicitly (e.g., an expression in the argument list that evaluates to NIL) and one that is passed as the result of omitting an argument.

Operators

Along with functions, constants, and variables, operators are the basic building blocks of expressions. An operator is like a function in that it performs a specific operation and returns a value. This section gives a general discussion of operators, and categorizes and describes all of the operators available in the *CA-Clipper* language.

Definitions

The following paragraphs discuss, briefly, several terms used regarding operators in order to help you understand the remaining material in this section.

Overloading

In *CA-Clipper*, the meaning of certain operators changes, depending on the data type of the operand(s). For example, the minus operator (-) can be used to concatenate character strings as well as subtract numeric or date values. Using the same symbol, or operator, for different operations is called *overloading*.

Unary and Binary Operators

All operators in *CA-Clipper* require either one or two arguments, called *operands*. Those requiring a single operand are called *unary operators*, and those requiring two operands are called *binary operators*.

Binary operators use *infix* notation which means that the operator is placed between its operands. Unary operators use *prefix* or *postfix* notation where the operator is placed either before or after the operand, depending on the operator and how you want to use it.

An example of a unary prefix operator is the logical negation (!) operator:

```
lTrue := .T.
? !lTrue              // Result: .F.
```

The postincrement operator (++) is an example of a unary postfix operator. This operator increments its operand by a value of one:

```
nCount := 1
nCount++              // Result: nCount is now 2
```

The multiplication operator (*) is an example of a binary operator which demonstrates infix notation:

```
? 12 * 12             // Result: 144
```

Precedence

Precedence rules determine the order in which different operators are evaluated within an expression. These rules define the hierarchy of all of the operators within an expression. Parentheses can be used to override the default evaluation order and to make complicated expressions more readable.

Error Handling

Operators are used to form expressions and must, therefore, adhere to certain rules. For example, you cannot use a unary operator that allows only prefix notation as a postfix operator and, except in a few well-defined circumstances, you cannot use any binary operator on operands with different data types. All of the rules that apply to operator usage are described in this section, and using any operator incorrectly results in a runtime error.

String Operators

The following table lists each string operator and the calculation that it performs. These operators are binary, requiring two character and/or memo type operands, and return a character value:

String Operators

Symbol	Operation
+	Concatenate
-	Concatenate without intervening spaces

The - concatenation operator moves the trailing spaces of the first operand to the end of the resulting string, so that there are no intervening spaces between the two original strings. The + operator leaves spaces intact.

Date Operators

The + and - mathematical operators discussed in the next section may be used to add or subtract a number of days from a date value. The result of such an operation is a date that is a number of days before (subtraction) or after (addition) the date value. The order of the operands in the case of subtraction is significant—the date value must come first. This is an exception to the rule against operations using mixed data types. In the precedence rules defined at the end of this section, this type of mixed operation is considered to be mathematical rather than date.

You may also subtract one date from another using the mathematical subtraction operator. The result of this type of operation is a numeric value that represents the number of days between the two dates. Subtraction of one date from another is the only true date operator.

Operators

Mathematical Operators

The following table lists each mathematical operator and the calculation it performs. As a general rule, these operators require numeric type operands (see Date Operators above for exceptions) and return a numeric value. Except where noted in the table, the mathematical operators are binary. The unary operations use prefix notation:

Mathematical Operators

Symbol	Operation
+	Addition or Unary Positive
-	Subtraction or Unary Negative
*	Multiplication
/	Division
%	Modulus
** or ^	Exponentiation

The modulus operator returns the remainder of a division operation. If the remainder is desired as the result of a division, use the modulus operator (%) instead of the division operator (/).

Relational Operators

Below is a list of relational operators and their purpose. All of them are binary operators requiring either two operands of the same data type or at least one NIL operand. The result of a relational operator is a logical value:

Relational Operators

Symbol	Operation
<	Less than
>	Greater than
=	Equal
==	Exactly equal for character; equal for other data types
<>, #, or !=	Not equal
<=	Less than or equal
>=	Greater than or equal
$	Is contained in the set or is a subset of

SET EXACT affects both the $ and the = operator when comparing strings. If SET EXACT is ON, both strings must be identical in content

2-44 CA-Clipper

and length (except trailing spaces) in order for these operations to return true (.T.).

The == operator compares strings (i.e., character and memo types) for exact equality in length and content, including trailing spaces. For arrays, it checks that both arrays refer to the same area of memory. For all other data types, the == operator is equivalent to the = operator.

Logical Operators

The following table lists the logical operators:

Logical Operators

Symbol	Operation
.AND.	Logical and
.OR.	Logical or
.NOT. or !	Logical negate

All of the logical operators require logical operands and, except for negate, they are binary. .NOT. and ! are unary prefix operators. The result of a logical operation is always a logical value.

The quick way to define these operators is to tell when they return true: .AND. returns true if both operands are true; .OR. returns true if either operand is true, and .NOT. returns true if its operand is false. Formal truth tables defining all of the logical operators are shown in the table below:

Truth Chart

.AND.	.T.	.F.	.OR.	.T.	.F.	.NOT.	.T.	.F.
.T.	.T.	.F.	.T.	.T.	.T.		.F.	.T.
.F.	.F.	.F.	.F.	.T.	.F.			

Assignment Operators

CA-Clipper has several operators that assign values to variables and these are summarized in the table below:

Assignment Operators

Symbol	Operation
=	Assign
:=	Inline assign
+=	Inline add (or concatenate) and assign
-=	Inline subtract (or concatenate) and assign
*=	Inline multiply and assign
/=	Inline divide and assign
**= or ^=	Inline exponentiate and assign
%=	Inline modulus and assign

All assignment operators are binary and require a single variable as the first operand. The second operand may be any valid expression, including NIL, provided that the data type is suitable for the operation. For the compound operators (i.e., all except = and :=), the variable used as the first operand must exist and must be the same data type as the second operand, except for some exceptions with date and numeric operations. These exceptions are described after the following table which summarizes the assignment operators by data type:

Assignment Operators By Data Type

Operator	Valid Data Types
=	All
:=	All
+=	Character, Date, Memo, Numeric
-=	Character, Date, Memo, Numeric
*=	Numeric
/=	Numeric
**= or ^=	Numeric
%=	Numeric

In addition to these data types, -= and += allow a date expression as the first operand and a numeric expression as the second operand. The result is a number of days added or subtracted from the date and the new value assigned.

The assignment operators in CA-Clipper always assume that the assigned variable is a MEMVAR unless it is explicitly declared as a different storage class or explicitly qualified with an alias. The "field first" rule does not apply when assigning. This is compatible with previous versions of *CA-Clipper*.

Note: With respect to the storage class of variables, there is no difference between := and =. (The only difference between := and = is that := can be used as part of an expression since it is not confused with an equality test.) Any assignment operator assumes MEMVAR if the variable reference is ambiguous. If an attempt is being made to assign to a field, the field variable must be either declared with the FIELD statement or referenced with an alias--either the assigned alias or the FIELD-> alias.

Simple Assignment

The simple assignment operator (=) assigns a value to a variable. It is identical in operation to the STORE command that initializes a single variable, and must be specified as a program statement. If used within an expression it is interpreted as the equality operator. For example:

```
nValue = 25
nNewValue = SQRT(nValue) ** 5
nOldValue = nValue
```

are all valid simple assignment statements. The first operand, as with all other assignment statements, can be a database field as well as any other variable. The result is that the field is REPLACEd with the new value. For example, the following two lines of code perform exactly the same function of replacing the current value in the *CustAge* field with the numeric value 20:

```
FIELD->CustAge = 20
REPLACE CustAge WITH 20
```

Inline Assignment

You can use the assignment operator (:=) interchangeably with the simple assignment operator. For example:

```
nValue := 25
nNewValue := SQRT(nValue) ** 5
nOldValue := nValue
```

duplicates the previous example. This operator, however, is an *inline* operator which means it can be used in any command or function whose syntax allows the use of an expression. The data type of an inline assignment operation is the same as the data type of the second operand.

Operators

Several examples of inline assignment follow:

```
LOCAL nValue := 10
//
IF (dDate := (DATE() - 1000)) = CTOD("12/20/79")
//
? SQRT(nValue := (nValue ** 2))
//
cTransNo := cSortNo := (CustId + DTOC(DATE()))
```

This last example demonstrates multiple assignments using the inline assignment operator as you would use the STORE command. When := is used in this way, the assignments are executed from right to left.

This feature is particularly useful when you need to store the same value to many different fields, possibly in different database files:

```
CustFile->CustId := TransFile->TransNo := ;
      (CustId + DTOC(DATE()))
```

Compound Assignments

The compound assignment operators perform an operation between the two operands and then assign the result to the first operand. These operators are summarized and defined in the following table:

Compound Assignment Operator Definitions

Operator	Example	Definition
+=	a += b	a := (a + b)
-=	a -= b	a := (a - b)
*=	a *= b	a := (a * b)
/=	a /= b	a := (a / b)
%=	a %= b	a := (a % b)
**= or ^=	a ^= b	a := (a ^ b)

Note that the definitions for these operators use the inline assignment operator. This means that all of the compound operators are also inline operators and can be used as such. The data type for these operations is determined by the second operand using the definitions in the table above. If the assignment operation is formed correctly, this should be the data type of the first operand in the original compound assignment statement.

Note: It is not advisable to use compound assignment operators with REPLACE or UPDATE since the WITH clause provides the assignment function as part of the command syntax.

Increment and Decrement Operators

The increment (++) and decrement (--) operators, new to *CA-Clipper*, are summarized in the table below:

Increment and Decrement Operators

Symbol	Operation
++	Prefix or Postfix Increment
--	Prefix or Postfix Decrement

Both are unary operators you can use with either a numeric or a date operand. Unlike other operators which can operate on more complicated expressions, the operand here must be a single, nonfield variable. The resulting data type is the same as that of the operand.

The ++ operator *increments*, or increases the value of, its operand by one, and the -- operator *decrements*, or decreases the value of, its operand by one. Thus, both operators perform an operation on, as well as an assignment to, the operand. The following shows how these operators might be defined in terms of addition (and subtraction) and assignment operators:

- *value++* is equivalent to *value := value + 1*
- *value--* is equivalent to *value := value - 1*

Both operators may be specified as prefix or postfix: the prefix form changes the value of the operand before the assignment is made, whereas the postfix form changes the value afterwards. In other words, the postfix form delays the assignment portion of the operation until the rest of the expression is evaluated, and the prefix form gives the assignment precedence over all other operations in the expression.

The following code illustrates the preincrement operator in an assignment statement:

```
nValue := 1
nNewValue := ++nValue
? nNewValue        // Result:   2
? nValue           // Result:   2
```

Since the increment occurs before the assignment takes place, both variables have the same value. The next example demonstrates the postdecrement operator:

```
nValue := 1
nNewValue := nValue--
? nNewValue        // Result:   1
? nValue           // Result:   0
```

Because the assignment takes place before the original variable is decremented, the two values are not the same. The next example further illustrates the difference in the prefix and postfix forms of these operators:

```
nValue := 10
? nValue++ * nValue     // Result: 110
? nValue                // Result:  11
```

Here, the postincrement operator is used to increase the first operand of the multiplication operation by one, making its value 11; however, the assignment of this new value to the *nValue* variable will not take place until the entire expression is evaluated. Thus, its value is still 10 when the multiplication operation occurs, and the result of 11 * 10 is 110. Finally, when *nValue* is queried again after the expression is evaluated, the postincrement assignment is reflected in its new value, 11.

In this last example, the predecrement operator is used to decrease the first operand of the multiplication by one, making its value 9. Additionally, the assignment of this new value to the *nValue* variable takes place before the rest of the expression is evaluated. Thus, the new value is used to perform the multiplication operation, and the result of 9 * 9 is 81:

```
nValue := 10
? --nValue * nValue     // Result:  81
? nValue                // Result:   9
```

Special Operators

The following table lists all other symbols that have special meaning in the *CA-Clipper* language. These are special operators that often appear in expressions:

Special Operators

Symbol	Operation
()	Function or grouping
[]	Array element
{}	Constant array definition
->	Alias identifier
&	Compile and execute
@	Pass by reference

Parentheses () are used in expressions either to group certain operations in order to force a particular evaluation order or to indicate a function call. When specifying the grouping operator, the item that falls within the parentheses must be a valid expression. Subexpressions may be further grouped. For function calls, a valid function name must precede the left parenthesis, and the function arguments, if any, must be contained within the parentheses.

The subscript operator ([]) is used to reference a single array element. The name of a previously declared array must precede the left bracket and the array element index must appear as a numeric expression within the brackets.

Curly braces ({}) are used to create and reference a literal array. The array elements must be within the braces and separated by commas. Curly braces are also used to create code blocks. The code block arguments are further delimited within the curly braces with vertical bars (| |), and the expressions defining the code block are separated by commas. Though the vertical bars are required delimiters, they need not contain an argument.

The alias identifier (->) is used to make an explicit reference to a field or variable. If the name of an alias precedes the operator, it may be followed either by a field name from that database file or any other expression enclosed in parentheses. Additionally, the keywords FIELD and MEMVAR may precede the operator followed by a valid field or variable identifier.

The macro symbol (&) is the compile-and-run operator. It is a unary prefix operator whose only valid operand is a character variable.

The pass-by-reference operator (@) is valid only in the argument lists of function or procedure calls using the function-calling syntax. It is a unary prefix operator whose operand may be any variable name. It works by forcing the argument to be passed by reference rather than by value to the function or procedure.

Operator Precedence

When evaluating expressions with two or more operations that are not explicitly grouped together with parentheses, *CA-Clipper* uses an established set of rules to determine the order in which the various operations are evaluated. These rules, called precedence rules, define the hierarchy of all of the operators discussed so far in this section.

Note: All function calls and other special operators in an expression are evaluated before any other operators

Operators

Precedence of Categories

For the most part, expressions in *CA-Clipper* are operations that manipulate a single data type. For example, an expression might concatenate several character strings or perform a series of mathematical operations on several numbers. There are, however, expressions in which the evaluation of several different types of operations is necessary. For example, a complex logical expression may involve several related operations, on different data types, that are connected with logical operators as shown in this example:

```
cString1 $ cString2 .AND. nVal1++ > nVal2 * 10
```

When more than one type of operator appears in an expression, all of the subexpressions are evaluated for each precedence level before subexpressions at the next level are evaluated. Except for multiple inline assignments, all operations at each level are performed in order from left to right; multiple inline assignments are performed from right to left. The order of precedence for the operators, by category, is as follows:

1. Preincrement and Predecrement
2. Mathematical
3. Relational
4. Logical
5. Assignment
6. Postincrement and Postdecrement

The nonassignment portion of the compound assignment operators (e.g., the multiplication portion of *=) exists at level 2, and the assignment portion exists at level 5.

Precedence within a Category

Within each category, the individual operators also have an established precedence which is critical, especially in mathematical operations. Take the following two expressions:

- 5 * 10 + 6 / 2
- 6 / 2 + 5 * 10

Algebraically, these are equivalent expressions, though worded differently, the result of both expressions is 53. If *CA-Clipper* did not establish an order of precedence for evaluating mathematical operators, you could not evaluate this expression correctly without using parentheses.

Without precedence among operators, all mathematical operators would be performed in order from left to right, and the two forms of the expression shown above would be evaluated as follows: the first would multiply 5 * 10, add the resulting 50 to 6, and divide the resulting 56 by 2 to obtain 28; the second would divide 6 by 2, add the resulting 3 to 5, and multiply the resulting 8 by 10 to obtain 80. Neither result is correct, and the two results are not the same.

To save you from having to use parentheses in complicated expressions and to insure that equivalent expressions give the same result, *CA-Clipper* uses an established order for evaluating operations in each category.

Preincrement and Predecrement

As mentioned earlier, the prefix and postfix forms of the increment and decrement operators are considered as separate categories because they have distinct precedence levels. This category refers to the prefix form of the increment and decrement operators.

Both operators in this category (++ and --) exist at the same precedence level and are performed in order from left to right.

Mathematical

When more than one mathematical operator appears in an expression, all of the subexpressions are evaluated for each precedence level before subexpressions at the next level are evaluated. All operations at each level are performed in order from left to right. The order of precedence for the mathematical operators is as follows:

1. Unary positive and negative (+, -)

2. Exponentiation (**, ^)

3. Multiplication, division, and modulus (*, /, %)

4. Addition and subtraction (+, -)

Relational

All of the relational operators exist at the same precedence level and are performed in order from left to right.

Logical

Like the mathematical operators, the logical operators also have an established order of precedence. When more than one logical operator appears in an expression, all of the subexpressions are evaluated for each precedence level before subexpressions at the next level are evaluated. All operations at each level are performed in order from left to right. The order of precedence for the logical operators is as follows:

1. Unary negate (.NOT.)
2. Logical and (.AND.)
3. Logical or (.OR.)

Assignment

All of the assignment operators exist at the same precedence level and are performed in order from right to left. For the compound operators, the nonassignment portion (e.g., addition or concatenation) of the operation is performed first, followed immediately by the assignment.

The increment and decrement assignment operations exist at their own level of precedence, and are not part of assignment category.

Postincrement and Postdecrement

Both operators in this category (++ and --) exist at the same precedence level and are performed in order from left to right.

Parentheses

You can override the order of precedence for expression evaluation using parentheses. When parentheses are present in an expression, all subexpressions within parentheses are evaluated first using the precedence rules described in this section, if necessary. If the parentheses are nested, the evaluation is done starting with the innermost pair and proceeding outward.

Note that although the *CA-Clipper* language provides a specific order of precedence for evaluating expressions, it is better programming practice to explicitly group operations for readability and to be certain that what executes meets your expectations.

The Macro Operator (&)

The macro operator in *CA-Clipper* is a special operator that allows runtime compilation of expressions and text substitution within strings. Whenever the macro operator (&) is encountered, the operand is submitted to a special runtime compiler referred to as the macro compiler that can compile expressions, but not statements or commands.

You can specify a macro using a variable containing a character string or an expression that returns a character string. If a character variable is used to specify a macro, it is referred to as a *macro variable* and specified like this:

```
&<macroVar>.
```

The period (.) is the macro terminator and indicates the end of the macro variable and distinguishes the macro variable from the adjacent text in the statement.

You can also apply the macro operator (&) to an expression, referred to as a *macro expression*, if the expression evaluates to a character value and is enclosed in parentheses, like this:

```
&(<macroExp>)
```

In this instance, the expression is evaluated first, and the macro operation is performed on the resulting character value. This allows the contents of fields and array elements to be compiled and run.

A macro operator (&) can perform text substitution or runtime compilation of an expression depending on how it is specified.

Text Substitution

Whenever a reference to a private or public macro variable is encountered embedded within a character string, like this:

```
cMacro := "there"
? "Hello &cMacro"              // Result: Hello there
```

the contents of the macro variable are substituted for the variable reference. If a macro expression is specified (e.g., &(cMacro1 + cMacro2), and the macro variable is a local, static, field variable, or an array element, it is treated as literal text and not expanded.

The Macro Operator (&)

Compile and Run

When *CA-Clipper* encounters a macro variable or macro expression within an expression, it treats it like an expression and the macro symbol behaves as the compile-and-run operator. If the macro is specified as a macro variable like this:

```
cMacro := "DTOC(DATE())"
? &cMacro
```

the contents of the macro variable is compiled by the macro compiler and then executed. The compiled code is then discarded. If an expression is specified enclosed in parentheses and prefaced by the macro operator (&) like this:

```
? &(INDEXKEY(0))
```

the expression is evaluated and the resulting character string is compiled and run just as a macro variable is.

One of the interesting effects of macro expressions is that you can compile a character string containing a code block definition, like this:

```
bBlock := &("{ |exp| QOUT(exp) }")
   .
   .
   .
EVAL(bBlock, DATE())
```

In this case, the character string containing the code block is compiled and the run portion of the operation returns the code block as a value. The resulting code block is saved and evaluated later with the EVAL() function.

Relationship To Commands

Because of the long history of the macro operator (&) in *CA-Clipper* and its antecedents, it is important to understand the precise nature of the relationship between commands and macros.

Using with Command Keywords

First, the macro operator (&) *cannot* be used to substitute or compile command keywords. In *CA-Clipper*, this is because commands are preprocessed into statements and expressions at compile time.

Note: Redefinition of command keywords, however, can be accomplished by either modifying the command definition in STD.CH, overriding an existing command definition with a new definition using

The Macro Operator (&)

the #command directive, or redefining a command keyword using the #translate directive. In any case, redefinition of a command keyword can only occur at compile time rather than at runtime.

Using with Command Arguments

Second, in prior versions of *CA-Clipper* as well as other dialects, macro variables were often used to specify the arguments of commands requiring literal text values. This included all file command arguments as well as SET commands with toggle arguments. In these instances, you can now use an *extended expression* in place of the literal argument if the expression is enclosed in parentheses. For example, the following:

```
xcDatabase = "Invoices"
USE &xcDatabase.
```

can be replaced with:

```
xcDatabase = "Invoices"
USE (xcDatabase)
```

It is important to use extended expressions, especially if you are using local and static variables. Commands are generally preprocessed into function calls with command arguments translated into function arguments as legal *CA-Clipper* values. With file commands, for instance, filenames are *stringified* using the smart stringify result marker and passed as arguments to the functions that actually perform the desired actions. If a literal or macro value is specified as the command argument, it is stringified. If, however, the argument is an extended expression, it is written to the result text exactly as specified. For example, the following specifications of the RENAME command:

```
#command RENAME <xcOld> TO <xcNew> =>;
      FRENAME( <(xcOld)>, <(xcNew)> )
//
RENAME &xcOld TO &xcNew
RENAME (xcOld) TO (xcNew)
```

are written to the result text as this:

```
FRENAME("&xcOld", "&xcNew")
FRENAME(xcOld, xcNew)
```

when preprocessed. When the macro variables are stringified, the macro variable names are hidden in the string and not compiled. Later, at runtime, they are substituted into the string and passed as arguments to the FRENAME() function. This precludes local and static macro variables since the names of the variables are not present at runtime to be substituted. Public and private variables, however, behave as you might expect. If this seems somewhat confusing, refer to the *Variables* section for more information about the difference between local and static variables, and private and public variables.

The Macro Operator (&)

Note: Extended expressions are denoted in command syntax with metasymbols prefaced by a lowercase x.

Using with Lists

The macro operator (&) will not fully substitute or compile a list as an argument of most commands, particularly those commands in which an argument list is preprocessed into an array or a code block. Instances of this are arguments of the FIELDS clause and SET INDEX. An exception is the SET COLOR command which preprocesses the list of colors into a single character string and passes it to the SETCOLOR() function.

In any case, list arguments should always be specified as extended expressions with each list argument specified:

```
LOCAL xcIndex := {"Ntx1", "Ntx2"}
SET INDEX TO (xcIndex[1]), (xcIndex[2])
```

Macros and Arrays

You can use the macro operator (&) in combination with arrays and array elements. However, because of the increased power of *CA-Clipper* arrays, you may find less need to use the macro operator (&) to make variable references to arrays. You can now assign array references to variables, return array references from user-defined functions, and nest array references within other arrays. In addition, arrays can be created by specifying literal arrays or using the ARRAY() function. For more information on arrays, refer to the Arrays section of this chapter.

You can then refer to arrays and array elements using both macro variables and macro expressions with the restriction that the subscript references cannot be made in a PRIVATE or PUBLIC statement. A further restriction is that the macro operator (&) cannot be specified in a LOCAL or STATIC declaration statement. Attempting this will generate a fatal compiler error. For example, valid references to array elements using macro variables include:

```
cName := "aArray"
nElements := 5
cNameElement := "aArray[1]"

//Creates "aArray" with 5 elements
PRIVATE &cName.[nElements]
&cNameElement. := 100        // Assigns 100 to element 1
&cName.[3] := "abc"          // Assigns "abc" to element 3
```

You may successfully apply a macro operator (&) to an array element if you use a macro expression. A macro variable reference, however, will generate a runtime error. For example, the following lists the values of all fields of the current record:

```
USE Customer NEW
aStruc := DBSTRUCT()
//
FOR nField := 1 TO LEN(aStruc)
    ? &(aStruc[nField, 1])
NEXT
```

Macros and Code Blocks

The macro operator (&) can be applied to a macro variable or expression in a code block in most cases. There is a restriction when the macro variable or macro expression contains a declared variable. You may not use a complex expression (an expression that contains an operator and one or more operands) that includes the macro operator within a code block. If this is attempted, a runtime error occurs.

Note that this restriction has important implications on the use of local and static variables in the conditional clauses of commands since these clauses are blockified as they are written to the result text when the command is preprocessed. This applies to all FOR and WHILE clauses, the SET FILTER command, and the SET RELATION linking expression. The general workaround is to gather the entire expression into a single macro variable and then apply the macro operator (&) to the variable.

Using with Database Command Conditions

When using the macro operator (&) to specify conditional clauses of database commands such as FOR or WHILE clauses, there are some restrictions based on the expression's complexity and size as follows:

- The maximum string size the macro compiler can process is 254 characters

- There is a limit to the complexity of conditions (the more complex, the less the number of conditions that can be specified)

Invoking Procedures and Functions

Procedure and function calls can be referenced using macro variables and expressions. With DO, the macro variable reference to the procedure can comprise all or part of the procedure name. With a call to a function (built-in or user-defined), the macro variable reference must include the function name and all of its arguments.

The Macro Operator (&)

With the advent of code blocks in *CA-Clipper*, this was no longer the preferred practice. Instead, all invocations of procedures and functions using the macro operator should be converted to the evaluation of code blocks. For example, the following code fragment:

```
cProc := "AcctsRpt"
.
.
.
DO &cProc
```

can be replaced with:

```
bProc := &( "{ || AcctsRpt() }" )
.
.
.
EVAL(bProc)
```

The clear advantage of a code block in place of a macro evaluation is that the compilation of a string containing a code block can be saved and therefore must only be compiled once. Macro evaluations compile each time referenced.

External References

Procedures and user-defined functions, used in macro expressions and variables but not referenced elsewhere must be declared external using the REQUEST statement. Otherwise, the linker will not include them in the executable file (.EXE). The only exception to this rule is a procedure or user-defined function prelinked into a prelink library (.PLL).

Nested Macro Definitions

The processing of macro variables and expressions in *CA-Clipper* allows nested macro definitions. For example, after assigning a macro variable to another macro variable, the original macro variable can be expanded resulting in the expansion of the second macro variable and evaluation of its contents. For example:

```
cOne = "&cTwo"
cTwo = "cThree"
cThree = "hello"
//
? &cOne                    // Result: "hello"
```

Arrays

An array is a collection of related data items that share the same name. Each value in an array is referred to as an element. Array elements can be of any data type except memo, which is an exclusive field type. For example, the first element can be a character string, the second a numeric value, the third a date value, and so on. Arrays can also contain other arrays and code blocks as elements.

Arrays are references. This means that a variable to which an array is assigned (or which is declared as an array) does not actually contain the array. Instead, it contains a reference to the array. In addition, if a variable containing an array is passed as an argument to a procedure or user-defined function, a copy of the reference is passed. The array itself is never duplicated.

TYPE() and VALTYPE() return "A" for an array, which is a distinct data type. In *CA-Clipper*, you can create complex array expressions because many of the existing array functions have been changed to return an array reference. The valid array operations are discussed in this section.

Creating Arrays

PRIVATE and PUBLIC operate as in previous versions of *CA-Clipper* with the exception that they let you define multidimensional arrays. In addition, LOCAL and STATIC statements have been added to the language to allow lexically scoped array names (see Variables). The syntax for specifying an array with any of these statements is either:

```
<identifier>[<nElements1>, <nElements2>,...]
```

or:

```
<identifier>[<nElements1>][<nElements2>]...
```

Unlike other syntax representations, the square brackets are part of the array definition and must be included. Other than the first dimension, <nElements1>, all other dimensions are optional.

With the exception of static array declarations whose dimensions must be completely specified at compile time, arrays in *CA-Clipper* are dynamic, allowing the size to be completely determined at runtime. Thus, you can create an array whose dimensions are specified as expressions. For example:

```
i := 12, j := 4
LOCAL myArray[i, j]
```

Arrays

You can also use the ARRAY() function to create an array. In this function, specify the dimensions as arguments. The return value is an array. Like other functions, ARRAY() can be used in expressions where statements cannot.

Each of the array declaration statements translates into two parts: the declaration of the array name and the subsequent creation of an array and assignment of a reference. For example, PUBLIC myArray[12][4] is the same as:

```
PUBLIC myArray := ARRAY(12, 4)
```

Other functions that create arrays are DBSTRUCT() and DIRECTORY().

Addressing Array Elements

After an array is created, its elements are accessed using an integer index, commonly referred to as a *subscript*. To address an array element, place the subscript of the element in square brackets following the array name. For example, suppose that *myArray* is a one-dimensional array with ten elements. To address the first element, you would use:

```
myArray[1]
```

Note that subscript numbering begins with one.

To specify more than one subscript (i.e., when using multidimensional arrays), you can either enclose each subscript in a separate set of square brackets, or separate the subscripts with commas and enclose the list in square brackets. For example, if *myArray* is a two-dimensional array, the following statements both address the second column element of the tenth row:

```
myArray[10][2]
myArray[10, 2]
```

It is illegal to address an element outside of the boundaries of the array. Attempting to do so will result in a runtime error.

When making reference to an array element using a subscript, you are actually applying the subscript operator ([]) to an array expression. An array expression, of course, is any expression that evaluates to an array. This includes function calls, variable references, subscripting operations, or any other expression that evaluates to an array. For example, the following are all valid:

```
{"a", "b", "c"}[2]
x[2]
ARRAY(3)[2]
&(<macro expression>)[2]
(<complex expression>)[2]
```

Assigning Values to Array Elements

When you create an array with one of the declaration statements or with ARRAY(), each element is set to NIL until the array is initialized to some other value. Array initialization is the assignment of values to its elements with the assignment (=) or inline assignment operator (:=).

You can create an array and initialize it with a single function. The DBSTRUCT() function creates an array and initializes it with database file structure information. Similarly, DIRECTORY() initializes the array that it creates with disk directory information. You can assign these function results to variables which can, in turn, be referenced as arrays.

It is also possible to initialize arrays at the time they are created by using declaration statements. To accomplish this, do not define the array with dimensions in the declaration statement. Instead, just give it a name. After the array name, place an inline assignment operator followed by an expression that returns an array. For example, the following statement creates a local array and initializes it with directory information:

```
LOCAL myArray := DIRECTORY("*.*", "D")
```

The only limitation on this ability is with the STATIC statement which allows only constants and simple expressions in assignments.

The AFILL() function initializes all the elements of an existing array to a single value. For example, the following example leaves you with a local numeric variable whose value is one:

```
LOCAL myArray[30]
myArray := 1
```

while the next example assigns a value of one to each element in *myArray*:

```
STATIC myArray[30]
AFILL(myArray, 1)
```

You can also use AFILL() to initialize a range of elements. For example, the following lines of code initialize the first ten elements to one, the second ten elements to two, and the last ten elements to three:

```
AFILL(myArray, 1, 1, 10)
AFILL(myArray, 2, 11, 10)
AFILL(myArray, 3, 21, 10)
```

Arrays

Array elements may also be initialized on an individual basis. For example:

```
myArray[1] := 1, myArray[2] := 2, myArray[3] := 3
myArray[4] = SQRT(4)
```

You may initialize array elements using any expression that results in a valid *CA-Clipper* data type, including code blocks and other arrays. Once initialized, array elements can be used as variables anywhere in the language that is appropriate for their data type.

Multidimensional Arrays

In *CA-Clipper*, you can create a multidimensional array by declaring it with more than one dimension parameter or by assigning an array to an existing array element.

You can create and maintain traditional multidimensional arrays that have a fixed number of elements in each dimension. For example, a two-dimensional array is often used to represent the rows and columns of a table. To declare a two-dimensional array with ten rows and two columns, you might use the following:

```
PUBLIC myArray[10][2]
```

Usually, arrays that are created in this manner are expected to adhere to certain rules. For example, each column contains the same type of information for each row in the array (e.g., column one may be a character string and column two a number). Since *CA-Clipper* does not enforce any rules on what can be stored in an array, you must implement this level of control programmatically.

The fact that you can assign an array to an existing array element allows you to dynamically change the structure of an array. For example, after an array is created, there is nothing to prevent you from doing something like this:

```
myArray[1][2] := {1, 2, 3, 4, 5}
```

This assignment changes *myArray* significantly so it can no longer be thought of as a two-dimensional array in the traditional sense because one of its elements is an array reference. In particular, a reference to myArray[1][2][1] is now valid (it returns 1) where it was not before. References to myArray[1][1][1] and myArray[2][1][1], however, result in runtime errors.

This feature of assigning an array reference to an array element is handy in many of applications. The thing to remember is that you as the programmer must exercise whatever control you think is necessary for storing and addressing array elements. You cannot make any assumptions about the structure of an array. You must enforce that structure.

Literal Arrays

Literal arrays, sometimes called constant arrays, can be created at runtime using the following syntax:

```
{ [<exp> [, <exp> ... ]] }
```

For example:

```
x := { 1, 2, 3 }
y := { "Hello", SQRT(x[2]), Myfunc(x) }
```

Creating an array like this is the same as declaring the array (e.g., with PUBLIC or LOCAL) and then assigning the values to each element individually. You can use literal arrays anywhere an array can be specified, including as literal array elements. For example, to create a two-dimensional literal array, you could do the following:

```
aTwoD := { {1, 2, 3}, {"a", "b", "c"}, {.t., .t., .f.} }
```

If expressed as a character string, a literal array can be compiled with the macro operator. For example:

```
cArray := "{1, 2, 3}"
aNew := &cArray
? VALTYPE(aNew)              // Returns: A
```

This is useful since you could not otherwise store arrays as database fields or as character values in text files. Note, however, that the macro compiler can only process strings up to 254 characters in length.

Arrays as Function Arguments and Return Values

Arrays can be passed as arguments to procedures and user-defined functions. When you specify an array as an argument to a routine, it is passed by value, as a default, if the routine is called using a function-calling convention. This means a copy of the reference to the array is passed to a receiving parameter variable. Although a copy of the reference is passed, any changes made to the referenced arrays are reflected in the original automatically. For example:

```
LOCAL myArray[10]         // All elements set to NIL
AFILL(myArray, 0)         // Elements initialized to 0
MyFill(myArray)           // Elements incremented by 1
//
? myArray[1]              // Result: 1
```

Arrays

```
FUNCTION MyFill( tempArray )
   FOR i = 1 TO LEN(tempArray)
      tempArray[i]++
   NEXT
   RETURN (NIL)
```

When the user-defined function MyFill() executes, a copy of the reference to *myArray* is passed to *tempArray*. Within MyFill() the values in the referenced array are incremented by one. When the function returns, the reference held by *tempArray* is discarded but the changes to the referenced array appear after the call.

Arrays can be returned as values from functions. For example:

```
myArray := MakeArray(10, "New Value")
//
FUNCTION MakeArray( nElements, fillValue )
   LOCAL tempArray[nElements]
   AFILL(tempArray, fillValue)
   RETURN (tempArray)
```

Traversing an Array

There are two ways to traverse an array in *CA-Clipper*.

FOR...NEXT

The first and most easily understood traverse method is a FOR...NEXT loop. This construct lets you step through the array one element at time using the control variable as the array subscript for the current dimension. In the following example, each element in *myArray* is assigned its array position as a value:

```
LOCAL myArray[10]
FOR i := 1 TO 10
   myArray[i] := i
NEXT
```

To traverse a multidimensional array, nest FOR...NEXT constructs, one level for each dimension. The following example displays each element in a two-dimensional array:

```
FUNCTION ArrDisp( myArray )
   LOCAL i, j
   //
   FOR i := 1 TO LEN(myArray)
      FOR j := 1 TO LEN(myArray[1])
         ? myArray[i][j]
      NEXT j
   NEXT i
   RETURN (NIL)
```

This method of array traversal is quite traditional and is probably familiar to you if have experience with a high-level language such as C, Pascal, or BASIC.

AEVAL()

The other method, AEVAL(), requires an understanding of code blocks. This function evaluates a code block for each element of an array, passing the element value as a block parameter, and returns a reference to the array. For example, the following invocation of AEVAL() displays each element in a one-dimensional array:

```
AEVAL(myArray, { |element| QOUT(element) })
```

For a multidimensional array, you must specify nested AEVAL() functions to traverse each dimension. The following example, initializes and displays a two-dimensional array:

```
LOCAL myArray[10][2], i := 1
// Fill each element with row number
AEVAL(myArray, { |element| AFILL(element, i++) })
// Display each element
AEVAL(myArray, { |element| AEVAL(element,;
      { |value| QOUT(value) }) })
```

Empty Arrays

An empty array is an array with zero elements and therefore no dimensions. To create an empty array, declare an array with zero elements or assign an empty literal array to a variable. For example, the following statements are equivalent:

```
LOCAL aEmpty[0]
LOCAL aEmpty := {}
```

You can use empty arrays whenever you do not know in advance what size array you will need. Later, you can add elements to the array with AADD() or ASIZE().

Arrays

To test for an empty array, use the EMPTY() function. For example:

```
FUNCTION ZeroArray( entity )
   IF VALTYPE(entity) = "A" .AND. EMPTY(entity)
      RETURN (.T.)
   ENDIF
   RETURN (.F.)
```

You cannot, on the other hand, test for an empty array by comparing it to NIL. This comparison will always evaluate to false (.F.).

Determining the Size of an Array

You can determine the number of elements in an array using the LEN() function. This returns the number of element slots in the array, or the last element position in the array. The following code fragment demonstrates:

```
LOCAL myArray[10][12]
? nArraySize := LEN(myArray)      // Returns: 10
```

Note that this example returns the number of elements in the first dimension of *myArray*. Remember that *CA-Clipper* creates multidimensional arrays by nesting subarrays within a containing array. LEN() returns the number of elements in the specified containing array.

To determine the number of elements in a subarray, you can take the LEN() of the first element in the containing array, which returns the number of elements in the next nested dimension:

```
? nNestedSize := LEN(myArray[1])  // Returns: 12
```

Note: LEN() returns zero for an empty array.

Comparing Arrays

The == operator compares two arrays for *equivalence*. Arrays are equivalent if they are references to the same array (i.e., they point to the same location in memory). For example:

```
LOCAL myArray := {1, 2, 3}
sameArray := myArray              // Create a new reference
                                  // to myArray.
? sameArray == myArray            // Result: .T.
newArray := ACLONE(myArray)       // Create a new array
? newArray == myArray             // Result: .F.
```

Note that the = operator is not valid for comparing arrays and that there is no single operator that can check whether all elements are the same in two distinct arrays. To accomplish this, you must traverse the arrays and compare them element by element. The following function accomplishes this for one-dimensional arrays:

```
FUNCTION ArrComp( aOne, aTwo )
   IF LEN(aOne) <> LEN(aTwo)      // Sizes are different
      RETURN(.F.)
   ENDIF
   FOR i := 1 TO LEN(aOne)
      IF (aOne[i] <> aTwo[i])     // At least one element
                                  // is different
         RETURN (.F.)
      ENDIF
   NEXT
   RETURN (.T.)                   // Sizes and elements
                                  // are identical
```

Changing the Size of an Array

There are two functions, ASIZE() and AADD(), that change the size of an existing array. ASIZE() specifies the array to change and the new size as function arguments. With this function, you can make an existing array either larger or smaller. For example:

```
LOCAL myArray[10]
ASIZE(myArray, 25)      // Increase myArray by 15 elements
? LEN(myArray)          // Returns: 25
ASIZE(myArray, 5)       // Decrease myArray by 20 elements
? LEN(myArray)          // Returns: 5
```

When you enlarge an array the new elements are added to the end of the array and are set to NIL. Shrinking an array (i.e., making it smaller) removes elements beyond the specified new array size. ASIZE() updates the array and returns a reference to it.

AADD() enlarges an array by one element and initializes the new element to a particular value. The new element is added to the end of the array. For example:

```
LOCAL myArray[10]
AADD(myArray, 500)      // Add one element
? LEN(myArray)          // Returns: 11
? myArray[11]           // Returns: 500
```

Inserting and Deleting Array Elements

The AINS() function inserts an element into an existing array at a specified location, pushing all subsequent elements down one position. The last element in the array is lost.

Arrays

ADEL() deletes an element at a specified location, and all subsequent elements are moved up one position. The last element becomes NIL.

Both ADEL() and AINS() update the original array and return a reference to it. Unlike AADD() and ASIZE(), these functions do not change the size of the array.

Copying Elements and Duplicating Arrays

ACOPY() copies elements from one array to another. The target array must exist prior to invoking the function. With this function, you can copy all elements or a particular range of elements. If the source array contains nested arrays, the target array will contain only references to the subarrays, not actual copies. ACOPY() returns a reference to the target array.

For multidimensional or nested arrays, use ACLONE() if you want to duplicate the entire array. If the source array contains a subarray, ACLONE() creates a matching subarray and fills it with copies of the values in the original subarray, as opposed to copying a reference. This function creates the target array and returns a reference to it.

The following example illustrates the difference between these two functions:

```
LOCAL aOne[4], aTwo[14]
aOne[1] := {1, 2, 3}
ACOPY(aOne, aTwo)           // Copies elements from aOne to aTwo
? aOne[1] == aTwo[1]        // Returns: .T.
aThree := ACLONE(aOne)      // Duplicate aOne in its entirety
? aOne[1] == aThree[1]      // Returns: .F.
```

The first element in *aOne* is an array. Thus, after ACOPY() is invoked to copy elements to *aTwo*, the first elements in these arrays are equivalent (i.e., are references to the same subarray). ACLONE() creates *aThree* as a duplicate, and the equivalence test fails because the subarrays being compared are not the same. If, however, aThree[1] and aOne[1] were compared on an element by element basis, all elements would be the same.

Sorting an Array

The ASORT() function sorts an array and returns a reference to the newly sorted array. With this function, you can sort an entire array or a specified portion. For example:

```
ASORT(myArray)
```

sorts the entire array.

```
ASORT(myArray, 10, 5)
```

sorts elements ten through fifteen, leaving all other elements in their original place. An additional feature has been added to the ASORT() function allowing you to specify the sort criteria as a code block. For example, the following statement does a descending sort of an entire array:

```
ASORT(myArray,,, { |x , y| x < y })
```

Searching an Array

ASCAN() searches an array for a particular value specified as an expression or a code block. This function operates on the entire array or a specified range of elements. If used with a simple search expression, the function searches the array until it finds a matching value and returns the subscript, or zero if the value was not found. For example:

```
LOCAL aArray := { "Tom", "Mary", "Sue", "Mary" }
? ASCAN(aArray, "Mary")           // Result: 2
? ASCAN(aArray, "mary")           // Result: 0
```

If the search criteria is specified as a code block, the function is much more powerful. In the example above, the second ASCAN() could not find the name because the search was case-sensitive. The following example uses a code block to perform a search that is not case-sensitive:

```
? ASCAN(aArray, { |x| UPPER(x) == "MARY" }) // Result: 2
```

Since the function lets you specify a range of subscripts to include in the search, it can resume the search with the next element as in the following example:

```
LOCAL myArray := { "Tom", "Mary", "Sue", "Mary" }, nStart := 1
nEnd := LEN(myArray)                    // Set boundary condition
DO WHILE (nPos := ASCAN(myArray, "Mary", nStart)) > 0
   ? nPos, myArray[nPos]
   IF (nStart := ++nPos) > nEnd         // Test boundary condition
      EXIT
   ENDIF
ENDDO
```

Code Blocks

Code blocks provide a means of exporting small pieces of executable program code from one place in a system to another. You can think of them as *assignable unnamed functions*. They are assignable because, except when executing them, CA-Clipper treats them as values. That is, they can be stored to variables, passed as arguments, and so forth.

Code blocks bear a strong resemblance to macros, but with a significant difference. Macros are character strings which are compiled *on the fly* at runtime and then immediately executed. Code blocks, on the other hand, are compiled at compile time along with the rest of the program. For this reason code blocks are more efficient than macros, while offering similar flexibility.

The difference between code blocks and macros becomes especially important with declared variables. Variable declarations act at compile time and, thus, have no effect within runtime macros. Since code blocks are compiled at compile time, declarations can be used to control access to variables in code blocks. Variables that require compile-time declarations (static and local variables, and declared parameters) are never visible within runtime macros whereas they may be accessed freely in blocks.

Defining a Code Block

The syntax of a code block is as follows:

```
{ | [<argument list>] | <exp list> }
```

Both <argument list> and <exp list> are comma-separated. The expressions must be valid CA-Clipper expressions, they may not contain commands or statements such as control structures and declarations.

Note: The vertical bars that delimit a code block argument list must be present, even if there are no arguments, to distinguish a code block from a literal array.

Some examples of code blocks follow:

```
{ || "just a string" }
{ |p| p + 1 }
{ |x, y| SQRT(x) + SQRT(y) }
{ |a, b, c| Myfunc(a) , Myfunc(b) , Myfunc(c) }
```

Operations

The following table lists all valid *CA-Clipper* language code block operations. These operations act on one or more code blocks to produce a result. The result is not necessarily a code block data type:

Code Block Operations

Operation	Description
=	Assign
:=	Inline assign
AEVAL()	Evaluate a block for each element in array
DBEVAL()	Evaluate a block for each record in a work area
EVAL()	Evaluate a block

Executing a Code Block

The EVAL() function executes a code block. EVAL() takes a block as its first argument followed by a list of parameters. The function then executes the code block, passing the specified parameters as arguments. The expressions are evaluated in order from left to right. The returned value is the result of the last (or only) expression in the list. For example:

```
bBlock := { |nValue| nValue + 1 }
? EVAL(bBlock, 1)                           // Result:  2
```

The AEVAL() and DBEVAL() functions also execute code blocks. These are iterator functions that process an array and a database file, respectively, executing a code block for each element or record.

Scope of Variables Within a Code Block

When a code block is executed, local and static variable references (other than arguments) are scoped to the procedure or function in which the code block was defined. A local variable referenced in a code block is still visible after the declaring function is no longer active as long as there is an active reference to the code block. For example:

Code Blocks

```
bBlock := MyFunc()
FOR i := 1 TO 10
? EVAL( bBlock )
NEXT
RETURN

FUNCTION MyFunc()
LOCAL nNumber
LOCAL bBlock
nNumber := 10
bBlock := {|| nNumber++}
RETURN bBlock
```

In this example, each time the code block returned by MyFunc() is evaluated, it increments *nNumber* and returns the new value.

nNumber is accessible through the code block as the function defining the code block's activation record becomes *detached* from the activation stack when the function returns. The function's local variables then become, in essence, a free-floating array. The detached activation lives in object memory and stays alive as long as any of the blocks referring to it are alive.

This facility has significant ramifications for the capabilities of a code block. In particular, it solves the problem in which a code block instance needs long term ownership of some values. It means that you must macro-compile a code block only if the actual code to be executed isn't known until runtime, not for example, because you wanted the block to always contain or operate on a value which isn't known until runtime.

Note also that every call to the function generates unique instances of the local variables. Those variables will continue to exist as long as there is a code block that refers to them.

Using this fact, along with the fact that a code block can be passed as a parameter to another program, you can export static and local variables. For example:

```
FUNCTION One()
   LOCAL myVar
   myVar := 10
   bBlock := { |number| number + myVar }
   //
   NextFunc(bBlock)
   //
   RETURN NIL

FUNCTION NextFunc( bBlock )
   RETURN (EVAL(bBlock, 200))
```

When *bBlock* is evaluated in NextFunc(), *myVar*, which is local to function One(), becomes visible even though it is not passed directly as a parameter.

Note that if the specified variable is a private or public, the variable is not exported from the defining routine. The currently visible copy of the variable is used instead.

Using Code Blocks

The block is a special data type that contains executable program code. A block can be assigned to a variable or passed as a parameter in exactly the same way as other data types.

Code blocks are particularly useful when creating *black box* functions which can operate without knowing what type of information passed to them. Consider the following:

```
FUNCTION SayData( dataPassed )
   // Display dataPassed as a string
   DO CASE
   CASE VALTYPE(dataPassed) = "C"           // Character
      @ 10,10 SAY dataPassed
   CASE VALTYPE(dataPassed) = "N"           // Numeric
      @ 10,10 SAY LTRIM(STR(dataPassed))
   CASE VALTYPE(dataPassed) = "L"           // Logical
      @ 10,10 SAY IF(dataPassed, "True", "False")
   ENDCASE
   RETURN (.T.)
```

In this example, an error will occur if data of any type other than those specified is passed to SayData(). If, for example, a date needs to be displayed, SayData() must be modified.

In the example below, the code to be executed is stored in a block:

```
charCode := { |data| data }
numCode  := { |data| LTRIM(STR(data)) }
logCode  := { |data| IF(data, "True", "False") }
dateCode := { |data| DTOC(data) }
```

SayData() can now be defined as a black box routine which can act on any type of data. The decision of which block to pass to SayData() is made in the calling program:

```
FUNCTION SayData( dataPassed, codeBlock )
   // Display dataPassed as a string
   @ 10,10 SAY EVAL(codeBlock, dataPassed)
   RETURN (.T.)
```

Code blocks are also used to pass an expression to a function or procedure that you do not want to execute until sometime later. A good example of this can be found in STD.CH with the definitions of the SET KEY and SET FUNCTION commands. Both commands translate into a call to the same function although the operation performed is not the same in both cases: SET KEY assigns the name of a procedure or

Code Blocks

user-defined function to call when the specified key is pressed, while SET FUNCTION assigns a string to stuff into the keyboard buffer.

Evaluation of Macros in Code Blocks

When a code block contains a macro, an ambiguity arises regarding when the macro is evaluated. There are two possibilities, early or late evaluation.

Early evaluation means that the macro is expanded at the time the code block is created, and the expanded value remains constant for all subsequent evaluations of the block.

CA-Clipper uses early evaluation by default, with an alternate method available for late evaluation.

As an example of early evaluation, consider a SET FILTER command in which the filter expression is a macro:

```
SET FILTER TO &cFilter
```

A filter expression, specified by a SET FILTER command, is executed each time a SKIP command is issued. The macro string contained in *cFilter* above, however, should be expanded only once, when the command is first issued. Otherwise, if the macro is re-expanded each time the filter block is evaluated, the macro variable (*cFilter* above) must be preserved for the duration of the filter.

In *CA-Clipper*, the SET FILTER command is implemented by use of a code block. The code for the above command, after preprocessing, is similar to the following example:

```
DBFILTER( { || &cFilter } )
```

Early evaluation of the macro within the code block results in the desired behavior: *cFilter* is evaluated when the block is defined (as part of the creation of the block). The block is created via the macro system and remains constant through all subsequent block evaluations, regardless of any change to the value of *cFilter*.

Late evaluation means that the macro is reevaluated each time the block is evaluated.

As an example of a situation where late evaluation is required, consider a block which must perform a macro operation on one of its own parameters.

The following block attempts to assign a value to a variable by use of a macro:

2-76 CA-Clipper

```
assignVarByNameBlock := { |vName, value| &vName := value }
```

Because *CA-Clipper* implements early evaluation, this code block would not have the desired effect: instead of evaluating the macro each time the block is run, *CA-Clipper* would evaluate the macro when the block was first created. The initial value of *vName* would become part of the block.

Late evaluation is available by using a macro expression. The above code block will function correctly if defined as follows:

```
assignVarByNameBlock := { |vName, value| &(vName) := value }
```

The use of a macro expression (the macro operator applied to a parenthesized expression) prevents the early evaluation from taking place, and the code block will reevaluate the macro string contained in *vName* during each block evaluation.

Storing and Compiling Code Blocks at Runtime

Although you cannot store code blocks directly to database fields or (.mem) files, you can store them as character strings. When you want to use a code block definition stored as a character string in a database field in a program, compile the field using the macro operator then assign the resulting code block to a variable. For example:

```
USE DataDict NEW
//
// Create a block string and assign it to a field variable
APPEND BLANK
REPLACE BlockField WITH "{|| Validate()}"
.
.
.
// Later compile and execute the block
bSomeBlock := &(BlockField)
EVAL(bSomeBlock)
```

Instead of storing the compiled code block to a variable and then evaluating it, you could compile and evaluate in a single step with EVAL(&(BlockField)). There is, however, a speed advantage to storing the compiled code block to a variable since the compile step only has to be performed once. Once compiled, the variable can be passed throughout the system and evaluated at the same speed as other *CA-Clipper-compiled* code.

Note: The macro operator is appropriate only for use with character expressions. Thus, you cannot compile a code block directly with the macro operator.

Objects and Messages

CA-Clipper offers a limited implementation of a special data type called *object*. The concept of an object is part of the object-oriented programming paradigm.

Note: *CA-Clipper*, is not an object-oriented language. The *CA-Clipper* implementation of objects represents only a small subset of the object-oriented paradigm.

CA-Clipper objects are complex data values which have a predefined structure and set of behaviors. A small number of predefined object types, called classes, are part of the system. These classes support particular operations in *CA-Clipper*. Creation of new classes or subclassing of existing classes is not available.

Classes

In *CA-Clipper*, there are several different types of objects. An object's type is known formally as its *class*. The information contained in an object, and the operations that can be applied to it, vary depending on the class of the object.

For each available class, there exists a special function called a *create function*. The create function facilitates the creation of new objects of the associated class.

Instances

You create a new object with a call to the create function for a particular class. The object will have the attributes and behaviors specified in the description of the class. The new object is formally known as an *instance* of the class.

Like *CA-Clipper* arrays, objects are handled referentially. This means that a program variable cannot actually contain an object. It may, however, contain a reference to an object. The variable is then said to *refer to* the object, and operations are performed on the object by applying the send operator (:) to the variable. Object references may be freely assigned, passed as parameters, and returned from functions. Two or more variables may contain a reference to the same object (if, for example, a variable containing an object reference is assigned to another variable). In this case, both variables refer to the object, and operations

Objects and Messages

are performed on the object by applying the send operator to either variable.

Objects continue to live as long as there are active references to them somewhere in the system. When all references to an object are eliminated, the system automatically reclaims the space occupied by the object.

When you create a new object, the create function returns a reference to the new object. You can then assign this reference to a variable, and access the object through that variable.

Instance Variables

An object contains within it all of the information necessary for it to perform the operations specified for its class. This information is stored inside the object in special storage locations called *instance variables*.

When you create a new object, it receives its own dedicated set of instance variables. The new instance variables are automatically assigned initial values.

Most instance variables are invisible to the programmer. However, some instance variables are accessible. These are known as *exported* instance variables. Exported instance variables can be inspected and, in some cases, assigned using the send operator.

Sending Messages

Each class defines a set of operations that can be performed on objects of that class. These operations are performed by *sending a message* to the object using the send operator (:).

The syntax for sending a message to an object is as follows:

```
<object>:<message> [(<parameters>)]
```

For example:

```
oGet:left()
```

In this example, oGet is the name of a variable that contains an object reference. left() is the message being sent; it specifies the operation to be performed. The available operations (and corresponding messages) vary depending on the class of the object.

Parentheses are optional if no parameters are supplied in the message send. By convention, however, parentheses are used to distinguish a

Programming and Utilities Guide   2-79

Objects and Messages

message send with no parameters from an access to an exported instance variable.

Executing a message send produces a return value, much like a function call. The return value varies depending on the operation performed. Often, this return value is a reference to the object that was just operated on (referred to as *self*). This allows multiple messages to be sent in a chain:

```
oGet:left():right()
```

Since this is very hard to read, this technique should be used in moderation.

In order to understand how an object can return a reference to self, one must understand how objects are handled in memory.

Accessing Exported Instance Variables

Some classes allow access to the instance variables within objects of that class. These variables are called *exported instance variables*. The definition for a class specifies which, if any, instance variables are exported. Some exported instance variables are said to be *assignable*. An assignable instance variable can be modified using a special variant of the message send operator.

You access exported instance variables using a variant of the normal message sending syntax. The syntax for retrieving the value of an instance variable is:

```
<object>:<instance variable>
```

The syntax for assigning a new value to an assignable exported instance variable is:

```
<object>:<instance variable> := <newValue>
```

For example:

```
nRow := oBrowse:rowPos
oBrowse:cargo := xNewCargo
```

Conceptually, it is easiest to consider the *<object:instanceVar>* as if it is a single variable; both are valid in all of the same contexts. The < > is much more powerful than a variable. Internally, assignments and retrievals are mapped to methods that set and retrieve the values of the variables, allowing those methods to also enforce valid values and perform other actions.

For example, in the Error class, assigning a value to the Error:genCode instance variable sets the Error:description instance variable to the default description for that type of error. This helps ensure consistency in the messages the user receives when errors occur.

Though this is useful, there is still a need to act upon the data stored in the object, independently of setting instance variables. This is accomplished by sending messages to the object to invoke the desired methods.

The Database System

The *CA-Clipper* database system is composed of work areas that are used to manage database and other related files, and operations designed to manipulate those files. A database file is a collection of related information that is stored in the form of a table. The design of the table, known as the file structure, is also stored in the database file.

The file structure is defined and added to the database file when the file is created. It consists of one or more field definitions describing the name, width, and data type attributes for each column in the table. The table rows, or records, are added to the file using append operations which understand and enforce the file structure on a field by field basis. Records are added to a database file in a particular physical order that is used, by default, when accessing the file.

An index file is an ancillary file that is created separately from the database file. It lets you define and maintain a logical ordering for a database file without affecting the physical file order.

Several database files that have related structures and data can be associated, along with their index files, using the SET RELATION command. SET RELATION lets you establish relationships between several files and operate on them as a single entity known as a database, or view.

The *CA-Clipper* package contains a database utility, DBU.EXE, that lets you perform most database and index operations using a menu-driven system. It is an excellent tool for designing database file structures for your applications and for entering test data. Refer to the *Database Utility* chapter in the *Programming and Utilities* guide for further information regarding DBU.

This section describes the various facets of the database system and gives a brief overview of all database and index operations.

Work Areas

The *CA-Clipper* database system defines 255 work areas, 250 of them are available for your use; the remaining five are reserved for internal use. A work area is essentially an area in memory in which you can manage a single database file along with an optional memo file and up to 15 index files. A work area is *occupied* or *unoccupied*, depending on whether or not it contains an open file. At application startup all work areas are unoccupied, and one is the current, or active, work area number.

Accessing Work Areas

To open a database file, you must first access the work area you want to use. The most common way to do this is with the SELECT command. SELECT is designed specifically to move between work areas by changing the current work area to the one you specify.

The general usage rules for SELECT are: to access a specific unoccupied work area, specify SELECT with a number between one and 250; to access the next available unoccupied work area, use SELECT 0; to access an occupied work area (i.e., an area in which the database file is in USE), specify SELECT with the alias name rather than the work area number.

In *CA-Clipper*, the USE command supports a NEW clause that SELECTs the next unoccupied work area before performing the open operation. This feature makes the SELECT 0 command obsolete, so that any code of the form:

```
SELECT 0
USE <xcDatabase>
```

can be replaced with:

```
USE <xcDatabase> NEW
```

After a USE, the selected work area remains active.

If no ALIAS clause is specified, as in the above example, the work area in which the database file is opened is assigned an alias name that is the same as the database filename. This name is used with SELECT to access the work area.

Aliased expressions are used to temporarily access a work area for the purpose of expression evaluation. To form an aliased expression, enclose the expression in parentheses and prefix it with the desired alias name and the operator (->) (e.g., <idAlias>->(<exp>)). The work area will be selected, the expression evaluated, and finally the original work area restored.

You may also evaluate simple database functions and more complicated expressions in unselected work areas this way. For example:

- Cust->(EOF()) evaluates the end of file status in the Cust work area.
- Cust->(HEADER() + (LASTREC() * RECSIZE())) computes the size of the file in the Cust work area.
- Cust->total + Cust->Amount adds the value of the *Amount* field in the Cust work area to the variable *total*.

Notice that in the final example above, *Amount* is referenced explicitly with its alias name, no parentheses are used within the actual aliased expression. This explicit field reference avoids any ambiguity that may exist among field and variable names and illustrates the following point.

Note: Using an aliased expression does not override any compiler assumptions about variables within the expression, whereas using an alias as an explicit field reference does.

In other words, the compiler does not assume that everything within an expression refers to a field simply because the expression is preceded by an alias. Instead, the standard precedence rules resolve ambiguities when the expression is evaluated at runtime. You can explicitly reference fields with an alias. Such fields' unambiguous references are resolved at compile time.

To illustrate, suppose you have the following code that uses a local variable *amount* and a field called *Amount* in the Cust work area:

```
LOCAL total := 0, amount := 0
total := Cust->(total + Amount)          // Field Cust->total + 0
```

This aliased expression accesses the local variable named *amount* because when the expression is evaluated, the variable is visible and takes precedence over the field with the same name. Changing the code to:

```
// Field Cust->total + Field Cust->Amount
total := Cust->total + Cust->amount
```

accesses the field *Amount* rather than the local variable, because the explicit reference to the field through its alias overrides the LOCAL declaration statement.

As a good programming practice, you should make all identifier references explicit and avoid naming conflicts whenever possible. Otherwise, depending on the compiler to resolve ambiguities as they occur, you may not agree with its decision.

Work Area Attributes

A work area provides a means for logically grouping attributes pertaining to a database file. These are called work area attributes and are listed in the table below:

Work Area Attributes

Command/Function	Attribute
SET DELETED	Delete record filter
SET FILTER	Logical record filter
ALIAS()	Database alias name
BOF()	Beginning of file flag
DBFILTER()	SET FILTER condition
DBRELATION(), DBRSELECT()	SET RELATION information
EOF()	End of file flag
FCOUNT()	Number of fields
FOUND()	Search flag
FLOCK(), RLOCK()	Lock status flags
INDEXKEY()	Index key expression
INDEXORD()	SET ORDER value
LOCATE, CONTINUE	LOCATE condition
RECCOUNT() I LASTREC()	Number of records
RECNO()	Record number
SELECT()	Work area number

Database Files

The database file is the main file type and has a default extension of (.dbf). A database file, often called a *table*, consists of one variable length header record that defines the file structure in terms of its field definitions, and zero or more fixed length records that contain actual data for nonmemo fields and pointer information for memo fields. Each record has one additional byte for the record delete status flag. The *CA-Clipper* database file format is compatible with dBASE III PLUS.

Database files can be thought of as tables where each field in the file structure represents a column of the table, and each record represents a row.

Memo Files

If a database file structure has definitions for one or more memo fields, there is one memo file associated with that database file. The memo file has a default extension of (.dbt) and the same root filename as the associated database file. By default, the database and memo files are created and maintained in the same directory.

The data for all memo fields is stored in the same memo file, no matter how many memo fields are defined in the database file structure. Each memo field in the database file contains a pointer into the memo file, so the database system can quickly locate the associated data.

The memo file is automatically opened and maintained in the same work area as the database file.

Database File Attributes

Like work areas, database files have associated attributes as listed in this table:

Database File Attributes

Function	Attribute
DELETED()	Record delete status flag
FIELD()	Field name
HEADER()	Header size in bytes
LUPDATE()	Date of last update
RECSIZE()	Record size in bytes

Database Operations

There are several operations (database operations) that can be performed on an open database file. The following table categorizes all of the database operations. Note that all of these operations require an open database file in the currently selected work area:

The Database System

Database Operations

Category	Command/Function
Add	APPEND BLANK, APPEND FROM, BROWSE()
Close	CLOSE, USE
Compute	AVERAGE, COUNT, SUM
Create	COPY STRUCTURE, COPY STRUCTURE EXTENDED, COPY TO, CREATE, CREATE FROM, JOIN, SORT, TOTAL
Delete	DELETE, PACK, RECALL, ZAP
Display	DISPLAY, LABEL FORM, LIST, REPORT FORM, BROWSE(), DBEDIT()
Filter	INDEX...UNIQUE, SET DELETED, SET FILTER
Information	AFIELDS()
Iterate	DBEVAL()
Lock	SET EXCLUSIVE, UNLOCK, FLOCK(), RLOCK()
Navigate	CONTINUE, FIND, GO, LOCATE, SEEK, SKIP
Open	USE
Order	INDEX, REINDEX, SET INDEX, SET ORDER, SORT
Relate	SET RELATION
Update	@...GET <idField>/READ, REPLACE, UPDATE, BROWSE(), DBEDIT()

Record Scoping

Many database operations can process subsets of records within a work area using scope and conditional clauses. For any database command that allows a <scope> as part of its syntax, the syntax of the <scope> is as follows:

[ALL | NEXT <nRecords> | RECORD <nRecord> | REST]

- ALL processes all records

- NEXT processes the current record and the specified number of records

- RECORD processes the specified record

- REST processes all records from the current record to the end of file

Commands specified without a scope default to the current record (NEXT 1) or ALL records, depending on the command. For example, DELETE and REPLACE typically process the current record by itself and, therefore, default to NEXT 1. Other commands, such as REPORT and LABEL FORM, typically process a set of records and, therefore, default to ALL records. Specifying a scope changes this default by indicating how many records to process and where to begin.

The set of records processed can also be restricted using a conditional clause which specifies a subset of records based on a logical condition. The two conditional clauses are FOR and WHILE.

A FOR clause defines a condition that each record within the scope must meet in order to be processed. If no other scope is specified, FOR changes the default scope to ALL records.

A WHILE clause defines another condition that each record processed must meet; as soon as a record is encountered that causes the condition to fail, the command terminates. If no other scope is specified, WHILE changes the default scope to REST.

Specifying a scope, a FOR clause, and a WHILE clause within the same command syntax can raise questions regarding the order in which the clauses are processed. The scope is evaluated first in order to position the record pointer. Then, the WHILE clause is evaluated and, if the condition is not met, the process terminates. If the WHILE condition is met, the FOR clause is evaluated. If the FOR condition is also met, the record is processed; otherwise, it is not. Either way, the record pointer is moved to the next record within the scope until the scope is exhausted.

The Record Processing Primitive—DBEVAL()

The iterate operation, DBEVAL(), is relatively new and, therefore, deserves further explanation. This function is the database primitive in STD.CH that defines database commands that allow record scoping, and it is sometimes called the record processing primitive.

DBEVAL() evaluates a code block for each record matching the specified scope and conditions. It creates user-defined database processing commands. The DBEVAL() entry in the *Language Reference* chapter of the *Reference* guide has more information and examples. For additional information on code blocks, see the Code Blocks section in this chapter.

Index Files

Database files are maintained and processed in a particular physical order identical to the order in which records are added to the file. Each record is numbered sequentially beginning with one, and the number of the current record may be accessed with the RECNO() function.

Indexing provides a means for logically ordering the database file according to the data it contains. In *CA-Clipper*, you can index by creating index files in a work area with a database file.

CA-Clipper's default method for creating and maintaining index files is unique and incompatible with other dialects. This method creates index files with a default extension of (.ntx).

Creating

Create index files for the database file in the current work area with the INDEX command in which you specify an index key expression. The maximum length of the key expression is 250 characters and its type can be character, date, or numeric.

Determine the index key value for each record in the database file by using the index key expression. This expression is stored in the index file in such a way that the database file can be processed in order according to the key values which can be searched very quickly.

The UNIQUE clause of the INDEX command and the SET UNIQUE ON flag create an index file in which only the first occurrence of any given key value is maintained in the index file. Uniqueness is an attribute that is stored in the index file.

Opening

You can create several index files for a single database file so the data can be accessed in many different ways. Once the necessary index files are created, you must open them along with the associated database file in order to use them. You can open up to 15 index files at the same time with the INDEX clause of the USE command or with SET INDEX.

Like all other database file opening operations in *CA-Clipper*, USE and SET INDEX first search the current SET DEFAULT drive and directory for the file to open. If the file is not found, the SET PATH list is searched in the order specified. Only after all paths are searched will an open operation be considered a failure.

Ordering

With both USE and SET INDEX you list the index files you want to open, and the first one in the list becomes the *controlling* index. The controlling index defines the order in which the database file is processed and is the only index that can be searched.

The SET ORDER command changes the controlling index by specifying a position in the index file list. SET ORDER TO 0 specifies the physical order as if no index files were in use.

Searching

SEEK and FIND locate a record based on its index key value in the controlling index file. SEEK is the preferred search command; FIND is an obsolete command that is provided only for compatibility. Searching an index file is much faster than performing a sequential search.

CA-Clipper uses one of two index search methods depending on a global flag setting, SET SOFTSEEK. If SET SOFTSEEK is OFF (the default), a failed index search moves the record pointer to the end of file, setting EOF() to true (.T.). If SET SOFTSEEK is ON, a failed index search moves the record pointer to the record with the next higher index key value.

The FOUND() function tests the success or failure of a search operation.

Updating

As long as an index file is open, any changes made to the database file are automatically reflected in the index file. This is true of all open index files not just the controlling index. Changes to a database file are made using any of the database update operations mentioned earlier in this section.

Additionally, database operations that move the record pointer update the index file before moving the record pointer if the index key value has changed. A COMMIT or an index close operation also updates the index file.

The REINDEX command recreates all open index files in the current work area by reindexing each record in the database file.

Closing

Close index files with SET INDEX TO with no parameters or CLOSE INDEX. Any of the database close operations also closes open index files in the same work area.

The Input/Output System

The *CA-Clipper* language provides a complete input/output system allowing data entry and data display on the screen and the printer. The system includes built-in data validation that is extensible through the use of user-defined functions. Included facilities let you create simple columnar reports and mailing labels, and, with the provided commands and library functions, you can create reporting mechanisms to suit your own needs. This section introduces you to *CA-Clipper's* input/output system, giving you basic information on controlling the console and printer.

Console Operations

Commands and functions that display output on the screen without reference to row and column position are called *console operations*. Used in conjunction with SET ALTERNATE and SET PRINTER, these operations can simultaneously send output to the screen, a text file, and the printer. However, the majority of the console commands have TO PRINTER and TO FILE clauses that supersede SET PRINTER and SET ALTERNATE. The following table lists all the console operations in the *CA-Clipper* language:

Console Operations

Operation	Output
?	??	The result of one or more expressions
ACCEPT	A prompt. Waits for character input.
DISPLAY	LIST	The contents of a database file
INPUT	A prompt. Waits for input of any type.
LABEL FORM	The contents of a database file as labels
REPORT FORM	The contents of a database file as a report
TEXT...ENDTEXT	A block of text
TYPE	The contents of a file
WAIT	A prompt. Waits for a single character input.
QOUT()	QQOUT()	The result of one or more expressions

SET CONSOLE is an important command to console operations since it lets you temporarily suppress the screen output without affecting file and printer output. For example, if you are printing a report, it may be disconcerting to see the report scrolling up the screen while it is printing. To control this, SET CONSOLE OFF before printing the report as in this code segment:

```
USE Accounts NEW
SET CONSOLE OFF
REPORT FORM Accounts TO PRINTER
SET CONSOLE ON
CLOSE Accounts
```

Full-screen Operations

The *full-screen operations* address the screen (and sometimes the printer) directly, usually by addressing a specific row and column. These operations are distinguished from console operations in that they ignore SET ALTERNATE, SET CONSOLE, and SET PRINTER:

Full-Screen Operations

Operation	Result
@...BOX	Draw a box
@...CLEAR	Erase the screen
@...SAY	Display the result of an expression
@...TO	Draw a box
CLEAR	Erase the screen
COL()	Return current column position
ROW()	Return current row position
MAXCOL()	Return the maximum column position
MAXROW()	Return the maximum row position
SAVESCREEN()	Save a screen region
RESTSCREEN()	Restore a screen region

Of all the full-screen operations, only @...SAY can be redirected to the printer with SET DEVICE and written to a file with SET PRINTER. Printing is discussed later on in this section.

Controlling Screen Color

SETCOLOR() saves the current color setting and optionally sets new colors for subsequent screen painting. A color string contains several color settings, each corresponding to a different logical region of the screen.

All console and full-screen operations (with the exception of @...GET) use the standard color setting when displaying to the screen. The GET area (@...GET) uses the enhanced setting and the unselected setting causes all GETs other than the current one to be displayed using a different color attribute.

The Input/Output System

For more information regarding color including a list of the available color codes, see SETCOLOR() in the *Language Reference* chapter of the *Reference* guide.

Controlling Output Destination

In *CA-Clipper*, the screen is always the default output destination, but it you may also direct output to the printer or a disk file.

Directing Output to the Printer

The way to direct output to the printer depends on what operation you are using. Many console commands have a TO PRINTER clause designed to echo output to the printer. For instance, to print a set of labels use:

```
LABEL FORM Ship TO PRINTER
```

For console commands that do not support this clause, use the SET PRINTER command. For instance:

```
SET PRINTER ON
DO WHILE .NOT. EOF()
   QOUT(Name)
   SKIP
ENDDO
SET PRINTER OFF
```

You can use SET PRINTER instead of the TO PRINTER clause, if you prefer. As stated earlier, you may want to SET CONSOLE OFF before directing console output to the printer.

For @...SAY, use the SET DEVICE command as in this example:

```
SET DEVICE TO PRINTER
DO WHILE .NOT. EOF()
   @ 1, 1 SAY Name
   SKIP
ENDDO
SET DEVICE TO SCREEN
```

Unlike console operations, @...SAY displays either to the screen or to the printer, but never to both devices simultaneously.

CA-Clipper allows other printer controls, including access to the current printhead position with PROW() and PCOL(). These functions allow relative printer addressing with @...SAY in much the same way as ROW() and COL() do relative screen addressing.

When printing with @...SAY, a page eject is issued anytime you address a row that is less than the current PROW() value. SETPRC() resets

PROW() and PCOL() to prevent an automatic page eject where it is not desired.

SET MARGIN sets the left printer margin. With console output, the margin setting indents output whenever there is a new line. @...SAY adds the margin setting to the specified column value.

The SET PRINTER command has a second form that is used to change the output device, allowing you to access more than one printer and to direct printed output to a file. SET PRINTER TO sets the output device for all printed output, including console and full-screen operations.

Directing Output to a File

You cannot send @...SAY output directly to a file. Instead, you must reroute its printed output to a file with SET PRINTER TO as in this example:

```
SET PRINTER TO AtOut.prn
SET DEVICE TO PRINTER
//
// <@...SAY commands go here>
//
SET DEVICE TO SCREEN
SET PRINTER TO
```

Redirecting printer output to a text file with SET PRINTER TO causes all printed output to go to the file, not just @...SAYs. Thus, it can be used to send @...SAY and console output to the same file, as long as both SET DEVICE TO PRINTER and SET PRINTER ON are in effect.

```
SET PRINTER TO AtOut.prn
SET DEVICE TO PRINTER
//
// <@...SAY commands go here>
//
SET PRINTER ON
//
// <console operations go here>
//
SET PRINTER OFF
SET DEVICE TO SCREEN
SET PRINTER TO
```

You may direct output to a file without rerouting printer output. The easiest way is to use the TO FILE option if it is supported. If not, use SET ALTERNATE:

```
SET ALTERNATE TO OutFile.prn
SET ALTERNATE ON
//
 // <console operations go here>
//
SET ALTERNATE OFF
CLOSE ALTERNATE
```

The Keyboard System

By default, *CA-Clipper* applications automatically save the characters typed on the keyboard in a buffer called the keyboard, or typeahead, buffer. The characters in the keyboard buffer are then removed on a first in, first out basis and used as input when required by a wait state or INKEY(). The wait states include ACCEPT, INPUT, READ, WAIT, ACHOICE(), DBEDIT(), and MEMOEDIT().

The buffering of keystrokes lets the user continue typing without having to wait for the application to catch up, and it is very convenient, especially for applications that are data entry intensive where the user happens to be a very fast typist. This feature may never affect the way you program your applications, but there may be cases when you want to control the keyboard buffer. This section describes the commands and functions available in *CA-Clipper* that affect the keyboard buffer.

Changing the Keyboard Buffer Size

SET TYPEAHEAD controls the number of characters the keyboard buffer can hold. The minimum size of the keyboard buffer is zero characters, a setting which essentially disables the keyboard buffer so the user may type only when an operation requiring input is actively executing. The maximum keyboard buffer size is 32,767 characters. Issuing a SET TYPEAHEAD command in an application clears the current contents of the keyboard buffer before changing the buffer size.

Putting Characters in the Keyboard Buffer

KEYBOARD is used to programmatically put characters into the keyboard buffer. This command clears all pending keystrokes before placing one or more ASCII characters in the keyboard buffer. You can use the CHR() function with KEYBOARD for nonprintable keystrokes such as a carriage return (CHR(13)).

Characters placed in the keyboard buffer this way are treated just as if they were typed from the keyboard, they are not executed until extracted by a wait state or by INKEY().

KEYBOARD is useful for developing self-running demonstration programs in *CA-Clipper*, and it is also commonly used with the ACHOICE() user function to force a particular menu selection.

Refer to the INKEY() Codes appendix in the *Error Messages and Appendices* guide for a complete list of key codes.

Reading Characters from the Keyboard Buffer

There are three functions that access the keyboard buffer. These are INKEY(), LASTKEY(), and NEXTKEY(). Each function returns a numeric value between -39 and 386, identifying the INKEY() code of keys including function, *Alt*-function, *Ctrl*-function, *Alt*-letter, and *Ctrl*-letter key combinations.

INKEY() is the only keyboard function that actually extracts a character from the keyboard buffer. In this sense, it is like a wait state, but it may or may not pause for input depending on its argument. INKEY() extracts the next pending keystroke from the keyboard buffer and returns the ASCII value of that key. This function is useful for polling the keyboard or pausing program execution.

NEXTKEY() returns the ASCII value of the next pending keystroke without removing it from the keyboard buffer.

LASTKEY() returns the ASCII value of the last keystroke extracted from the keyboard buffer. Keystrokes are extracted from the buffer either by INKEY() or by a wait state. This function has several applications including determining the key used to terminate a READ.

Controlling Predefined Keys

There are several commands and functions that control the action of specific predefined keys. These are summarized in the following table:

Commands and Functions That Control Specific Keys

Command/Function	Purpose
SET ESCAPE	Toggle ability to terminate READ with Esc
SET WRAP	Toggle circular menu navigation with Uparrow and Dnarrow
ALTD()	Toggle ability to invoke the Debugger with Alt-D
READEXIT()	Toggle ability to terminate READ with Uparrow or Dnarrow
SETCANCEL()	Toggle ability to terminate application with Alt-C

The Keyboard System

In addition to these limited controls, all predefined keys can be completely redefined with SET KEY.

Reassigning Key Definitions

SET KEY lets you redefine any key on the keyboard by specifying a procedure to be executed when the key is pressed during a wait state. Three parameters, PROCNAME(), PROCLINE(), and READVAR(), are automatically passed to the procedure. As with INKEY(), keys are identified by a numeric ASCII code.

Another command, SET FUNCTION, redefines function keys just as SET KEY redefines other keys. This command, however, is preprocessed into SET KEY and KEYBOARD commands, causing SET FUNCTION to clear a SET KEY for the same key number. This is incompatible with previous releases of *CA-Clipper* which maintained a separate list for SET FUNCTION and SET KEY definitions.

A total of 32 keys may be SET at one time. At startup, the system assumes that *F1* is set to Help. If you link a procedure called Help into the current program, pressing *F1* from a wait state will execute it.

All SET KEY definitions take precedence over the standard key definitions. This includes predefined keys such as *Ins* and *Del* that are used in editing, as well as keys such as *Esc* and *Alt-C* that you can control with built-in commands and functions.

Clearing the Keyboard Buffer

You can clear the keyboard buffer of all pending keystrokes with CLEAR TYPEAHEAD. This command is useful when you want to be sure the user does not bypass important messages on the screen. Simply displaying the message and waiting for a response quickly flashes the message and uses the next pending character in the keyboard buffer for the response. However, issuing a CLEAR TYPEAHEAD before displaying the message insures that there are no pending characters and forces the user to read the message before entering a response.

The Low-level File System

Several functions are provided that allow you to manipulate DOS binary files directly from a *CA-Clipper* application. These include the capability to open existing binary files, to create new files, and to read from and write to open files. This section describes these functions and their capabilities. Note that the functions described herein allow low-level access to DOS files and devices. They require a thorough knowledge of the operating system and should be used with care. Refer to the *Language Reference* chapter of the *Reference* guide for more information on a particular function.

Opening a File

There are two functions available to open a binary file. The first, FOPEN(), opens an existing file while the second, FCREATE(), creates a new file, leaving it open for use. Both functions return a number representing the DOS file handle for the open file. The file handle must be saved in a variable to use it with other functions to identify the file. FCREATE() lets you specify a DOS file attribute, and FOPEN() lets you specify the DOS open mode.

Since these functions are dealing with files at the operating system level, they do not respect the SET DEFAULT and SET PATH settings. Instead, unless a filename is unambiguously identified with a path, the current DOS drive and directory are assumed. Also, no file extension is assumed unless one is explicitly specified.

Reading from a File

Once a binary file is open, you can read its contents with one of two functions, FREAD() or FREADSTR(). Both functions require that you identify the file using its file handle number. Specifying a particular file is necessary since more than one binary file may be open at a time.

FREAD() and FREADSTR() are very similar in functionality. The main difference is that FREAD() requires you to specify where to save the characters read from the file, and FREADSTR() returns the characters read as a character string.

In either case, the data read from a binary file is in binary form. Thus, several functions are provided to convert binary data to numeric so that you can manipulate the information in your *CA-Clipper* application.

All of the conversion functions require that the argument be represented as a character string, which makes them work easily with FREAD() and FREADSTR().

Writing to a File

Once a binary file is open, you can write to it with FWRITE(). This function requires that you identify the file using its file handle number. It is necessary to specify a particular file since more than one binary file may be open at a time. You will not be able to use FWRITE() on a file that was opened in read-only mode.

With FWRITE() you specify a character string to write to the file. Functions are provided to convert *CA-Clipper* numeric data to binary form.

The I2BIN() and L2BIN() functions are provided to convert *CA-Clipper* numeric values to binary form and return the results as character strings. These character strings can then be written to a binary file.

Manipulating the File Pointer

When a binary file is initially opened for use, the file pointer is positioned to the beginning of the file. The read and write functions move the file pointer, as needed, to perform their operation, and leave the pointer in its new position in the file.

Another function, FSEEK(), lets you move the pointer directly to a specific position in the file. This function is particularly useful if you must read or write characters somewhere in the middle of a file but are not interested in the information preceding those characters. Instead of performing a read operation to bypass the unwanted data, you would perform an FSEEK()—which is much more efficient—to move the file pointer to the correct position in the file. Then, you would read or write the necessary characters.

FSEEK() is very flexible in its ability to position the file pointer. You specify an argument that tells the function to begin from the beginning of file, from the current file position, or from the end of file, and another argument that indicates the number of positions to move.

Closing a File

FCLOSE() closes an open binary file and writes the associated DOS buffers to disk. This function requires that you identify the file you want to close by its file handle number. Specifying a particular file is necessary since more than one binary file may be open at a time.

Error Detection

When using the low-level file functions, you can test for errors with the individual function return value or with the FERROR() function. Which method you use depends on what level of detail you want regarding the error. The following table summarizes the return values of the low-level file functions that indicate an error condition:

Low-Level File Function Error Conditions

Function	Error Condition
FCLOSE()	Returns false (.F.)
FCREATE()	Returns -1
FOPEN()	Returns -1
FREAD()	Returns zero or a value less than the number of bytes to read
FREADSTR()	Returns a null string
FSEEK()	No error condition
FWRITE()	Returns zero or a value less than the number of bytes to write

FERROR() returns a DOS error number for the last low-level file function executed. If there is no error, FERROR() returns zero. After determining that a function has failed by checking its return value, you can use FERROR() to narrow down the cause of the error. This function retains its value until the next execution of a low-level file function.

Chapter 3
The Runtime Environment

After you have designed, written, debugged, and configured a *CA-Clipper* application program, the next step is to install the program on the end-user's machine. This process consists of preparing the user's workstation. Before doing this there are a number of issues to understand in order to guarantee that your application program will run correctly in the user's computing environment.

This primarily consists of setting the workstation operating system environment so the program will have enough resource to run, configuring the application program's environment, and configuring the user's network environment to guarantee that software installed on the network will operate correctly and have access to shared resources.

In This Chapter

In this chapter, the following major topics are discussed:

- Setting the workstation environment
- Setting the application program environment
- Setting the network environment
- The application batch file

Setting the Workstation Environment

The workstation's operating environment is configured by two facilities at system initialization: settings defined in CONFIG.SYS and commands executed in AUTOEXEC.BAT.

CONFIG.SYS is an ASCII text file that DOS processes during system initialization. It is used to configure the workstation's operating system environment by setting the number of internal buffers and open files and installing device drivers to enable RAM disks, terminal emulation, extended or expanded memory, and other peripheral devices. Of particular interest when executing *CA-Clipper* programs are the FILES and BUFFERS settings, although there are others.

When the user's computer is started, CONFIG.SYS must be located in the root directory of the user's startup disk. After the operating system is configured the system batch file AUTOEXEC.BAT is called.

AUTOEXEC.BAT is a special batch file located in the root directory of the workstation startup disk. After the command processor (COMMAND.COM) is loaded, it looks for AUTOEXEC.BAT and executes all of the commands found in the file. Typically, an AUTOEXEC.BAT would perform one or more of the following operations:

- Define the PATH environment variable to tell COMMAND.COM where to look for executable files if they are not located in the current drive/directory

- Execute a series of SET commands to define common environment variables

- Use the MODE command to configure the serial port and redirect printed output to it

- Load TSR programs such as the network shell or a local print spooler

- Invoke a default application program such as a DOS shell or user menu

For more information on the full range of configurations that can be made in CONFIG.SYS and AUTOEXEC.BAT, refer to your DOS manual.

Files and Buffers

In order for a *CA-Clipper* application to run, it must have enough file handles and buffers. You can configure these items using the FILES and BUFFERS commands in the CONFIG.SYS file.

The FILES command uses *file handles* to set the number of files/or devices that can be open at one time. The default value is eight, but FILES can be set as high as 255 if the user's workstation is running DOS 3.3 or later. If the number of file handles is greater than 20, you must use the CLIPPER environment variable (discussed later on in this chapter) to tell *CA-Clipper* how many handles to use.

The actual number of file handles to specify depends on your application. As a general principle, you should allow one handle for each database and index file in your application. Note, however, that

Setting the Workstation Environment

some operations such as indexing and sorting create temporary files. This increases the number of file handles required for the application to run.

If the user's workstation is running a version of DOS earlier than version 3.3, the maximum number of files that can be open at one time is 20. If this is the case, you may want to upgrade to a more recent version of DOS.

The BUFFERS command allocates the number of buffers (disk cache) DOS uses to keep copies of the sectors most recently read from or written to disk. If not specified, the default value is two for IBM PC/XTs and three for IBM ATs and beyond. The recommended number of buffers for *CA-Clipper* is eight.

Example

The CONFIG.SYS file should contain the following two statements for versions of DOS prior to 3.3:

```
FILES=20
BUFFERS=8
```

Changing the DOS Environment Size

When setting up your application on a user's machine, you may add more environment variables than will fit in the default environment space (160 bytes in version 3.2). Whenever this happens, a DOS error message is displayed and the variable is not assigned.

In DOS versions 3.2 and above, you can change the default environment size by specifying a SHELL command in CONFIG.SYS, like this:

```
SHELL=COMMAND.COM /E:<nBytes> /P
```

<nBytes> is the initial allocation of environment space and can be any value between 160 and 32,768. The recommended value is 1024 bytes.

Specifying the Location of Executable Files

With a *CA-Clipper* program, there are several types of files associated with the application's executable image. These include the following:

- Executable files (.EXE)
- Overlay files (.OVL)
- Prelinked library files (.PLL)

Setting the Workstation Environment

In the *runtime environment*, there are some specific rules as to how these files are located when invoked or requested.

The executable file (.EXE) is the program file that can be invoked directly from the DOS command prompt and is searched for in the current directory. If not found, the current PATH is searched.

Overlay files (.OVL) are files associated with the application .EXE file and contain either dynamic or static overlays. They are searched for only in the directory where the .EXE is located. No other directories are searched. If not found, a runtime error is generated. This means you must install all overlay files in the same location as the executable (.EXE) file.

Prelinked library (.PLL) files contain the executable code loaded into memory whenever you execute the associated .EXE file. This means that each time the .EXE is executed, the .PLL file must be available and must, therefore, be distributed with your system. Prelinked libraries can be shared between multiple .EXE files and therefore must have a wider search rule than .OVL files. The order in which .PLL files are searched for at runtime is as follows:

- The directory in which the .EXE is located
- The path specified in the PLL environment variable
- The path specified in the LIB environment variable

If the file is not found, a runtime error is generated.

To assure executable files (.EXE) and associated files can be located, add the location of the .EXE file to the PATH environment variable and assign a separate PLL variable for .PLL files in the user's AUTOEXEC.BAT file:

```
SET PATH=C:\APPS\MYAPP
SET PLL=C:\APPS\PLL
```

Warning! *When running under any DOS version prior to 3.0, do not rename an .EXE file created using .RTLink or you will not be able to execute the file.*

Temporary Files

In the process of some operations such as sorting and indexing, *CA-Clipper* creates temporary files. These files are created either in the current directory or in the directory specified by the environment variable TMP. To assign a TMP variable, place a command line like the following in the user's AUTOEXEC.BAT file:

```
SET TMP=C:\TEMP
```

Note: In contrast to *CA-Clipper*, Summer '87 created temporary files in the directory specified by SET DEFAULT or in the DOS current directory if no SET DEFAULT was issued.

Specifying the Location of COMMAND.COM

When you use the RUN command to execute external programs, the application runtime system needs to load another copy of COMMAND.COM to execute the specified program. To do this successfully, it requires that the environment variable COMSPEC point to the location of a copy of COMMAND.COM. Normally, DOS sets the value of this variable to the root directory of the startup disk.

If for some reason this is not the case, add an explicit assignment of COMSPEC in the user's AUTOEXEC.BAT or application batch file, as follows:

```
SET COMSPEC=C:\DOS
```

The application batch file is discussed later on in this chapter.

Configuring the Serial Port

If the workstation is printing to a local printer attached to a serial port, you must configure the port with the MODE command. MODE is used to configure the serial port with the appropriate baud rate, parity, word length, and stop bits for the attached printer. A subsequent MODE command can also be used to redirect output from the default parallel port (LPT1, LPT2, or LPT3) to a serial port (COM1 or COM2).

For *CA-Clipper* programs running on a workstation with a serial port and a local printer attached, the recommended procedure is to first run MODE with the configuration options within the application batch file. To redirect printed output from the default parallel port, you have two choices: you can either redirect output using the MODE command in the application batch file, or you can use the SET PRINTER command within the application program.

For example, the following application batch file shows how to use the MODE command to configure, as well as redirect output to serial port:

```
REM Application batch file
MODE COM1:=1200,8,1,P
MODE LPT1:=COM1
MYAPP.EXE
MODE LPT1
```

If the SET PRINTER command is used, the reference to the port can be a literal or variable value. A literal value designation is somewhat restrictive and environment specific. The printer port can be specified to the SET PRINTER command as an extended expression. This allows you to pass the print destination to an application program via an environment variable or as a command line argument of the application program.

For example, this application batch file configures the serial port and then defines an environment variable designating the printer port as COM1:

```
REM Application batch file
MODE COM1:=1200,8,1,P
SET CLIPPORT=COM1
MYAPP.EXE
MODE LPT1
```

In the *CA-Clipper* application program, the GETENV() function is used to obtain the port. If the environment variable exists, SET PRINTER directs output to the specified port:

```
IF !EMPTY( (cDest := GETENV("CLIPPORT")) )
   SET PRINTER TO (cDest)
ELSE
   SET PRINTER TO LPT1
ENDIF
```

Setting the Application Program Environment

Once you have set the workstation environment, you may need to set the application program environment. *CA-Clipper* provides runtime configuration control of *CA-Clipper* compiled programs with a number of settings that can be specified in a DOS environment variable named CLIPPER or on the application program command line.

The following table lists the syntax and a brief description of each setting. Settings marked with an asterisk (*) do not have any meaning in the application command line context and should, therefore, only be specified using the CLIPPER environment variable.

Setting the Application Program Environment

Application Program Environment Settings

Setting	Meaning
BADCACHE	Save/restore EMM page frame on each EMM access
CGACURS	Prevent use of EGA/VGA extended cursor capability
DYNF:<*nHandles*>	Specify number of file handles for dynamic overlay system use
E:<*nExpandedKbytes*>	Configure amount of expanded memory
F:<*nHandles*>	Set maximum number of file handles
INFO	Display memory configuration details at startup
NOIDLE	Prevent detection of idle time during execution of compiled applications
SWAPK:<*nBytes*>	Specify maximum size of disk swap file used for VM system
SWAPPATH:'<*path*>'	Specify location of VM swap file
TEMPPATH:'<*path*>'	Control placement of temporary sort and index files
X:<*nKbytes*>	Exclude available memory

You should adhere to the following rules when specifying these environment settings regardless of whether they are specified on the application command line or in the CLIPPER environment variable:

- Preface each setting with a double-slash
- Place a single blank space between settings
- Place a colon between setting and argument with no intervening space

Application Program Command line

To specify environment settings on the application command line, use the following syntax:

```
C><app> [//<setting> ... ] [<app arguments>]
```

Note that all environment settings must come before any application arguments on the command line. If you wish to use the command line method for specifying environment settings but also wish to hide the command line complexity from your end user, use an application batch file to invoke the application with the configuration settings.

Programming and Utilities Guide   3–7

The CLIPPER Environment Variable

In addition to the command line method for specifying environment settings, *CA-Clipper* provides a special DOS environment variable named CLIPPER to specify configuration information. When a *CA-Clipper* program is invoked, it looks for the environment variable CLIPPER and then processes the environment settings defined therein. The CLIPPER variable can be entered at the DOS prompt or included in the user's AUTOEXEC.BAT file. If the user will be accessing the application from a network, you can also add the SET CLIPPER command to the user's login script.

The CLIPPER environment variable is specified as follows:

```
SET CLIPPER=[//<setting> ...]
```

Saving/Restoring EMM Page Frame—BADCACHE

BADCACHE causes the Virtual Memory Manager (VMM) to preserve and restore the state of the Expanded Memory Manager (EMM) page frame before and after every EMM access (the EMM page frame is an area in real address space through which EMM data is accessed). This setting can be used to correct problems when there are conflicts with other programs that use EMM.

Note that on some EMM systems, the BADCACHE setting may adversely affect VMM performance. It should only be used if you experience disk or file corruption because of a conflict with a disk cache or other resident software.

Preventing Extended Cursor Use—CGACURS

CGACURS prevents the use of the extended cursor capability of the EGA/VGA. Specifying this setting may preclude some cursor modes by some display adapters. The CGACURS prevents the unusual cursor behavior when a *CA-Clipper* program is executed in some multitasking and TSR environments.

Specifying Number of Dynamic Overlay File Handles—DYNF

DYNF specifies the number of file handles the dynamic overlay system is allowed to use. Valid settings range from 1 to 8 inclusive. If not specified, the default is 2.

Configuring Expanded Memory—E

CA-Clipper-compiled and linked programs can use expanded memory to speed up processing as well as some disk-based operations. *CA-Clipper* can use memory configured as expanded memory according to the Lotus-Intel-Microsoft (LIM) Expanded Memory Specification (EMS) version 4.0 or higher.

Expanded memory is automatically allocated in its entirety at startup—there is no dynamic allocation as execution proceeds. The maximum amount allocated is limited by four factors:

- The amount available
- The SET CLIPPER=//E:nnn environment setting
- A theoretical maximum of 32MB
- The amount of conventional memory available

The E parameter can be used to restrict the amount of expanded memory automatically allocated to <*nExpandedKbytes*>. For example, specifying the following E parameter restricts the expanded memory *CA-Clipper* will allocate to 2MB:

```
SET CLIPPER=//E:2000
```

Note that a certain amount of conventional memory must be used to contain management tables for the virtual memory system—the more total memory (both conventional and expanded), the more space taken up by this control information. The amount of expanded memory used may be less than the amount available if there is insufficient conventional memory to hold the tables.

Specifying the Number of Files—F

If you have an application program that uses more than 20 files and are running under DOS 3.3 or greater, you must use the F parameter to inform *CA-Clipper* of the maximum number of file handles to use. This parameter is used in combination with the value you specified with the FILES command in CONFIG.SYS.

When specified, *CA-Clipper* determines the number of files that can be opened using the smaller of the F parameter and the CONFIG.SYS FILE value. For example, if the FILES command is set to 120 and the F parameter is set to 50, the maximum number of files that can be opened is 50. The ideal <*nHandles*> is an odd number and 5 less than the number specified with the FILES command.

Keep in mind, the F parameter should be set judiciously. It must be large enough to allow the use of all needed files and yet not so large that valuable memory space is unnecessarily used up.

Displaying Memory Configuration Details at Startup—INFO

INFO provides the following information about *CA-Clipper*'s memory usage at startup of an application.

- The first line describes the general product version, revision, and international version.

- *DS=<offset>:0000* is the address for the data segment or DGROUP. This value has no application functionality.

- *DS avail=<memory>KB* reflects the amount of DGROUP available which is used for the allocation of the processor stack, *CA-Clipper* statics, *CA-Clipper* environment variables and the evaluation stack.

- *OS avail=<memory>KB* represents the amount of conventional memory available for VMM swap space. If the reported value is too low, there is a strong possibility the program will terminate with a "Conventional Memory Exhausted" unrecoverable error.

- *EMM avail=<EMM memory>KB* shows the amount of EMM (expanded memory) allocated to the current application. A "P" following the *EMM avail* value appears when the application is loaded with the //BADCACHE configuration setting.

- *Fixed Heap=<fixed heap>KB/<number of fixed segments>* shows the size of the fixed heap in kilobyte increments in addition to the number of fixed segments. Fixed heap is used for symbols at runtime (macros), static allocations (extend system), the EMM table and the symbol table.

Preventing Detection of Idle Time—NOIDLE

NOIDLE prevents *CA-Clipper* from detecting and taking advantage of idle time during execution of compiled applications. *CA-Clipper* detects idle states (e.g., keyboard wait states) during execution of compiled applications. When an idle condition is detected, the system uses the slack time to perform garbage collection, file updates, and other routine housekeeping duties. This increases system performance by doing this work while the application is waiting for user input.

Note: NOIDLE is provided for applications in which idle time processing is unacceptable. Since it reduces overall system performance, its use is generally not recommended.

Specifying Maximum Swap File Size—SWAPK

SWAPK specifies the maximum allowable size of the disk swap file used for the virtual memory (VM) system. Settings are specified in kilobyte increments. Valid settings range from 256 to 65,535, inclusive. If this setting is not specified, the default is 16,384 (16 MB).

Note: Swap space is allocated only as needed—a particular setting does not guarantee that the swap file will get that big. Suppressing or restricting disk swapping may cause an application to fail.

Specifying Swap File Location—SWAPPATH

SWAPPATH specifies the location of the virtual memory swap file. If not specified, the swap file is created in the current DOS drive and directory.

Specifying Temporary File Location—TEMPPATH

TEMPPATH controls the placement of temporary files created during sorting and indexing. By default these files are placed in the current DOS directory.

Note: Temporary files created during sorting and indexing can be quite large. Setting TEMPPATH to a small volume (e.g., a ram disk) may cause these operations to fail. In general, the volume where these temporary files will be written should have an available capacity at least twice the size of the largest index to be created or database file to be sorted.

Excluding Available Memory—X

In certain situations, especially debugging, you may need to test how a program operates in a memory environment less than the current machine offers. You can do this using the X parameter to exclude <nKbytes> of memory from being allocated, except for the RUN command. For example, if you specify //X:64 with a RAM configuration of 640K memory is allocated as though the computer contains only 556K of memory.

Valid values for <nKbytes> range from zero to 64, inclusive. The default value is zero.

Note that memory is excluded before other memory is allocated. The amounts allocated for other purposes are diminished accordingly.

Setting the Network Environment

If your application program is being installed on a network, there are some basic requirements that must be fulfilled for the application program to operate properly.

Network Hardware Requirements

In order for a *CA-Clipper*-compiled program to operate on a network, there are some basic hardware and operating system requirements:

- The workstation operating system must be DOS 3.1 compatible
- The workstation must have enough memory to run the network operating system shell, the workstation operating system, and the *CA-Clipper* application program
- The network operating system must support the DOS 3.1 networking protocol

Note that the amount of memory used by the network shell varies from one network operating system to another.

Assigning Rights

Most network environments govern the access of users to various locations on the file server disk as a part of the network security system. Network security is necessary in a shared environment to provide suitable levels of protection to other people's work as well as privacy. This access control is maintained through a system of rights to directories and files. Rights are granted to users by a network supervisor.

With most PC-based network operating systems, rights are pessimistically granted. Users are given a minimal set of rights as a default and must be explicitly granted subsequent rights. If the user attempts to access a file and perform some operation without the required rights for that operation, an error is generated from the network operating system to the application program. This means that for an application program to operate properly in a network environment, the user must be granted the appropriate level of rights.

With a *CA-Clipper*-compiled program, the minimum set of rights required to guarantee that the program will successfully run is as follows:

- *Open rights:* The user has the right to open existing files
- *Read rights:* The user has the right to read from or execute files
- *Write rights:* The user has the right to write to existing files
- *Search rights:* The user has the right to search the directory's file list

Remember when the application program attempts to open a file in read-write mode and the user does not have write privileges to that directory or file, an error is generated. As a default, *CA-Clipper* opens database and index files as read-write unless specified with the READONLY clause of the USE command.

Note also that if the application program creates files for any reason, the user will also require create rights. This could include indexing, copying, or writing output to a print file.

Setting Up Network Devices

Part of the resource a network offers is shared access to peripherals and devices. For most application programs these are network printers.

In a network environment, output to a network printer is captured by the network shell (resident in the workstation's memory) and redirected to the network's printing system. Typically, captured output is spooled to a file until the print job terminates. The spool file is then placed in a *print queue* and printed when the network printer becomes ready.

With a *CA-Clipper* program, one of the workstation's ports must be redirected to the specified network printer or queue. This is usually done by executing a network utility prior to executing the application. This can be done either in the user's login script or in the application batch file.

Refer to the *Network Programming* chapter in this book for more information on programming considerations for network printing.

Refer to your network documentation for more information about accessing and setting up network printers.

The Application Batch File

If your application program has specific invocation requirements, you can create a batch file to invoke it. This batch file can be quite simple, changing the directory to the application home directory and then invoking the application program. For example:

```
ECHO OFF
REM Accounts Receivable Module
CD \APPS\ACCOUNT
AR
CD \
```

Additionally, you can set the *CA-Clipper*-specific environment variables for the duration of the run and then release them when the application program quits. For example:

```
ECHO OFF
REM Accounts Receivable Module
SET CLIPPER=//F:54 //E:2000
SET TMP=C:\TEMP
CD \APPS\ACCOUNT
AR
CD \
SET CLIPPER=
SET TMP=
```

Additionally, the application batch file can run network configuration utilities to redirect output to the network printer among other workstation specific actions.

Depending on the standards of the user's site, such as corporate or company standards for invoking programs, you can put the batch file in the user's root directory, batch file directory, or \DOS directory. Most importantly, the batch file must be in the path so it can be invoked from anywhere on the user's disk.

Summary

In this chapter, you have been exposed to most of the major issues that you need to know about in order to install and run *CA-Clipper* application programs both in a single-user and in a network environment.

Chapter 4
Network Programming

This chapter contains conceptual information as well as detailed instructions for developing application programs in *CA-Clipper* to run in a Local Area Network environment. For more specific reference material on any of the commands and functions mentioned here, see the *Reference* guide.

If your application does not require access to a shared database file, even though your computer system has a Local Area Network, the information in this chapter does not apply to you. By default, a *CA-Clipper* application assumes a single user and opens all files in exclusive mode (SET EXCLUSIVE ON); therefore, no file sharing takes place and no locking (file or record) is necessary.

In This Chapter

The following topics are covered:

- Definition of a Local Area Network
- Local Area Network requirements for *CA-Clipper*
- *CA-Clipper* network features
- Network commands
- Network functions
- Programming in a network environment
- Update visibility
- Network printing

Definition of a Local Area Network

A Local Area Network, or LAN, is two or more personal computers connected to one or more file servers that share data and peripheral equipment in a controlled environment. The computers, the file servers, the peripheral devices, the hardware connecting these items, and the software controlling the communication are collectively referred to as a LAN.

Connecting the computers and enabling them to share data and peripheral devices requires additional equipment and software. Usually, the required hardware consists of a board installed in each of the computers and a cable that connects all of the boards to one main computer, called the *file server*. In a simple network, there is a single file server that contains the hard disk for the shared files and has attached to it all of the shared peripherals. The network software must be installed on each computer to identify the file server, the shared peripherals, and the computer workstations that are part of the LAN.

Sharing data is possible because of a method for opening files (called shared mode) provided by the network operating system that lets two or more users open the same file. *CA-Clipper* utilizes this capability and allows you to control it with functions that temporarily override the shared mode in favor of private control by a single-user.

Local Area Network Requirements for CA-Clipper

The only requirement for using *CA-Clipper* with a Local Area Network is that the LAN must adhere to DOS 3.1 or greater function calls. *CA-Clipper* uses DOS calls exclusively for all network related operations and, consequently, applications that are compiled with *CA-Clipper* can run on any LAN designed to the DOS standard.

You should be familiar with the nature and design of your LAN before attempting to develop applications that use the *CA-Clipper* network features described in this chapter.

Note: Each network workstation must be running under PC/MS-DOS version 3.1 or greater in order to use files in shared mode.

CA-Clipper Network Features

CA-Clipper provides the following features that allow you to take advantage of LAN capabilities to develop shared applications:

- **Shared access** to allow two or more users to open the same file simultaneously (USE...SHARED)

- **Exclusive access** to prevent other users from opening the same file at the same time (USE...EXCLUSIVE)

- **Logical file locking** to prevent two or more users from updating the same file at the same time (FLOCK() function)

- **Logical record locking** to prevent two or more users from updating the same record at the same time (RLOCK() function)

- **Lock status checking** built into both file and record locking mechanisms (FLOCK() and RLOCK() return the current lock status as a logical value)

- **Record and file lock release** mechanism to return to shared access mode (UNLOCK command)

- **Printer redirection** to allow use of a shared network printer (SET PRINTER TO command)

- **Network error checking** to test the success or failure of certain commands (NETERR() function)

The *CA-Clipper* networking features offer extensive flexibility and optimal computer usage. It should *not* be assumed, however, that *CA-Clipper* applications written to take advantage of these features will be compatible with other network software products, such as dBASE III PLUS.

Network Commands

SET PRINTER TO

```
SET PRINTER TO [<xcDevice> | <xcFile>]
```

Establishes the destination of the printed output, allowing the use of shared and local printers.

UNLOCK

```
UNLOCK [ALL]
```

Releases file and record locks previously set by the lock functions.

USE

```
USE [<xcDatabase>
   [INDEX <xcIndex list>]
   [ALIAS <xcAlias>] [EXCLUSIVE | SHARED]
   [NEW] [READONLY]
   [VIA <cDriver>]]
```

The EXCLUSIVE clause opens the database file and its associated index files in exclusive mode, and the SHARED clause opens the files in shared mode. Both of these clauses operate without regard to SET EXCLUSIVE, allowing you to operate in a shared or exclusive mode while selectively opening files in whatever mode you need for a particular operation.

Network Functions

FLOCK()

Attempts to place a lock on the current database file. If the lock is successful, the function returns true (.T.); otherwise, it returns false (.F.) To lock a file in an unselected work area, precede FLOCK() with an alias expression (e.g., *<alias>*->(FLOCK())). A successful FLOCK() releases any previous file or record lock placed on the file by the same user.

NETERR()

Returns true (.T.) if certain *CA-Clipper* network operations fail. These operations and the circumstances of their failure are listed in the table below.

NETERR() Returns .T.

Command	Cause
USE	Another user has exclusive use of the file
USE...EXCLUSIVE	Another user has the file open
APPEND BLANK	Another user has an FLOCK() in place or has issued a simultaneous APPEND BLANK

NETNAME()

Gives the identification of the current workstation.

RLOCK()

Attempts to place a record lock on the current record in the active database file. If the lock is successful, the function returns true (.T.); otherwise, it returns false (.F.). To lock the current record in an unselected work area, precede RLOCK() with an alias expression (e.g., <alias>->(RLOCK())). A successful RLOCK() releases any previous file or record lock placed on the file by the same user.

Programming in a Network Environment

This section gives you tips on network programming in *CA-Clipper*. To help you write effective network applications, the following issues are discussed:

- Attempting a lock
- How to open files for sharing
- Other commands that open files
- Resolving a failed lock
- The mechanics of locking
- When to lock records and files
- When to obtain exclusive use
- Overlays on a network

By default, a *CA-Clipper* application assumes a single-user and opens all files in exclusive mode (SET EXCLUSIVE ON). Network operation requires you to include SET EXCLUSIVE OFF in your application in order to share the information in database files.

Attempting a Lock

When developing a strategy for obtaining file and record locks, an issue of persistency arises. If you cannot lock what you want, when is giving up premature, and when is it appropriate? Several approaches are possible:

Programming in a Network Environment

- Try once
- Retry for a fixed time limit
- Retry until interrupted (e.g., by pressing *Esc*)
- Retry for a fixed time limit or until interrupted
- Retry forever

The *try once* approach is not persistent enough, whereas *retry forever* is too persistent. Both are impractical for most applications. To make your network programs both flexible and practical, you need a compromise. There are several alternatives for deciding when to terminate the retry effort: set a time limit, let the user terminate it, or use a combination of the two.

The user-defined functions RecLock() and FilLock() in the Locks.prg sample program provide for the time limit. A slight modification permits user termination. The source code for these functions is listed below. You can either use them as is, or as a starting point that you will adapt to suit your application.

The FilLock() Function

FilLock() attempts to lock the current shared file. The only argument is *nSeconds*, a numeric value defining the number of seconds to retry the file lock—a value of zero retries forever. Like FLOCK(), FilLock() returns the logical value, *lvalue* which is true (.T.) if the file lock was obtained and false (.F.) if it was not.

```
FilLock( <nSeconds> ) → lValue
```

The RecLock() Function

RecLock() attempts to lock the current shared record. The only argument is *nSeconds,* a numeric value defining the number of seconds to retry the record lock—a value of zero retries forever. Like RLOCK(), RecLock() returns the logical value, *lvalue* which is true (.T.) if the record lock was obtained and false (.F.) if it was not.

RecLock(<nSeconds>) → *lValue*

How to Open Files for Sharing

Network programming presupposes that you will USE database files in shared mode. There are a few simple requirements for doing so.

In network programming, the failure of a USE becomes a normal possibility; therefore, you should always check NETERR() immediately after a USE to make sure the file was successfully opened.

For this reason, you should never include the INDEX clause in a USE command when programming in a network environment. Although USE...INDEX is a single command, it performs several distinct open operations. Instead, open the index files with SET INDEX after checking NETERR() to determine that the USE succeeded. Using this method, you can prevent possible runtime errors and be guaranteed that the index files will be opened successfully and in the same mode as the associated database file. SET INDEX automatically opens the index files in the same mode (i.e., shared or exclusive) as the current database file.

The following code illustrates the correct way to open database and index files in a network program:

```
USE MyFile SHARED
IF NETERR()                     // Returns true (.T.) if USE failed
    ? "File not available in shared mode."
    BREAK
ENDIF
SET INDEX TO MyIndex1, MyIndex2
```

Programming in a Network Environment

The next example is very similar to the one above except that it uses the NetUse() function described in the next section to open the database file:

```
#define DB_SHARED     .F.
#define DB_EXCLUSIVE  .T.

IF NetUse("File", DB_SHARED, 5)
   SET INDEX TO MyIndex1, MyIndex2
ELSE
   ? "File not available in shared mode."
   BREAK
ENDIF
```

The NetUse() Function

NetUse() is a user-defined function found in the sample program Locks.prg. The function demonstrates a flexible and comprehensive file opening scheme that attempts to open a database file in either shared or exclusive mode and continues trying to open the file for a given amount of time or until it succeeds.

The function operation depends on the following arguments: *cDatabase* is a character string representing the name of the database file to open; *lOpenMode* is a logical value describing the open mode; and *nSeconds* is a numeric value defining the number of seconds to retry the operation—a value of zero retries forever. NetUse() returns the logical value, *lvalue* which is true (.T.) if the file was opened successfully and false (.F.) if it was not.

```
NetUse( <cDatabase>, <lOpenMode>, <nSeconds> ) → lValue
```

Notice that your application must anticipate that the file may not be available for USE (even in the more permissive shared mode) since another user could have the file open in the exclusive mode already. Possible failure of a USE to open the file is a normal event in a network environment.

Other Commands that Open Files

In addition to USE and SET INDEX, several other *CA-Clipper* commands and functions automatically open one or more files in the course of operation and assign the open mode based on their operation.

There are two general rules that will help you to decide how a given command or function operates: if it writes to the file, the open mode is exclusive; if it only reads the file, the open mode is shared. As a programmer you do not have control over this, but knowing it helps you make the proper allowance in your network applications. For example, SAVE writes to a (.mem) file and so opens it exclusively;

Programming in a Network Environment

RESTORE reads a (.mem) file, and so opens it shared. Therefore, your programs must anticipate possible simultaneous SAVEs by providing mechanisms for avoidance (e.g., using interworkstation communication) or recovery (e.g., using the error system).

The table on the next page shows the open modes for all commands and functions that automatically open files. In cases where there are two possible file open operations, both files are shown; the ordering of the files refers to the order in which they appear in the command syntax.

Open Modes For Automatic File Opening

Command/Function	File1	File2
APPEND FROM	shared	
COPY FILE	shared	exclusive
COPY STRUCTURE	exclusive	
COPY STRUCTURE EXTENDED	exclusive	
COPY TO	exclusive	
CREATE	exclusive	
CREATE FROM	exclusive	shared
DISPLAY...TO FILE	exclusive	
INDEX	exclusive	
JOIN	exclusive	
LABEL FORM...TO FILE	shared	exclusive
LIST...TO FILE	exclusive	
REPORT FORM...TO FILE	shared	exclusive
RESTORE	shared	
SAVE	exclusive	
SET ALTERNATE	exclusive	
SET PRINTER	exclusive	
SORT	exclusive	
TEXT...TO FILE	exclusive	
TOTAL	exclusive	
TYPE...TO FILE	shared	exclusive
UPDATE	shared	
MEMOREAD()	shared	
MEMOWRIT()	exclusive	

Note: In addition to these commands and functions, the low-level file open function, FOPEN(), allows you to directly specify the open mode. The operation of FOPEN() is unaffected by SET EXCLUSIVE and is mentioned here for completeness.

Resolving a Failed Lock

As stated earlier, you can develop a locking strategy that continues to retry a failed lock until it is successful, but this is not recommended since you have no idea how long it will take to obtain the lock. Thus, no matter what other locking strategy you devise, you have to consider that it might fail to obtain the lock. RecLock() and FilLock(), for example, both fail if the lock is not obtained within the indicated number of seconds. If you find that your target data is unavailable, it is time for contingency plans. Here lies the real difference between network code and familiar single-user code.

The code adaptations for opening files and attempting locks are straightforward. If, however, the needed record or file lock cannot be obtained, the program has to abandon its original intentions and come up with substitute plans. You have choices at this point:

- Communicate the lock failure to the user
- Make a branching choice

You can make the communication transient (a two second duration, for example) or follow it by an INKEY(0) pause to be sure the user has seen it. It can be a small message displayed in the corner of the screen or a whole screen displayed between SAVESCREEN() and RESTSCREEN().

The branching can be hard-coded or can provide for multiple-way branching with user intervention at a small menu. In that case the menu itself constitutes the user advice. A typical menu might offer three choices:

- Retry
- Try same activity, different data
- Abort current activity and go back to higher menu

The first choice allows the user to extend the lock effort. It does not genuinely seek to resolve the failure.

The second choice makes sense in many applications. If the target record is not available, use another instead. For example, a data entry operator working on a stack of credit adjustment slips will not mind putting Smith's slip at the bottom of the pile and moving on to Jones and Brown. When the operator gets back to Smith an hour later, that record will probably be free. Therefore, make the code branch to the record selection entry point.

The third choice implies abandoning not only the record in question, but the activity itself. For example, the user leaves the credit adjustment section of the program altogether and goes back to the main menu.

The Mechanics of Locking

If you are operating in shared mode, you must obtain exclusive control of the shared file to perform certain operations, as mentioned earlier. You do this with the lock functions, FLOCK() and RLOCK(). Each of these functions serves two purposes: to attempt the lock and report back the result of the attempt.

If a user executes a successful lock function, the function returns a logical true (.T.) and the appropriate lock is put into place. FLOCK() places a lock on the entire file in the current work area, and RLOCK() locks the current record only.

Once the lock is in place, the user is allowed to write to the file. Other users' attempts to lock the same record or file will fail, but they will still have read access to the file. Attempting to write to an unlocked shared file normally returns an error message.

The lock remains in place until the user who obtained it releases it by:

- Issuing an UNLOCK from the work area of the locked file
- Issuing UNLOCK ALL from any work area
- Closing the locked file
- Terminating the program normally
- Issuing another lock function for the same file

Because efficient network applications seek to minimize the duration of the locks they impose, you should always remember to release your lock as soon as possible after it has served its purpose.

Note: Attempting a lock releases any existing lock before trying to achieve the new one, regardless of whether or not the attempt is successful

When to Lock Records and Files

If you choose to allow global file sharing in your application, you must determine when file or record access by more than one user must be prevented. Then, you can incorporate lock functions into your application with accompanying logic to control execution in accordance with the lock results. Use these basic guidelines to determine when locking is necessary:

- Locking is *required* whenever you are going to write to a database file
- Locking is *optional* at all other times

Commands that write to a database file include @...GET, DELETE, RECALL, and REPLACE and, as stated in the first guideline above, locking is a hard and fast requirement for these commands. *CA-Clipper* enforces it with an error message if you execute any of them without a record lock when the active file is open in shared mode.

A file lock must be obtained for processes that update multiple records in a database file, that is, whenever you DELETE, RECALL, REPLACE, or UPDATE with a multiple record scope.

CA-Clipper does not require a lock for operations that only read files such as COUNT, LIST, REPORT FORM, and SUM—even if they have multiple record scopes. The automatic buffer management technique that is built into *CA-Clipper* ensures that these commands have access to the most up-to-date information in the database files; however, you may wish to obtain a file lock before performing these operations to make sure that none of the data changes during the process. Whether to lock the file depends on your application requirements.

The following table summarizes the commands that require locks in a network environment:

Commands Requiring Locks

Command	Requirement
@...SAY...GET	RLOCK()
DELETE (single record)	RLOCK()
DELETE (multiple records)	USE...EXCLUSIVE or FLOCK()
RECALL (single record)	RLOCK()
RECALL (multiple records)	USE...EXCLUSIVE or FLOCK()
REPLACE (single record)	RLOCK()
REPLACE (multiple records)	USE...EXCLUSIVE or FLOCK()
UPDATE ON	USE...EXCLUSIVE or FLOCK()

The AddRec() Function

CA-Clipper allows you to APPEND BLANK records to a shared database without locking it. APPEND BLANK attempts to add and automatically lock the new record and sets NETERR() to indicate its success or failure.

AddRec() is a user-defined function found in the sample program Locks.prg that illustrates how to use APPEND BLANK on a shared database. The only argument is *nWaitSeconds*, a numeric value defining the number of seconds to retry the append—a value of zero retries forever. AddRec() returns true (.T.) if the new record is added and locked and false (.F.) if it is not.

```
AddRec( <nWaitSeconds> ) → lValue
```

When to Obtain Exclusive Use

Certain operations cannot be performed in the shared mode; they must have EXCLUSIVE USE of the file. *CA-Clipper* enforces this requirement with an error message if you attempt to use any of the following commands with a database file opened in shared mode:

- PACK
- REINDEX
- ZAP

If SET EXCLUSIVE is OFF, attempt to open the file with USE...EXCLUSIVE before performing an operation. Because exclusive use is precluded if any other users are sharing the file, it may be necessary to provide logic for communication among workstations to request another user to close the file.

Overlays on a Network

.RTLink, the *CA-Clipper* linker, produces overlays containing compiled program code. By default, all compiled *CA-Clipper* code is placed in dynamic overlays. Additionally, static overlaying may be requested for C or Assembler code.

By default, overlays are appended to the end of the application's .EXE file (the code may be written to a separate overlay file by using the INTO clause in the overlaying directive). During execution, the runtime system opens the .EXE file and reads overlays as required.

When the runtime system needs to read overlays, it opens the .EXE file for shared, read-only access. On a network, this allows multiple processes to read the file at the same time.

On some networks, however, a problem can arise if an attempt is made to run the .EXE when another process has opened it to read overlays. The error occurs because the DOS EXEC function (used by

COMMAND.COM to execute programs) opens .EXE files using a sharing mode called "compatibility mode." On some networks, this causes a sharing violation if the file is already in use by another process, even if the other process has specified that it wishes to share the file.

If this conflict occurs, it can usually be resolved by marking the affected file read-only on the disk. A small utility program RO.COM, located in \CLIPPER5\BIN, is supplied for this purpose. RO toggles the read-only status of a specified file.

Update Visibility

When programming in a network environment, it is important to determine when a database update actually becomes visible to other processes. A process is another application program running on the network. The visibility of an update differs depending on the process or entity observing the update. In this discussion, there are three levels of observation of an update with each process or entity referred to as an *observer*.

The Observers

The major *observers* are as follows:

- The initiator
- The operating system and other applications
- The physical disk

The Initiator

The initiator of an operation is the process which causes an update to occur, for example, a program which performs a REPLACE. The term *initiating machine* refers to the physical machine on which the initiator is executing.

The Operating System and Other Applications

Database updates performed within an application program are not visible to other processes (other applications running on the network) until they are written to the operating system on the initiating machine. Note that if the file being updated does not reside on the initiating machine, the operating system sends the update to the machine on which the file resides (the *target machine*).

Once the update has been received by the operating system on the target machine, it becomes visible to other processes. This communication between machines is only visible to the operating system, so for all practical purposes an update becomes visible to other processes when the initiator writes the update to the operating system.

The Physical Disk

Writing an update to the operating system does not guarantee that a physical disk write will be attempted. The operating system on the target machine may hold recently written records in memory. These records appear to other processes to be on the disk; however, if a failure occurs on the target machine, the records may never be physically written to disk. In this case, updates are lost, and processes to which the updates were visible may be proceeding with erroneous data.

DOS provides an entry point which can be called to ensure that all pending writes for a particular file (whether local or remote) have been physically written to disk. This is the known as the *commit* request. If a process requests that DOS commit a file located on a remote machine, DOS sends this request to the target machine.

Update Rules

CA-Clipper adds another level of complexity to the update process because its database system often defers write operations for the sake of speed.

This section explains the three major categories of database update, which are:

- Update on an exclusive file
- Update on a shared file with FLOCK()
- Update on a shared file with RLOCK()

These categories all follow certain update rules that govern the point where an update is *guaranteed* to be visible to other processes.

Note: The rules specified below are for the DBFNTX and DBFNDX database drivers. Rules for other drivers may differ

Update Visibility

Updating an Exclusive File

- *Initiator:* The update appears to occur immediately.

- *DOS/other application:* The update is not guaranteed to appear until the file is closed or a COMMIT is performed.

- *Disk:* The update is not guaranteed to appear until the file is closed or a COMMIT is performed.

Updating a Shared File with FLOCK()

- *Initiator:* The update appears to occur immediately.

- *DOS/other applications:* The update is not guaranteed to appear until the file is closed or one of the following operations is performed: UNLOCK, RLOCK(), or COMMIT.

- *Disk:* The update is not guaranteed to appear on disk until a COMMIT is performed or the file is closed.

Updating a Shared File with RLOCK()

- *Initiator:* The update appears to occur immediately.

- *DOS/other applications:* The update is not guaranteed to appear until the file is closed or one of the following operations is performed: UNLOCK, COMMIT, or any record movement operation (e.g., SKIP, GOTO).

- *Disk:* The update is not guaranteed to appear on disk until a COMMIT is performed or the file is closed.

When Does a Commit Write to Disk?

When a process requests a commit, there is no way for it to determine whether a physical disk write actually occurred. In particular, some cache programs and network server software (underlying DOS on the target machine) may postpone physical writes even after a commit. This is a violation of DOS protocol unless the underlying system can guarantee that its method of committing updates is as reliable as writing them to the disk controller hardware.

Additional Visibility Issues

The update rules described above only guarantee that updates will be visible at some particular point. There is no guarantee that they won't ever be visible *before* that point. In particular, updates under FLOCK() will often become visible before the guaranteed time. Since the system doesn't guarantee this behavior, it is unwise to depend on it.

Abnormal Termination

The phrase *the file is closed*, refers in the update rules to all types of operations which would close the file, including exiting the application.

If the application terminates abnormally (for example, with an unrecoverable database error), then some updates may be lost. The missing updates may potentially include every update to a file since updates to that file last became visible to DOS.

If an application causes DOS to fail, missing updates may potentially include every update to a file since the last point at which updates to that file appeared on disk (i.e., were committed).

Index Updates

Record updates and index updates appear to occur simultaneously, from the point of view of other *CA-Clipper* processes. That is, no *CA-Clipper* application can ever *see* an update to a record without also being able to *see* any associated index updates.

Network Printing

A network by nature is a system that provides shared access to common resources. Printers are among several hardware devices that a network can provide shared access to. From a general point of view, network operating systems provide services to make access to shared printers as transparent as possible to the workstation so that printing to a local printer appears no different to an application program than does printing to a network printer.

The actual difference between printing to a local printer and a network printer is that the output to the printer port is *captured* by the network shell and *redirected* to the network file server. As the output is received by the file server, it is *spooled* to a file stored on the file server. When a print job is complete, the file is closed. If the network printer is busy, the spooled file is placed in a *print queue* and sent to the printer whenever it becomes ready. Network print queues are generally first in, first out, although most network operating systems provide print queue management facilities to change the order of print jobs.

When sending printed output to a network printer from a *CA-Clipper* program, there are some considerations and approaches to consider.

Setting Up the Network Printer

In order to print to a network printer, one of the workstation ports must be redirected to the network printer. This is usually done with a network utility that configures the network shell.

This configuration can be accomplished in several ways:

- In the user's login script
- In the workstation AUTOEXEC.BAT file, but after the user has logged into the network
- In the application batch file
- Running the configuration utility under program control

Program Design Considerations

With most PC-based networks, a print job begins when the application program opens the PRN device. Printed output is captured and redirected to the spooler until the PRN device is closed by the application. At that point, the print job is submitted to the print queue and is printed when the printer becomes available.

In *CA-Clipper*, you can explicitly control this process with the SET PRINTER TO command. SET PRINTER TO specified with the device name begins the print job. After all of the printed output has been sent to the printer, SET PRINTER TO with no argument closes the print job and places the output in the print queue.

For example, the following code fragment starts and ends a print job with SET PRINTER TO:

```
USE Customer NEW
SET PRINTER TO LPT1
SET PRINTER ON
//
DO WHILE !EOF()
    ? Customer, Amount
    SKIP
ENDDO
//
SET PRINTER OFF
SET PRINTER TO
CLOSE Customer
```

Note: Some network operating systems support a timeout feature that terminates a print job if the network shell has not received output for a specified number of seconds. A timeout value of zero requires that the print job be explicitly terminated. Refer to your network documentation for more information.

Printing to a File

One of the great problems of printing is the strong likelihood that something will happen during the printing process that could abort a critical print job. In large scale installations with many users and a large volume of transactions, a printer problem can have system consequences. Accounting systems may, for example, print a report of transactions as they are posted at the end of an accounting period. If the report is sent directly to the printer, a print job failure could compromise database integrity. In this instance, reprinting the report would require a rollback of all posted transactions, and all print jobs would then have to be resubmitted to the printer.

A better approach is to print everything to a file, and then sometime later submit the print file to the network printer. If there is a printer or network failure before the report is successfully printed, the print file need only be resubmitted to the network printer. Once the report has been successfully printed, the print file can be deleted.

After printing to a file, you need to place the file in the print queue. There are several approaches to this. Perhaps the easiest method is to use the DOS COPY command to copy the file to the PRN device.

The one caveat to this technique is that it uses disk space over and above the amount of disk space on the server required to spool the same and other print jobs. When considering the issue of disk space, you may want to measure the benefits of database integrity and flexibility against the costs related to problems that occur when a report must be generated completely from scratch.

Summary

In this chapter, you learned the basic concepts of data sharing and locking in a network environment and how to use the *CA-Clipper* commands and functions to write successful network applications. You were also introduced to the sample functions in Locks.prg. To use these functions, you must compile and link Locks.prg with your application.

Chapter 5
Introduction to TBrowse

TBrowse is one of four predefined object classes included in *CA-Clipper*. It allows the programmer a much greater degree of control and flexibility in the retrieval, display, and editing , the *browsing*, of tabular views of data. It provides a robust, open-architecture browsing mechanism that is not tied to the (.dbf), or any other, file format.

This chapter establishes criteria for effective use of the TBrowse class. It starts with creation of a TBrowse object and a description of the TBrowse class, continues with information about the most common uses of TBrowse objects, and concludes by outlining problems and establishing guidelines for their resolution. This text assumes that you are familiar with the concepts and terminology associated with *CA-Clipper*'s predefined classes as discussed in the *Basic Concepts* chapter of this guide.

In This Chapter

This chapter discusses the following topics:

- Creating TBrowse objects
- Basic browse operations
- Optimization of the browse
- Browsing with Get
- Adding color to the browse
- Controlling the scope

TBrowse

To understand TBrowse, you must understand what it *doesn't* do as well as what it *does*. TBrowse is a passive object in the sense that it never initiates an action. It never takes control from the programmer for more than a few moments, never, without explicit direction, handles a single keystroke, and never "ends." You simply stop using it when you are done with it. TBrowse is a data source access tool that can browse any data that is expressible in a table form; is subject to its calling activation for control and order; can exist as multiple browses.

TBrowse differs from the *CA-Clipper* functions DBEDIT(), MEMOEDIT(), and ACHOICE(), in its degree of *exclusive* control. The functions control all screen and data manipulation irrespective of the activation that calls them. TBrowse, on the other hand, operates as a tool of the calling activation. That activation controls and orders the action of the TBrowse object by sending messages to it.

TBrowse is not a replacement for DBEDIT() (though, internally, DBEDIT() is written entirely in *CA-Clipper* code, using TBrowse). TBrowse is less limiting and more powerful than any of the earlier functions. Due to the nature of objects under TBrowse, you can activate several simultaneous browses which can be suspended and restarted at will. TBrowse may be read-only or interactive.

Basic Browse Operations

Regardless of the nature of the data on which you use TBrowse in your application, you should use the same general approach. The first example is the simplest of all cases, a read-only browse of a database. It starts with creation of the TBrowse object and design of the main TBrowse loop, then continues through stabilization and keystroke handling on a monochrome display.

Creating TBrowse Objects

The browse is composed of two different object classes: TBrowse, the main class, which controls the overall browse, and TBColumn, of which each column is an instance, and which is 'owned' by the TBrowse object.

First, create the TBrowse object. There are two functions for doing this: TBrowseNew() and TBrowseDB().

TBrowseNew() creates a completely generic TBrowse object. The following code creates a new TBrowse object and stores it in the variable, *browse*. The coordinates define the rectangular area of the TBrowse object.

```
obrowse := TBrowseNew(<nTop>, <nLeft>, <nBottom>, <nRight>)
```

TBrowseDB() creates a TBrowse object customized for (.dbf) browsing (it does this by defaulting some of the exported instance variables: the objects are of exactly the same class). It has the same syntax as TBrowseNew() but is designed for use while browsing a database file. TBrowseDB() creates a TBrowse object with default code blocks to locate and position the data source within a database.

Next, add columns to the browse by creating TBColumn objects. This is a two-step process. First create the TBColumn object, then add the column to the TBrowse object.

```
oCol := TBColumnNew ( "Record #", { || RECNO() })
obrowse:addColumn(oCol).
```

This example, using TBrowseDB(), creates one object for every field in the database in the currently selected area.

```
//   Creating a TBrowse Object
//
//     StockBrowseNew()
//     Create a "stock" TBrowse object for the current work area.
//
FUNCTION StockBrowseNew(nTop, nLeft, nBottom, nRight)
   LOCAL oBrowse
   LOCAL n, oCol

   // Start with a new browse object from TBrowseDB()
   oBrowse := TBrowseDB(nTop, nLeft, nBottom, nRight)

   // Add a column for each field in the current work area
   FOR n := 1 TO FCOUNT()
      // Make a new column
      oCol := TBColumnNew( FIELDNAME(n), ;
         FIELDWBLOCK(FIELDNAME(n), SELECT()) )

      // Add a column to the browse object
      oBrowse:addColumn(oCol)

   NEXT
RETURN (oBrowse)
```

Main Loop

The main loop of the browse is very straightforward; stabilize the browse (explained below), get a keystroke, act on the keystroke, then repeat the process. If the keystroke happens to be an exit key (in this case, *Esc*), exit the loop and return. The reference to the TBrowse object, since it was held in a local variable, falls out of scope and the system automatically reclaims the space used by the object.

The relevant code looks like this.

Basic Browse Operations

```
//  Main loop of the browse object
//
lMore := .T.
DO WHILE (lMore)

   // Stabilize the browse
   oBrowse:ForceStable()
   // The new Method, oBrowse:ForceStable(), works
   // like earlier, user-coded stabilization
   // procedures, but is faster

   // Positioned at top or bottom of browse
   IF oBrowse:hitTop .OR.  oBrowse:hitBottom
      TONE(125, 0)
   ENDIF

   // Get a new keystroke
   nKey := INKEY(0)

   IF nKey == K_ESC
      // Exit the loop
      lMore := .F.
   ELSE
      // Act on the keystroke
      ApplyKey(oBrowse, nKey)
   ENDIF

ENDDO
```

Stabilization

To permit greater control, TBrowse does not update the screen immediately when messages are sent to it. Instead, it waits until the object is *stabilized* with the stabilize() method. When that message is received, the browse is stabilized incrementally, displaying one change at a time. For example, when the TBrowse object is created, it is not immediately displayed on the screen. The first time a TBrowse object receives a stabilize() message, it displays the headers; then every time stabilize() is received, it displays one record until everything is finally displayed. (When everything is displayed, the stabilize() method returns true (.T.)) Multiple browses are therefore displayed "in sync" and/or the redraw of the screen can be interrupted by a keystroke.

In this example, we do nothing during the stabilization process except maintain a simple loop that forces complete stabilization.

```
//   Stabilization
//
//    ForceStable()
//    Stabilize the browse
//
oBrowse:ForceStable()
```

Handle Keystrokes

Keystrokes are handled by sending messages to the TBrowse object based on the reaction we desire to a specific keystroke. There are no default keystroke mappings in TBrowse; everything is under your control. Usually, keystrokes are mapped so the browse "reacts" like DBEDIT(). This example displays that type of mapping.

```
// Keystroke handling
//   Apply one keystroke to the browse.
//
PROCEDURE ApplyKey(oBrowse, nKey)
   DO CASE
   CASE nKey == K_DOWN
      oBrowse:down()

   CASE nKey == K_PGDN
      oBrowse:pageDown()

   CASE nKey == K_CTRL_PGDN
      oBrowse:goBottom()

   CASE nKey == K_UP
      oBrowse:up()

   CASE nKey == K_PGUP
      oBrowse:pageUp()

   CASE nKey == K_CTRL_PGUP
      oBrowse:goTop()

   CASE nKey == K_RIGHT
      oBrowse:right()

   CASE nKey == K_LEFT
      oBrowse:left()

   CASE nKey == K_HOME
      oBrowse:home()

   CASE nKey == K_END
      oBrowse:end()

   CASE nKey == K_CTRL_LEFT
      oBrowse:panLeft()

   CASE nKey == K_CTRL_RIGHT
      oBrowse:panRight()

   CASE nKey == K_CTRL_HOME
      oBrowse:panHome()

   CASE nKey == K_CTRL_END
      oBrowse:panEnd()

   OTHERWISE
      // Invalid-keystroke warning
      TONE(125, 0)

   ENDCASE
RETURN
```

Optimization of the Browse

A minimum browse like the example is barely useful and you should consider the values of constructing a more robust 'production' browse. These additional operations optimize a browse for single or multi-user environments and include reduction in response time, creation of calculated fields, picture clauses and custom headers, explicit record pointer control, and use of the TBrowse:cargo and TBColumn:cargo instance variables, Calculated Fields, Picture Clauses and Custom Headers.

Using TRANSFORM() in a data retrieval block can apply a picture to a cell or even calculate a field.

This function adds a Record Number column to the passed browse and formats it with commas.

```
//  Adding a RecordNumber Column to a TBrowse Object
//
//    AddRecNo()
//    Insert a frozen column at the left that shows current
//    record number formatted with commas
//
//
PROCEDURE AddRecNo(oBrowse)
   LOCAL oCol

   // Create the column object with custom header.  Use
   // TRANSFORM() to apply picture clause
   oCol := TBColumnNew("Record #", ;
      {|| TRANSFORM(RECNO(), "999,999,999")})

   // Insert it as the leftmost column
   oBrowse:insColumn(1, oCol)

   RETURN
```

Quicker Response Time

Often, you want to let the user page through a database or other table quickly. Previous examples have forced the user to wait until the screen redraws before they can move again. This significantly slows user operations. Fortunately, you can easily reduce the response time: you can interrupt the stabilize loop. Do this with care throughout, to ensure that you do nothing that is only valid on a stable browse. Check this by using the TBrowse:stable exported instance variable.

This code represents a main loop, modified to allow a stabilize loop interrupt.

Optimization of the Browse

```
// Interrupting the stabilize loop
//
lMore := .T.

DO WHILE (lMore)
   // Stabilize the browse until it's stable or a key is pressed
   nKey := 0
   DO WHILE nKey == 0 .AND. .NOT. oBrowse:stable
      oBrowse:stabilize()
      nKey := INKEY()
   ENDDO

   IF oBrowse:stable
      // positioned at top or bottom of browse (these values
      // only valid if stable)
      IF oBrowse:hitTop .OR.  oBrowse:hitBottom
         TONE(125, 0)
      ENDIF

      // Get a new keystroke
      nKey := INKEY(0)
   ENDIF

   IF nKey == K_ESC
      // Exit the loop
      lMore := .F.
   ELSE
      // Act on the keystroke
      ApplyKey(oBrowse, nKey)
   ENDIF
ENDDO
```

Multi-user Usage

For performance reasons, TBrowse buffers all cells for the currently displayed row (whether the cells are visible or not). While this gives a good performance boost, it means that what is on the screen may not match what is in the database. In a multi-user application, another user may have deleted one of the displayed records, modified records, etc. Rather than verifying that everything on the screen is current (which might be impossible and would certainly be slow), modify the browse to ensure that the current record is up-to-date. Use the TBrowse:refreshCurrent() method, and immediately stabilize. You must ensure that the browse is stable before you send the message, otherwise, the current row is undefined.

Combining this immediate update and stabilize with the quicker response time modification, we now have the following.

Optimization of the Browse

```
// Permit Multi-user Access
//
// Main loop
lMore := .T.
DO WHILE (lMore)
   // Increment Stabilize() method
   // or react to keystrokenKey := 0
   DO WHILE nKey == 0 .AND. .NOT. oBrowse:stable
      oBrowse:stabilize()
      nKey := INKEY()
   ENDDO

   IF oBrowse:stable
      // positioned at top or bottom of browse (these values
      // only valid if stable)
      IF oBrowse:hitTop .OR. oBrowse:hitBottom
         TONE(125, 0)
      ENDIF

      // Ensure that the current record is showing
      // up-to-date data (e.g., on a network).
      oBrowse:refreshCurrent()
      oBrowse:ForceStable()

      // Get a new keystroke
      nKey := INKEY(0)
   ENDIF

   IF nKey == K_ESC
      // Exit the loop
      lMore := .F.
   ELSE
      // Act on the keystroke
      ApplyKey(oBrowse, nKey)
   ENDIF
ENDDO
```

Repositioning the Record Pointer

TBrowse is a general-purpose browser unaffected by the underlying data format. While this is useful in many cases, there are times when it means that a little more work is necessary on the programmer's part. One of these cases is when the record pointer is repositioned outside the control of TBrowse (when seeking a new value, for example).

When a TBrowse stabilizes, it tries to leave the same cell highlighted as before. That is, it tries to keep the highlight at the same position within the browse window unless it is explicitly moved by an up or down message. The TBrowse positions the data source correspondingly. If there aren't enough rows left in the data source (i.e., end of file is encountered while trying to adjust the database to match the window), the TBrowse moves the cursor upward, leaving it on the correct record but with part of the window unfilled.

This behavior is appropriate for logical end of file, but a problem occurs when something outside of TBrowse moves the record pointer so close to the logical beginning of file that it is impossible to highlight the correct record while leaving the highlight at its previous position within the window. In this case, TBrowse leaves the highlight in the same position within the window, even though that position no longer corresponds to the same record as before. That is, it repositions the database as far as it will go, then leaves the highlight where it was. The result is the highlight on a different record than the one just edited.

Correct this behavior by forcing a complete refresh (through TBrowse:refreshAll()), and a full stabilization. *CA-Clipper* then checks for repositioning of TBrowse to a different record than before. If so, *CA-Clipper* assumes the old record is somewhere above the current record, and a series of up() messages are issued to the browse to move the highlight to the proper position. Determination of proper positioning of the highlight varies depending on what is being browsed. You must have some way of determining the exact row position. In the case of databases, use the record number, or the index key, if it is unique.

This function is called from a keystroke handler that prompts for a key value for which to search when you type *Alt-S*. It correctly repositions the highlight by using the record number.

```
//   Repositioning the Record Pointer
//
//   Prompting for a Key Value to Seek
//
//   Search browse on key field with SEEK and adjust highlight
//
PROCEDURE SearchBrowse(oBrowse)
   LOCAL cSaveScreen, xSearch
   LOCAL nRec
   LOCAL GetList := {}

   // Ensure browse is stable before getting info about
   // current record
   oBrowse:ForceStable()

   // Get current key value and record number
   xSearch := & (INDEXKEY(0))
   nRec := RECNO()

   // Put prompt online surrounding browse
   cSaveScreen := SAVESCREEN( oBrowse:nBottom + 1, 0, ;
      oBrowse:nBottom + 1, MAXCOL() )

   SETPOS(oBrowse:nBottom + 1, oBrowse:nLeft + 2)
   DISPOUT("[Search For ")
   @ ROW(), COL() GET xSearch PICTURE "@K"
   DISPOUT("]")
   READ

   RESTSCREEN(oBrowse:nBottom + 1, 0, oBrowse:nBottom + 1, ;
      MAXCOL(), cSaveScreen)
```

Optimization of the Browse

```
        IF UPDATED() .AND.  (LASTKEY() != K_ESC)
           SEEK xSearch

           IF (! SET(_SET_SOFTSEEK)) .AND. (! FOUND())
              // Don't allow search to throw me to end of file
              GOTO nRec
              ALERT(xSearch + " not found.", {"OK"})
           ELSE
              // Record pointer was moved.  Save new location,
              // refresh browse,  then move highlight to proper
              // position

              // Ensure I have a valid record
              IF EOF()
                 GO BOTTOM
              ENDIF

              nRec := RECNO()
              oBrowse:refreshAll()
              oBrowse:ForceStable()

              DO WHILE (RECNO() != nRec)
                 oBrowse:up()
                 oBrowse:ForceStable()
              ENDDO
           ENDIF
        ENDIF

        RETURN
```

Using TBrowse:cargo and TBColumn:cargo

Sometimes it is desirable to track more information about the browse or its columns than the browse itself tracks. To let you do this, *CA-Clipper* defines the *cargo* exported instance variable in all of its classes. It is called 'cargo' because it is just "along for the ride." Cargo can hold data of any type, including an array.

If you have complete control over the objects, you can use cargo in any way you desire.

This example makes the preprocessor code easier to understand.

```
// These #defines use the browse's "cargo" slot to hold the
// "append mode" flag for the browse.  Having #defines for
// these just makes it easy to change later.

#define TURN_ON_APPEND_MODE(b)      (b:cargo := .T.)
#define TURN_OFF_APPEND_MODE(b)     (b:cargo := .F.)
#define IS_APPEND_MODE(b)           (b:cargo)
```

If, on the other hand, you are not in complete control of the objects (i.e., you are writing general purpose routines for others to use), we **strongly** recommend that you use cargo as a dictionary, as implemented in the Dict.prg sample program supplied with *CA-Clipper* (explained below).

This allows several programmers to use cargo without conflict, and without any change in the source code.

Dictionaries

Dictionaries provide a way of associating a key with a value, as in a real dictionary, where the word you look up (the key) is associated with a particular definition (the value). Note that the entries in data dictionaries have no particular order.

This example uses a data dictionary instead of the cargo instance variable to store the append mode. Because the dictionary is created and evaluated at runtime, any number of modules can use the dictionary without knowledge of how other modules are using it.

```
//   Using A Dictionary
//
//   Use a dictionary to hold cargo information.  Requires that
//   Dict.obj be linked in to work.

// Create new dictionary
oBrowse:cargo := DictNew()

// Turn on append mode
DictPut(oBrowse:cargo, "APPEND", .T.)

// Turn off append mode
DictPut(oBrowse:cargo, "APPEND", .F.)

// Check append mode
DictAt(oBrowse:cargo, "APPEND")
```

Browsing with Get

Browsing is more than read-only screens and pulldown menus. You can, for instance, create a popup edit screen; you can edit the fields in place. Using the Get system, you have unprecedented access to your data, but you are responsible for integrating the system into an application and controlling movement through the data source.

Though, at first, editing the browse fields "in-place" may seem an impossible task, it is actually quite easy. This discussion covers two different methods.

The first, and probably simplest method, employs cargo.

This example, a modification of the basic TBrowse object creation code, places the field name for the created objects into TBColumn:cargo instance variable:

Browsing with Get

```
//    Inserting a Column's FieldName
//    into oBrowse:cargo.
//
#define SET_FIELD_NAME(c, f)  (c:cargo := f)
#define GET_FIELD_NAME(c)     (c:cargo)

//    StockBrowseNew()
//    Create a "stock" Tbrowse object for the current work area.
//
FUNCTION StockBrowseNew(nTop, nLeft, nBottom, nRight)
   LOCAL oBrowse
   LOCAL n, oCol

   // Start with a new browse object from TBrowseDB()
   oBrowse := TBrowseDB(nTop, nLeft, nBottom, nRight)

   // Add a column for each field in the current work area
   FOR n := 1 TO FCOUNT()
      // Make a new column
      oCol := TBColumnNew( FIELDNAME(n), ;
         FIELDWBLOCK(FIELDNAME(n), SELECT()) )

      // Store the field name with the column
      SET_FIELD_NAME( oCol, FIELDNAME(n) )

      // Add the column to the browse object
      oBrowse:addColumn(oCol)

   NEXT

   RETURN (oBrowse)
```

which makes the question of when to edit the field a simple matter

```
//    Editing the current Call using
//    oBrowse:cargo and a Macro
//
//    DoGet()
//    Get current cell using cargo macro method
//
PROCEDURE DoGet(oBrowse)
   LOCAL oCol, cField
   LOCAL nCursSave
   MEMVAR GetList

   // Ensure browse is stable
   oBrowse:ForceStable()

   // Get current column
   oCol   := oBrowse:getColumn(oBrowse:colPos)

   // Get field name from column
   cField := GET_FIELD_NAME(oCol)

   // Turn cursor on
   nCursSave := SETCURSOR(SC_NORMAL)

   @ ROW(), COL() GET &cField
   READ

   // Restore cursor
   SETCURSOR(nCursSave)
   RETURN
```

Browsing with Get

A more efficient way to accomplish the same thing takes advantage of the fact that the TBColumn:block and the Get class's Get:block can be compatible. You can, therefore, manipulate the Get directly.

```
// Editing the current Call using GetNew()
//
//   DoGet()
//   Get current cell using GetNew() method
//
PROCEDURE DoGet(oBrowse)
   LOCAL oCol
   LOCAL oGet
   LOCAL nCursSave

   // Ensure browse is stable
   oBrowse:ForceStable()

   // Get current column
   oCol := oBrowse:getColumn(oBrowse:colPos)

   // Create Get at current position
   oGet := GetNew(ROW(), COL(), oCol:block, oCol:heading)

   // Turn cursor on
   nCursSave := SETCURSOR(SC_NORMAL)

   // And do READ
   ReadModal({oGet})

   // Restore cursor
   SETCURSOR(nCursSave)
   RETURN
```

Determining Whether the Record Has Moved

As discussed under the section, Repositioning the Record Pointer, indexes can cause problems in simple code examples. If you change the key field of a particular record, its relative position in the database might have changed, requiring that the display be refreshed.

It is relatively easy to determine if this is necessary. By evaluating the index key before and after the Read, determine whether it has changed. If it has, refresh the screen.

Here is the previous example, with the proper code added.

```
//   Determine whether the Record has Moved
//   DoGet()
//   Get current cell using GetNew() method and updating
//   TBrowse correctly when key field is changed
//
PROCEDURE DoGet(oBrowse)
   LOCAL oCol
   LOCAL oGet
   LOCAL nCursSave
   LOCAL xOldKey, nRec
```

```
// Ensure browse is stable
oBrowse:ForceStable()

// Indexes key expression, macroing returns the value
// of the expression for the current record
xOldKey := &(INDEXKEY(0))
nRec    := RECNO()

// Get current column
oCol := oBrowse:getColumn(oBrowse:colPos)

// Create Get at current position
oGet := GetNew(ROW(), COL(), oCol:block, oCol:heading)

// Turn cursor on
nCursSave := SETCURSOR(SC_NORMAL)

// And do READ
ReadModal({oGet})

// See if index key has changed
IF ! (&(INDEXKEY(0)) == xOldKey)
   // Force refresh, and reposition highlight if necessary
   oBrowse:refreshAll()
   oBrowse:ForceStable()
   DO WHILE (RECNO() != nRec)
      oBrowse:up()
      oBrowse:ForceStable()
   ENDDO
ENDIF

// Restore cursor
SETCURSOR(nCursSave)
RETURN
```

To adapt this code to a multi-user environment, apply a record lock immediately after the first call to ForceStable() and unlock after the call to READMODAL().

Adding Color

Though the previous examples have been for monochrome displays, the TBrowse class also gives more control over color. The simplest way to add color to a TBrowse object is to SET COLOR TO the desired standard and enhanced colors. The headers, footers and columns will be displayed in the standard color; the highlight will be in the enhanced color.

TBrowse:colorSpec

More complex use of color requires an understanding of the relationships between the various exported instance variables that control color. Just as *CA-Clipper* maintains a color table using SETCOLOR() to determine the colors of display objects, TBrowse maintains a color table using TBrowse:colorSpec for its own use. TBrowse:colorSpec, unlike SETCOLOR(), implies no specific meaning by the order in which colors are listed. When unspecified, TBrowse:colorSpec defaults to the value of SETCOLOR() at the time the TBrowse object is created.

Typical TBrowse:colorspec assignments look like:

```
oBrowse:colorSpec := "N/W, N/BG, B/W, B/BG, R/W, R/BG"

oBrowse:colorSpec += ",G/W, G/BG"
```

All of the other color-related instance variables and methods refer to colors as indexes into TBrowse:colorSpec. In the example above, color one is N/W, color two is N/BG, etc. In actual usage, colors are always expressed as an array of two such indexes. These are the normal and highlighted colors for the particular cells you are working with. The array { 3, 4 }, assuming the TBrowse:colorSpec listed above, represents a normal color of B/W and a highlighted color of B/BG. You can call this a *color pair array*.

TBColumn:defColor

TBrowse lets you set column colors individually through the TBColumn:defColor instance variable. The first, or normal, color in the TBColumn:defColor color pair array determines the color of the data values, headers and footers. The highlighted color determines the color of the highlight when it falls on this column.

This code displays every data type in a different color.

```
// Displaying Each Data Type in a Different Color
//
//    FancyColors()
//    Set up colors for the browse.
PROCEDURE FancyColors(oBrowse)
   LOCAL nCol, oCol
   LOCAL xValue

   // Set up a list of colors for the browse to use
   oBrowse:colorSpec := "N/W, N/BG, B/W, B/BG, R/W, R/BG, G/W"

   oBrowse:colorSpec += "G/BG"

   // Loop through the columns, choosing a different color
   // for each data type
   FOR nCol := 1 TO oBrowse:colCount

      // Get (a reference to) the column
      oCol := oBrowse:getColumn(nCol)
```

Adding Color

```
      // Get a sample of the underlying data by evaluating
      // the code block
      xValue := EVAL(oCol:block)

      DO CASE
      CASE VALTYPE(xValue) == "C"
         // Characters in black
         oCol:defColor := {1, 2}

      CASE VALTYPE(xValue) == "N"
         // Numbers in blue
         oCol:defColor := {3, 4}

      CASE VALTYPE(xValue) == "L"
         // Logicals in red
         oCol:defColor := {5, 6}

      OTHERWISE
         // Dates in green
         oCol:defColor := {7, 8}
      ENDCASE
   NEXT
   RETURN

//    TBColumn:defColor defaults to { 1, 2 }
```

TBColumn:colorBlock

TBrowse even lets you highlight the color of individual cells, based on the value retrieved by TBColumn:block (the data retrieval block). Use TBColumn:colorBlock; it is passed the result of the data retrieval block and returns a color pair array. TBColumn:colorBlock overrides the TBColumn:defColor for the column. This is valuable in a number of situations. For example, negative numbers can be displayed in one color, column values can be another color, different from the one used for the column header, and rows can be displayed in alternating colors.

```
//   Displaying negative numbers in a different color
//
//     FancyColors()
//     Set up colors for the browse.
//
PROCEDURE FancyColors(oBrowse)
   LOCAL nCol, oCol
   LOCAL xValue

   // Set up a list of colors for the browse to use
   oBrowse:colorSpec := "N/W, N/BG, R/W, W+/R"

   // Loop through the columns, set up default, and
   // force columns to display negative numbers in red
   FOR nCol := 1 TO oBrowse:colCount

      // Get (a reference to) the column
      oCol := oBrowse:getColumn(nCol)
```

Adding Color

```
            // Get a sample of the underlying data by evaluating
            // the code block
            xValue := EVAL(oCol:block)

            // Default color
            oCol:defColor := {1, 2}

            IF VALTYPE(xValue) == "N"
               // Negative numbers in red
               // The block is passed the current cell value.  If
               // it is less than 0, colors 3 and 4 are used;
               // otherwise, colors 1 and 2 are used
               oCol:colorBlock := {|n|   IIF(n < 0, {3,4}, {1,2})}
            ENDIF
         NEXT
         RETURN
```

This example shows column values in a different color than the column headers

```
      //   Displaying Column Values and Headers in Different Colors
      //
      //   FancyColors()
      //   Set up some colors for the browse.
      //
      PROCEDURE FancyColors(oBrowse)

         // Set up a list of colors for the browse to use
         oBrowse:colorSpec := "N/W, W/N, B+/W, W+/B"

         // Loop through the columns, set default and force
         // columns to display other than the default color
         FOR nCol := 1 TO oBrowse:colCount

      //   Get (a reference to) the column
      oCol := oBrowse:getColumn(nCol)

      //   Default color (this will be the color of the headers
      //   and the footers)
      oCol:defColor :={1,2}

      //   Use the color block to have all columns change their
      //   color to something other than the default color
      oCol:colorBlock := {|xJunk| {3, 4}}

NEXT
RETURN
```

This example displays cells in random colors.

```
      //   Displaying Cells in Random Colors
      //
      //   FancyColors()
      //   Set up some colors for the browse.
      //
      PROCEDURE FancyColors(oBrowse)
         LOCAL nCol, oCol

         // Set up a list of colors for the browse to use
         oBrowse:colorSpec := "B/W, W/B, R/W, W/R, G/W, W/G, " + ;
            "GR/W, W/GR, W/BG, BG/W, N/W, W/N"
```

Adding Color

```
    // Loop through the columns, setting up lines to display
    // in alternating colors
    FOR nCol := 1 TO oBrowse:colCount

       // Get (a reference to) the column
       oCol := oBrowse:getColumn(nCol)

       // Default color (this will be the color of the headers
       // and footers)
       oCol:defColor := {1, 2}

       // Ignore the value of the cell and choose color for
       // column based on the time at which the block is
       // executed
       oCol:colorBlock := {|x| RandColor(6)}
    NEXT
    RETURN

//
//  RandColor()
//  Return color pair based on time
//
FUNCTION RandColor(nSets)
   LOCAL nSelector

   // nSets is the number of color pair arrays available
   nSelector := (SECONDS() * 10) % nSets

   RETURN ({(nSelector * 2) + 1, (nSelector * 2) + 2})
```

Controlling the Highlight

By default, TBrowse turns the highlight on and off. TBrowse gives you complete control over when and where the highlight appears through TBrowse:autoLite, TBrowse:hilite(), and TBrowse:deHilite(). If TBrowse:autoLite is true (.T.), the default setting, TBrowse automatically turns the highlight on and off. In the example, under Repositioning the Record Pointer, you can see the highlight flicker as the necessary adjustment takes place. You can correct this by changing the searching function to turn off automatic highlighting before and turn it on after the movement is done.

```
//  Eliminating Flicker During Searches
//
//  SearchBrowse()
//  Search browse on key field with SEEK and adjust highlight
//
PROCEDURE SearchBrowse(oBrowse)
   LOCAL cSaveScreen, xSearch
   LOCAL nRec
   LOCAL GetList := {}

   // Ensure browse is stable before getting info about
   // current record
   oBrowse:ForceStable()

   // Get current key value and record number
```

Adding Color

```
xSearch := & (INDEXKEY(0))
nRec := RECNO()

// Put prompt online surrounding browse
cSaveScreen := SAVESCREEN(oBrowse:nBottom + 1, 0, ;
   oBrowse:nBottom + 1, MAXCOL())

SETPOS(oBrowse:nBottom + 1, oBrowse:nLeft + 2)
DISPOUT("[Search For ")
@ ROW(), COL() GET xSearch PICTURE "@K"
DISPOUT("]")
READ

RESTSCREEN(oBrowse:nBottom + 1, 0, ;
   oBrowse:nBottom + 1, MAXCOL(), cSaveScreen)

IF UPDATED() .AND. (LASTKEY() != K_ESC)
   SEEK xSearch

   IF (! SET(_SET_SOFTSEEK)) .AND. (! FOUND())
      // Don't allow search to throw me to end of file
      GOTO nRec
      ALERT(xSearch + " not found.", {"OK"})
   ELSE
      // Record pointer was moved.  Save new location,
      // refresh browse, and move highlight to proper
      // position

      // Check for valid record
      IF EOF()
         GO BOTTOM
      ENDIF

      nRec := RECNO()
      oBrowse:refreshAll()

      // Turn off highlighting while movement is
      // taking place
      oBrowse:autoLite := .F.
      oBrowse:deHilite()
      oBrowse:ForceStable()
      DO WHILE (RECNO() != nRec)
         oBrowse:up()
         oBrowse:ForceStable()
      ENDDO

      // Turn it back on now that we are done
      oBrowse:autoLite := .T.
   ENDIF
ENDIF

RETURN
```

TBrowse:colorRect()

TBrowse also includes a method, TBrowse:colorRect(), for coloring any rectangular region. This method takes as parameters the four corners of the box (top, left, bottom, and right) in an array, and a *color pair array*. The coordinates of the box are relative to the browse window, so that 1,1 is the first cell of the first row currently visible. The box "1,1, oBrowse:rowCount, oBrowse:colCount" refers to the entirety of the rows currently visible, though the column numbers could refer to columns that are not currently visible. Colors set with TBrowse:colorRect() remain in effect on a particular row until that row's data is refreshed, either because the row scrolled off the screen or due to one of the refresh methods. TBrowse:colorRect() overrides any TBColumn:colorBlock and TBColumn:defColor.

You may use TBrowse:colorRect() to highlight an entire row rather than an individual column. This function uses the passed color pair array to highlight a row. If you add a call to the function immediately after the stabilize portion of the main loop (in an area where the browse is guaranteed stable), the row maintains that color until refreshed, immediately before applying the key to the browse.

```
//   Highlighting the Current Row
//
//   HiliteRow()
//   Highlight current row with passed color pair array
//
PROCEDURE HiliteRow(oBrowse, aColorPair)
   LOCAL nCurRow

   nCurRow := oBrowse:rowPos

   // Color row
   oBrowse:colorRect({nCurRow, 1, nCurRow, oBrowse:colCount},;
      aColorPair)

   // And make it happen
   oBrowse:ForceStable()

   RETURN
```

Controlling the Scope

TBrowse lets you restrict a browse to a specific set of records without the speed penalty inflicted by a filter, or the need to create a temporary database. In *CA-Clipper*, you restrict the browse by defining the limits of cursor movement within the data source by using TBrowse:goTopBlock, TBrowse:goBottomBlock, and TBrowse:skipBlock instance variables. Restricting movement this way is, effectively, the same as defining a filter or redefining the scope of your search, only much, much faster.

To define movements within your redefined scope to TBrowse, you must specify how to move directly to the top, how to move directly to the bottom, and how to move forward and backward through records.

TBrowse:goTopBlock

TBrowse uses TBrowse:goTopBlock to go to the top of the data set being browsed. TBrowseDB() creates a default TBrowse:goTopBlock equivalent to:

```
oBrowse:goTopBlock := {|| DBGOTOP()}
```

TBrowse:goBottomBlock

TBrowse uses TBrowse:goBottomBlock to go to the bottom of the data set being browsed. TBrowseDB() creates a default TBrowse:goBottomBlock equivalent to:

```
oBrowse:goBottomBlock := {|| DBGOBOTTOM()}
```

TBrowse:skipBlock

TBrowse uses TBrowse:skipBlock to move forward and backward the specified number of records through the data set. TBrowse passes a positive number to TBrowse:skipBlock to move forward, and a negative number to move backward.

TBrowse:skipBlock returns only the number of rows it actually moved. Thus, if the current row is two rows from the end of the data set, and TBrowse:skipBlock is passed 4, it returns 2, indicating it could only move forward two rows. TBrowseDB() creates a default TBrowse:skipBlock equivalent to:

```
TBrowse:skipBlock := {|nMove| Skipper(nMove)}


//   TBrowse:skipBlock created by TBrowseDB()
//
//   Skipper()
//   Handle record movement requests from the TBrowse object.
//
STATIC FUNCTION Skipper(nMove)
   LOCAL nMoved

   nMoved := 0

   IF nMove == 0 .OR.  LASTREC() == 0
      // Skip 0 (significant on a network)
      SKIP 0

   ELSEIF nMove > 0 .AND.  RECNO() != LASTREC() + 1
      // Skip forward
      DO WHILE nMoved < nMove
         SKIP 1
         IF (EOF())
            SKIP -1
            EXIT
         ENDIF
```

Controlling the Scope

```
            nMoved++
        ENDDO

    ELSEIF nMove < 0
        // Skip backward
        DO WHILE nMoved > nMove
            SKIP -1
            IF (BOF())
                EXIT
            ENDIF

            nMoved--
        ENDDO
    ENDIF

    RETURN (nMoved)
```

Viewing a Specific Key Value

You can look at all records associated with a particular key value by defining the movements within the data set, then forcing the TBrowse:skipBlock to remain within the key value. Define movement to the top of the data set as seeking the particular key and movement to the bottom of the data set as seeking the last item matching the current key.

First, write a function to create a block for TBrowse:goTopBlock. Note that this function creates a code block that can only refer to *xKey*, relieving us of the burden of ensuring that the key value remains available.

```
//  TBrowse:goTopBlock
//
//      FirstKeyBlock()
//      Create block that seeks the first occurrence
//      of the passed key.  Use scoping nature of blocks
//
FUNCTION FirstKeyBlock(xKey)
    RETURN ({|| DBSEEK(xKey)})
```

Next, write a function to create a TBrowse:goBottomBlock. This is more complex because the browse should move almost instantaneously to the last matching key. Do this by seeking the next key value, then skip back one record. Calculate the next key value using the NextKey() function, as listed in the following example. LastKeyBlock() returns a block that implements TBrowse:goBottomBlock.

```
//  TBrowse:goBottomBlock
//
//      LastKeyBlock()
//      Create block that seeks the last occurrence
//      of the passed key.  Use scoping nature of blocks
FUNCTION LastKeyBlock( xKey )
    RETURN ({|| DBSEEK(NextKey(xKey), .T.), DBSKIP(-1)})
```

```
//    NextKey()
//
//    Return value of next key in database
//
//    WARNING! Does not work on logical values or
//    floating point numbers
//
FUNCTION NextKey(xValue)
   LOCAL cType := VALTYPE(xValue)
   LOCAL xNext

   DO CASE
   CASE cType == "C"
      // Find next value by incrementing the ASCII value of
      // the last character in the string.
      //
      // NOTE: Assume last char is not 255 for simplicity
      xValue := STUFF(xValue, LEN(xValue), 1, ;
         CHR(ASC(RIGHT(xValue,1)) + 1))

   CASE cType == "N"
      // Since we can only handle integers, increment value
      // by one
      xValue++

   CASE cType == "D"
      // Increment date by one day
      xValue++

   OTHERWISE
      // Couldn't handle it

   ENDCASE

   RETURN (xNext)
```

Finally, write a function that returns the necessary TBrowse:skipBlock. This function is like the default TBrowse:skipBlock described earlier, except that it checks for moving out of range as well as for beginning and end of file. One typical use of this function is zooming to child records from a browse of the parent database.

```
//   New TBrowse:skipBlock
//
//   KeySkipBlock()
//   Create skipBlock to employ block scoping
//
FUNCTION KeySkipBlock(xKey)
   RETURN ({|nMove| KeySkipper(nMove, xKey)})


//   KeySkipper()
//
//   Handle record movement requests from the TBrowse object
//   while staying within one key.
//
FUNCTION KeySkipper(nMove, xKey)
   LOCAL nMoved

   nMoved := 0
```

Controlling the Scope

```
      IF nMove == 0 .OR.  LASTREC() == 0
         // Skip 0 (significant on a network)
         SKIP 0

      ELSEIF nMove > 0 .AND.  RECNO() != LASTREC() + 1
         // Skip forward
         DO WHILE nMoved <= nMove .AND.  ! EOF() .AND. ;
               &(INDEXKEY(0)) == xKey
            SKIP 1
            nMoved++
         ENDDO

         // All above cases lead to skipping one extra
         SKIP -1
         nMoved--

      ELSEIF nMove < 0
         // Skip backward
         DO WHILE nMoved >= nMove .AND.  ! BOF() .AND. ;
               &(INDEXKEY(0)) == xKey
            SKIP -1
            nMoved--
         ENDDO

         // No phantom BOF() record, otherwise, above loop
         // leads to skipping back one extra record
         IF ! BOF()
            SKIP 1
         ENDIF

         nMoved++
      ENDIF

      RETURN (nMoved)

//  A Scoped Browse
//
//    BrowseWhile()
//
//    Browse database while the key field matches xKey
// PROCEDURE BrowseWhile(nTop, nLeft, nBottom, nRight, xKey)
      LOCAL oBrowse       // The TBrowse object
      LOCAL nKey          // Keystroke
      LOCAL lMore         // Loop control
      LOCAL nCursSave     // Save state

      // Create another generic browse
      oBrowse := StockBrowseNew(nTop + 1, nLeft + 1, ;
         nBottom - 1, nRight - 1)

      // Change the heading and column separators
      oBrowse:headSep := MY_HEADSEP
      oBrowse:colSep := MY_COLSEP

      // Change the movement methods
      oBrowse:goTopBlock := FirstKeyBlock(xKey)
      oBrowse:goBottomBlock := LastKeyBlock(xKey)
      oBrowse:skipBlock := KeySkipBlock(xKey)

      // Draw a box around the browse
      @ nTop, nLeft CLEAR TO nBottom, nRight
      @ nTop, nLeft TO nBottom, nRight

      // Save cursor shape, turn the cursor off while browsing
      nCursSave := SETCURSOR(SC_NONE)
```

Controlling the Scope

```
// Main loop
lMore := .T.
DO WHILE (lMore)

   // Stabilize the browse
   oBrowse:ForceStable()

   // Positioned at top or bottom of browse
   IF oBrowse:hitTop .OR.  oBrowse:hitBottom
      TONE(125, 0)
   ENDIF

   // Get a new keystroke
   nKey := INKEY(0)

   IF nKey == K_ESC
      // Exit the loop
      lMore := .F.
   ELSE
      // Act on the keystroke
      ApplyKey(oBrowse, nKey)
   ENDIF
ENDDO

SETCURSOR(nCursSave)

RETURN
```

Browsing Search Results

Before the implementation of TBrowse, when browsing a random collection of records (such as the results of a user's query), it was necessary to create temporary databases or deal with the slow display times of SET FILTER. Because TBrowse gives you complete control over movement through the database, you can execute fast browses of any arbitrary subset of the database.

Accomplish this by running the search through the database once, creating an array of the record numbers of records that satisfy the query. Use this array to control the browse movement. While this imposes a limit of 4,096 matching records, this is usually sufficient.

Under *CA-Clipper*, the fastest way to create the control array is to use the DBEVAL() function. This function receives a code block equivalent to the FOR condition in commands, and returns a control array:

Controlling the Scope

```
//  Creating an Array of Record Numbers
//     ArrayFor()
//
//     Create array of record numbers that match passed
//     condition
//
FUNCTION ArrayFor(bFor)
   LOCAL aRecs := {}

   DBEVAL({|| AADD(aRecs, RECNO())}, bFor)

   RETURN (aRecs)
```

which makes the TBrowse:goTopBlock and TBrowse:goBottomBlock easily specified (it assumes that nCurrRow is the current row of the browse):

```
oBrowse:goTopBlock := {|| DBGOTO(aRecs[nCurrRow := 1])}
oBrowse:goBottomBlock := {|| DBGOTO(aRecs[nCurrRow := ;
   LEN(aRecs)])}
oBrowse:skipBlock := {|nMove| SkipFor(nMove, aRecs, @nCurrRow)}
```

The function that TBrowse:skipBlock then calls to move through the database is unusual because it uses the end and beginning of the array, rather than the end and beginning of the database.

```
//  Moving Through the Database
//
//     SkipFor()
//     Move through array-controlled database
//
FUNCTION SkipFor(nMove, aRecs, nCurrRow)
   LOCAL nMoved := 0

   // Put move into an acceptable range
   IF nCurrRow + nMove < 1
      // would skip past top...
      nMove := - nCurrRow + 1

   ELSEIF nCurrRow + nMove > LEN(aRecs)
      // Would skip past bottom...
      nMove := LEN(aRecs) - nCurrRow

   ENDIF
   nCurrRow += nMove
   nMoved := nMove

   DBGOTO(aRecs[nCurrRow])

   RETURN (nMoved)
```

Summary

TBrowse, as the primary data-handling tool of *CA-Clipper*, is, in many respects, the core of the *CA-Clipper* programming language. This chapter has discussed the character and functioning of the major elements of TBrowse: the Tbrowse and TBColumn objects, their exported instance variables, and the most common ways they are used. Along with discussions of the obvious aspects of TBrowse, the chapter also discussed implications of the proper handling of TBrowse, on response-time, multi-user applications, and movement within a data source. These discussions and examples, though not exhaustive, are representative of, and should lead you to effective use of TBrowse in your own applications.

Chapter 6
Introduction to the Get System

The *CA-Clipper* Get System is an integral interface tool consisting of commands, objects (and their related data and methods), environments, and behaviors that implement data entry and retrieval. It gives the developer direct and specific control over the interaction between the user and data, while making the retrieval, display, manipulation, and storage of data easy and flexible. This chapter lays the groundwork required to take advantage of this new open architecture, after which, the reader should understand and be able to extend the standard Read system. The text assumes knowledge of the @...GET and READ commands, so the focus lies instead on more advanced uses of the system.

This chapter starts with a discussion of the Get class, continuing with the layers of the Get System. It concludes with a discussion of possible modifications to the behavior of the Get System. The text contains many examples and recommendations. It is assumed that you are familiar with the concepts and terminology associated with *CA-Clipper's* predefined classes as discussed in the *Basics Concepts* chapter of this guide.

In This Chapter

This chapter discusses the following topics:

- The Get class
- Creating Get Objects
- The Get System structure
    The Read Layer
    The Reader Layer
    The Get Function Layer
- Using code blocks in the Get system
- Extending the Read Layer
- Creating a New Read Layer

Objects and the Get System

To understand the Get System, you must be conversant with the vocabulary and concepts surrounding *CA-Clipper's* predefined objects. This terminology is the same as that of many object-oriented programming languages. It includes the concepts of Class, method, instance variables, the constructor, the selector, and the send operator. It includes the sending and receiving of messages, and the concepts of inheritance and persistence, and of focus. It includes the Get system provisions for preedit, postedit, and in-edit validation.

Using Objects

Once you master the vocabulary, you must understand some basic mechanics. The *Basic Concepts* chapter explains the details of using *CA-Clipper* objects, but the principles are fairly simple:

- **Create an object**: Of the four predefined *CA-Clipper* object classes, the Get system creates, of course, Get objects.

- **Evaluate or assign values:** Since an object is both data and methods, part of its power resides in the characteristics of the exported instance variable values. These are affected by sending one or more messages to the object.

- **Send a message:** Send the object an appropriate message, chosen from the list of methods. Messages may be simple or complex and may affect appearance or function or many other characteristics of the Get object.

This chapter discusses expansion of the basic principles, as described in Objects and Messages, relative to the Get system.

Get Class

The Get class implements the formatting behavior associated with *CA-Clipper's* Gets.

A Get object has several exported instance variables, including:

Get Class

Get Class Exported Instance Variables

Variable	Description
badDate	logical value indicating validity of buffer content data type
block	code block associative to the Get object variable
buffer	character value representing the edit-buffer
cargo	value of any data type
changed	logical value indicating change status of Get object
col	numeric value indicating screen-column of Get display
colorSpec	character string of color attributes
decPos	numeric value indicating decimal position in edit-buffer
hasFocus	logical value showing status of input focus
name	character string representing the name of the Get variable
original	value of the Get object as it acquired input focus
picture	character value defining the PICTURE string
pos	numeric value showing position of cursor within the edit-buffer
postBlock	code block to validate a newly-entered value
preBlock	code block to permit or deny access to edit-buffer
rejected	logical value showing status of Get:insert or Get:overStrike
row	numeric value indicating screen-row of Get display
type	character representing data type of the Get variable
typeOut	logical value indicating attempted movement beyond buffer. Reset by cursor movement

The *Language Reference* chapter of the *Reference* guide defines each of the Get Class exported instance variables. Here, we only discuss some important behavior and uses of get:block and get:cargo in preparation for deeper discussions of the Get System itself. We review get:block because of its importance to the functioning of any Get object. We discuss get:cargo because of its value to the programmer as an information source about a particular Get object. GetNew() creates a new Get object. You must supply parameters to GetNew() that specify the row and column position of the Get, as well as a block that defines how the value of the Get should be retrieved and stored, and the name to associate with the Get (usually the name of a variable or field being edited by the Get). These are stored in the instance variables Get:row, Get:col, Get:block, and Get:name, respectively:

```
LOCAL oGet
oGet := GetNew(1, 2, FIELDBLOCK("Zip"), "Zip")
```

Get:block

The Get:block instance variable holds a code block that stores the value passed, or retrieves the value associated with the Get if no parameter is passed. If the Get refers to a simple variable (*cVar* in this example), the code block is straightforward:

```
oGet := GetNew( 1, 1, ;
   {|cNew| IF(PCOUNT() == 0, cVar, cVar := cNew)}, "cVar")
```

The block may also refer to a function call. For example, to create a Get object to modify the _SET_WRAP setting, one would do the following:

```
cName := "_SET_WRAP"
oGet := GetNew(1, 1, ;
   {|lNew| IF(PCOUNT() == 0, SET(_SET_WRAP), ;
   SET(_SET_WRAP, lNew)}, cName)
```

Due to the nature of code blocks, arrays create a particular challenge. An array with a constant index works reliably:

```
cName := "aVar[1]"
oGet := GetNew(1, 1, ;
   {|cNew| IF(PCOUNT() == 0, aVar[1], aVar[1] := cNew)},;
   cName)
```

But when the index is a variable (e.g., aVar[x]), it doesn't work because the code block then refers to a variable, x, that can change before the block is executed. At the time the block was created, x may have a value of 3, making the block refer to aVar[3]. If x is later changed to 1, the code block refers to aVar[1].

```
// This doesn't work
cName := "aVar[" + LTRIM(STR(x)) + "]"
oGet := GetNew(1, 1, ;
   {|cNew| IF(PCOUNT() == 0, aVar[x], aVar[x] := cNew)}, ;
   cName)
```

To solve this problem, the Get class makes special provisions for arrays. Normally, a Get:block for an array simply returns the array reference. However, since the code block must access the actual array element rather than the reference to the entire array, and the subscript instance variable is not assignable, you must create the code block *within a function*.

```
cName := "aVar[" + LTRIM(STR(x)) + "]"
oGet := GetNew(1, 1, AGetBlock(aVar, x), cName)
...
STATIC FUNCTION AGetBlock(aArray, x)
   RETURN ({|xVal| IIF(PCOUNT() == 0, aArray[x], ;
      aArray[x] := xVal)})
```

This effectively hard-codes the array subscript, because the local variable x has gone out of scope and can no longer be changed. For more information, refer to the Arrays and Code Blocks sections of the *Basic Concepts* chapter in this guide.

Setting and Retrieving Values

Note that while it seems logical to evaluate the Get:block to set and retrieve the value of a Get object, the way arrays are handled makes this impractical. Use the Get:varGet() and Get:varPut() methods instead. These methods work correctly for both normal Get:blocks and array Get:blocks.

```
// Retrieve the current value of the Get
nVal := oGet:varGet()

//...and assign back one greater
oGet:varPut(nVal + 1)
```

Get:cargo

You may track more information about a Get than the Get object itself tracks. To do this, use the *cargo* exported instance variable defined in all of *CA-Clipper's* classes. It is called cargo because it is just "along for the ride." Cargo can hold data of any type, including an array.

If you have complete control over the objects, you can use cargo in any way you desire (e.g., to use the preprocessor to make the code easier to understand):

```
// These #defines use the GET's "cargo" slot to hold the
// message displayed at the bottom of the screen.  Having
// #defines for these just makes it easy to change later.

#define SET_MESSAGE( g, msg )   (g:cargo := msg)
#define GET_MESSAGE( g )        (g:cargo)
```

If, on the other hand, you are not in complete control of the objects it is **strongly** recommended that you use cargo as a dictionary, as implemented in the Dict.prg sample program supplied with *CA-Clipper* (explained below). This allows several programmers to use cargo without conflict and without any change in the source code.

Dictionaries

Dictionaries are a way of associating a key with a value, as in a real dictionary where the word you are looking up (the key) is associated with a particular definition (the value). Note that the entries in such a dictionary, unlike its literary counterpart, have no particular order. The storing of the help text could be done with dictionaries, as follows:

```
// Use a dictionary to hold cargo information.  Requires
// the sample program Dict.obj to be linked in to work.
//
// Create new dictionary
oGet:cargo := DictNew()
// Assign message to Get object
DictPut(oGet:cargo, "MESSAGE", "Help Text")

// Print message from Get object at the bottom of
// the screen
@ MAXROW(), 0 SAY DictAt(oGet:cargo, "MESSAGE")
```

Because the dictionary is created and evaluated at runtime, any number of modules can use the dictionary without knowledge of how other modules are using it.

Get Class Protocol

Once the object is created and the appropriate instance variables are set, it is ready for use. You must first display the Get on the screen using either of two methods, Get:display() and Get:colorDisp(), to accomplish this. Get:display() simply displays the Get in its current color (set through the Get:colorSpec instance variable), while Get:colorDisp() sets the Get:colorSpec then sends the Get:display() message to self. The Get is initially displayed in its unselected color.

When you perform actual data entry into the Get, you must first prepare the Get to accept characters. Do this by sending the Get:setFocus() message. Upon receiving this message, the Get object creates and initializes its internal state information, including the exported instance variables: Get:buffer, Get:pos, Get:decPos, Get:clear, and Get:original. The contents of the editing buffer are then displayed in the Get's selected color.

Keystrokes are intercepted externally to the object. Data is put into the Get using the Get:insert() and/or the Get:overStrike() methods, and movement within the Get is accomplished using any of the movement messages.

When editing is completed, the new value in the buffer is either accepted by Get:assign() or discarded by Get:undo(). The Get:killFocus message removes input focus from the Get. The Get is then redisplayed in its unselected color and internal state information is discarded.

This description supplies all the information necessary to follow the examples given below. Understand that the Get class is simply a tool used by the Get System to accomplish its goals—there is nothing about the Get class that ties the two together. This will become clearer in the following discussion of the Get System.

Get System

The Get System is an effective way to accept data input from the user. It is flexible and open enough to allow modifications that greatly enhance the power of the system.

The Get System is composed of several layers. Moving down the chart, the layers increase in both complexity of use and in programmer control. As with all areas of programming, *the highest possible layer should always be used.* This ensures maximum programmer productivity and system reliability.

READ Layer — READ

READER Layer — GETREADER()

GET Function Layer — GETAPPLYKEY(), GETDOSETKEY(), GETPREVALIDATE(), GETPOSTVALIDATE()

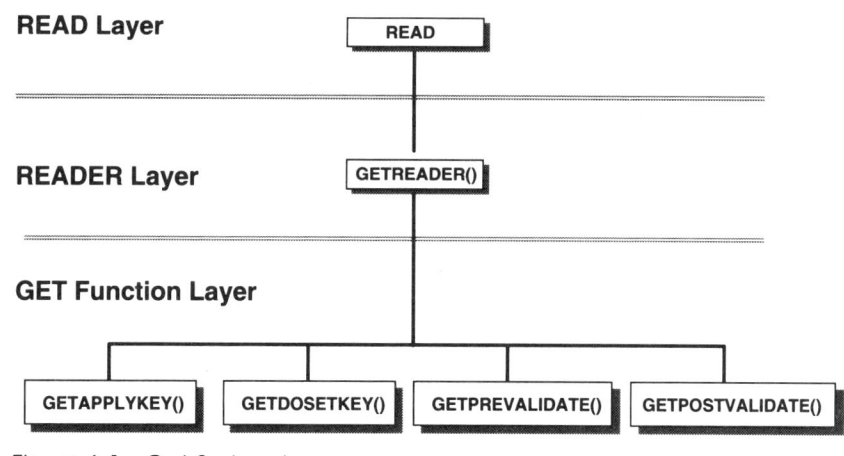

Figure 6-1—Get System Layers

The top layer, the Read Layer, is composed of the familiar @...GET and READ commands. Previous versions of *CA-Clipper* only accessed this layer, and this is where the bulk of *CA-Clipper* programming is done.

The next layer, the Reader Layer, controls the behavior of one Get at a time. This layer is implemented by default through the function GetReader(), but can be extended through additional custom *readers*.

The bottom layer, the Get Function Layer, separates the behavior of a single Get down into four distinct actions: prevalidating the Get, handling SET KEYs, applying keystrokes to the Get, and postvalidating the Get. The GetPreValidate(), GetDoSetKey(), GetApplyKey(), and GetPostValidate() functions perform the default implementation, respectively. This layer can also be extended through custom Get functions.

The Read Layer

One way to see the relationship between layers is to analyze the implementation of the Read Layer. This system is functionally equivalent to the Clipper, Summer '87 version which was written in C, but now it is written almost entirely in *CA-Clipper*. The file Getsys.prg contains the source code for the system.

A quick look at this source, or at the commands in STD.CH, show that a screen of Gets is represented, internally, as an array of Get objects. These Gets are stored in the public array *GetList* by default. This array is always present, just as though the first line of all your programs were:

```
PUBLIC GetList := {}
```

When you issue an @...GET command, a new Get object is created, added to the GetList array, and displayed on the screen. Behind the scenes, *CA-Clipper* implements this in a manner that is conceptually equivalent to:

```
AADD(GetList, GetNew(...))
ATAIL(GetList):display()         // Display just-created
                                 // Get object
```

This is repeated for each @...GET command.

READ is implemented through the function READMODAL(). READMODAL() is passed the GetList array to process, and returns a logical value that indicates whether or not any of the Gets were updated.

READ first initializes all of its variables then implements some Clipper Summer '87 compatibility behaviors. It then enters a loop to process all of the Gets. Within this loop, READ notifies the system that this is the "active" Get object (making it available through the GetActive() function) and passes control to the Reader Layer. The Reader Layer may be the default reader, GetReader(), or a custom reader supplied by the developer, to provide a new type of Get.

The Read Layer

This code demonstrates implementation of ReadModal()

```
//  Implementing ReadModal()
//
// ReadModal()
// Standard modal READ on an array of Gets.
 FUNCTION ReadModal( aGetList )
    LOCAL oGet
    LOCAL nPos
    LOCAL aSavedGetSysVars

    // If a format file (.fmt), execute the format
    // code  then continue
    IF (VALTYPE(Format) == "B")
       EVAL(Format)
    ENDIF

    // If there aren't any Gets, exit
    IF (EMPTY(aGetList))
       // S87 compatibility
       SETPOS(MAXROW() - 1, 0)
       RETURN (.F.) // NOTE
    ENDIF

    // Preserve state vars
    aSavedGetSysVars := ClearGetSysVars()

    // Set these for use in SET KEYs
    ReadProcName := PROCNAME(1)
    ReadProcLine := PROCLINE(1)

    // Set initial Get to be read
    nPos := Settle(aGetList, 0)

    DO WHILE (nPos <> 0)
       // Get next Get from list and post it as the
       // active Get
       oGet := aGetList[nPos]
       PostActiveGet(oGet)

       // Read the Get
       IF (VALTYPE(oGet:reader) == "B")
          // Use custom reader block if one was supplied
          EVAL(oGet:reader, oGet)
       ELSE
          // Use standard reader
          GetReader(oGet)
       ENDIF

       // Move to next Get based on exit condition
       nPos := Settle(aGetList, nPos)
    ENDDO

    // Restore state vars
    RestoreGetSysVars(aSavedGetSysVars)

    // S87 compatibility
    SETPOS(MAXROW() - 1, 0)

    // Updated is a filewide static set by the reader
    RETURN (Updated)
```

Extending the Read Layer

With understanding of the basic Read Layer, it is possible to make significant enhancements to the system without modifying it.

Nested Reads

The ability to nest reads, a most desired enhancement, is easy with *CA-Clipper*. Because the READ command simply passes an array named GetList to READMODAL(), you may use a different variable, GetList, in other functions. If you call a function from a VALID or WHEN clause, or a SET KEY procedure, then define a new array, GetList, local to that function, @...GETS and READs there will refer to the local GetList. This GetList falls out of scope when the function ends, returning control to the previous Read and its GetList.

```
// Creating nested reads
//PROCEDURE Main()
   LOCAL cName := SPACE(10)
   @ 10, 10 GET cName VALID SubForm( cName )
   READ
   RETURN

FUNCTION SubForm( cLookup )
   LOCAL GetList := {}        // Create new GetList
   USE Sales INDEX Salesman NEW
   SEEK cLookup
   IF FOUND()
      // Add Get objects to new GetList
      @ 15, 10 GET Salesman
      @ 16, 10 GET Amount

      // READ from new GetList
      READ
   ENDIF
   CLOSE Sales
   // Release new GetList
   RETURN (.T.)
```

Using GetActive()

Much of the power of the new Get System lies in its direct manipulation of a Get object after its retrieval with the GetActive() function. This function lets you not only retrieve and change the current value of the Get, but access all the information stored in it.

Extending the Read Layer

Example: Changing Gets Through SET KEYs

A perfect example of direct object manipulation is changing the value of a Get through a SET KEY. While you can do this by using the KEYBOARD command to stuff the keyboard buffer, or by macroing the result of ReadVar() (if the Get is being done on a field, private or public variable), it is much cleaner to just send messages to the current Get object.

This example examines the current content of the Get. If the Get is of character data type, it converts the contents to mixed upper and lowercase using the sample function Proper(), located in String.prg, when the user types *Ctrl-K*.

```
//  Converting Character Input to mixed case
#include "Inkey.ch"
.
. .<statements>
.
SET KEY K_CTRL_K TO GetCapFirst

. .<statements>
.
@ 1, 1 GET cTest
READ

SET KEY K_CTRL_K TO
RETURN


//  GetCapFirst()
//  Capitalize the first letter of every word
//

STATIC PROCEDURE GetCapFirst()
   LOCAL oGet, cString

   // Grab the current Get object
   oGet := GETACTIVE()

   // Get the current value of the Get
   cString := oGet:varGet()

   IF VALTYPE(cString) == "C"
      // Convert and put the value back
      oGet:varPut(Proper(cString))
   ENDIF

   RETURN
```

Example: Improving Help Routines

As mentioned earlier, you can also improve help systems by storing the help text (or an index to the help text) in the Get:cargo instance variable. Notice how dictionaries are used to reduce potential conflicts between this and any other modules that implement enhancements to the Get System.

```
// Dict.obj is necessary for the following example.
// Storing Help Text in Get:cargo
//
PROCEDURE ReadTest()
   LOCAL cVar1
   LOCAL GetList := {}

   cVar1 := SPACE(30)

   CLS
   @ 3,0 SAY PADC("GETs w/ Help Text in Get:cargo", ;
      MAXCOL())
   @ 4,0 SAY PADC("(<F1> for help)", MAXCOL())

   // Put the help text in the cargo dictionary manually
   @ 7,0 SAY "Variable 1" GET cVar1
   IF (ATAIL(GetList):cargo == NIL)
      ATAIL(GetList):cargo := DictNew()
   ENDIF

   DictPut(ATAIL(GetList):cargo, "Help Text", ;
      "Help for variable one.")
   READ

   RETURN


// Help()
// The system automatically assigns the F1 key to this
// procedure which will return help from wherever you are
PROCEDURE Help(cProc, nLine, cVar)
   LOCAL oGet
   LOCAL cMsg := "No help available for this item"

   IF (oGet := GETACTIVE()) != NIL
      // Retrieve help text from Get:cargo dictionary
      // In a more "robust" version of this system, some
      // sort of index into a help file would be stored
      // rather than the help text itself. This would
      // allow easier editing of the help text and permit
      // runtime modification of it.
      //
      cMsg := DictAt(oGet:cargo, "Help Text")
   ENDIF
   DispMsg(cMsg)

   RETURN
```

The first concern many people have when seeing this is the difficulty of adding the help text. Fortunately, *CA-Clipper* also provides a graceful way to do this. By defining a command that adds a HELP clause to the standard @...GET command, the interface is made much cleaner. A more complete discussion of commands can be found in the *Language Reference* chapter of the *Reference* guide.

Extending the Read Layer

```
// Adding a help clause to the @...GET Command
// Helpget.ch

#command @ <row>, <col> GET <var> [<clauses, ...>] ;
   HELP <help> [<moreClauses, ...>] ;

   => @ <row>, <col> GET <var> [<clauses>] ;
      [<moreClauses>];
      ;IF (ATAIL(GetList):cargo == NIL)
      ;ATAIL(GetList):cargo := DictNew();
      ;END
   ;DictPut(ATAIL(GetList):cargo, "Help Text", <help>)
```

Note that the command checks to see if Get:cargo is NIL before making it a dictionary because this command's definition lets you use it with other extensions to the GET command. It is very plausible that some other extension has already made Get:cargo a dictionary. If so, the command just inserts the new entry into it.

Use the command by including the header file and simply adding the HELP clause to any existing Gets.

If Get:cargo isn't NIL, and it isn't a dictionary, the command will crash at runtime.

```
//   Using the Help Clause
//

#include "HelpGet.ch"

PROCEDURE ReadTest()
   LOCAL cVar1
   LOCAL GetList := {}

   cVar1 := SPACE(30)

   CLS
   @ 3,0 SAY PADC("GETs w/ Help Text in Get:cargo", ;
      MAXCOL())
   @ 4,0 SAY PADC("(<F1> for help)", MAXCOL())

   // Put the help text using new command
   @ 7,0 SAY "Variable 1" GET cVar1 PICTURE "@!" ;

      HELP "Help for variable one."

   READ

   RETURN
```

Using Code Blocks for WHEN/VALID

Because the @...GET command accepts a code block as a legal parameter for either clause, it is easier to manipulate a GET from within a WHEN or VALID clause than from a SET KEY.

Extending the Read Layer

Example: Dynamic Pictures

The following example changes the PICTURE clause of a Get, 'on the fly,' in the WHEN clause:

```
// Dynamically Changing the PICTURE Clause of a Get

PROCEDURE Main()
   LOCAL GetList := {}
   LOCAL cName := SPACE(20)
   LOCAL cIDType := "I"
   LOCAL cSSNo := SPACE(9)

   SET CONFIRM ON
   SET SCOREBOARD OFF

   CLS
   @ 3,0 SAY PADC("Member Information")

   @ 5,0 SAY "Name                    " GET cName
   @ 6,0 SAY "C)ompany or I)ndividual" GET cIDType ;
       PICTURE "!" VALID cIDType $ "CI"

   // Change picture based on whether this is for a person
   // or company
   @ 7,0 SAY "SS# or Tax ID#          " GET cSSNo ;
       PICTURE StartPic(cIDType) ;
       WHEN {|oGet| ChangePic(oGet, cIDType)}
   READ

   RETURN


//   StartPic()
//   Determine what the initial value for the picture
//   should be.
//

STATIC FUNCTION StartPic(cIDType)
   LOCAL cPic

   IF cIDType == "C"
      cPic := "@R 99-99999-99"
   ELSE
      // Default value
      cPic := "@R 999-99-9999"
   ENDIF
   RETURN (cPic)

//   ChangePic()
//   Change picture on-the-fly from SS# to Tax ID or
//   vice versa (to be called from WHEN clause).
STATIC FUNCTION ChangePic(oGet, cIDType)
   IF cIDType == "C"
      oGet:picture := "@R 99-99999-99"
   ELSE

      // Default value
      oGet:picture := "@R 999-99-9999"
   ENDIF
   RETURN (.T.)
```

Processing the Entire GetList

While modifying a single Get is useful, there are times when it is desirable to modify all of the Gets. One might want to change their colors or move them as a unit across the screen. Because you can dynamically modify almost everything about a Get, such changes are surprisingly easy.

First write the functions to process the entire GetList. The functions below move the Gets a specified number of rows and columns, and change the colors of the Gets. Both force the Gets to redisplay themselves.

```
//  Moving Rows and Columns and Changing Colors
//
//  MoveGets( <aGets>, <nRows>, <nCols> ) --> <aGets>
//  Move Get array to new location and redisplay

FUNCTION MoveGets(aGets, nRows, nCols)
   LOCAL nGet

   FOR nGet := 1 TO LEN(aGets)
      aGets[nGet]:row += nRows
      aGets[nGet]:col += nCols
      aGets[nGet]:display()
   NEXT

   RETURN (aGets)



//  ColorGets( <aGets>, <cColor> ) --> <aGets>
//  Change colors of all Gets in Get array and redisplay

FUNCTION ColorGets(aGets, cColor)
   AEVAL(aGets, {|oGet| oGet:colorDisp(cColor)})
   RETURN (aGets)
```

Creating a New Read Layer

While the previous examples demonstrate that many additions can be made without modifying the Read Layer, sometimes it is necessary or cleaner to replace the default Read Layer.

Basic Guidelines

How do you know whether to extend the Read Layer or to create a new one? There is no absolute rule, but there are some basic guidelines. Create a new Read Layer under the following circumstances:

- The new feature would, otherwise, require adding a function to the VALID or WHEN clause of every Get.
- The new feature requires actions between each Get.

Extend the default Read Layer, if either of the following is true:

- The new feature only requires modifying the Get at the time it is created. For example, storing information in Get:cargo or one of the other instance variables.
- The new feature doesn't affect the order in which Gets are traversed, or what tasks are carried out between Gets. This type of feature is usually added by making a modification at one of the lower layers (leaving the Read Layer intact).

Note that one set of modifications to the Read Layer is generally incompatible with any other set (i.e., both can exist in the same application but cannot be used simultaneously). Combining them requires creating a third Read Layer that incorporates ideas from both implementations.

Important Implementation Rules

If you determine that a new Read Layer is necessary, be mindful of several things. First and foremost is the following rule, important in all aspects of *CA-Clipper* programming:

Avoid changing the default behavior of the system.

In the case of the Get System, this means:

- Do not modify Getsys.prg or redefine any of the functions contained within it.
- Do not redefine the basic READ command. You may add clauses and map the new command to a different Read Layer.
- Do not redefine the basic @...GET command; add clauses through a new command definition.
- If you add a clause to the @...GET command, map the new command to the basic @...GET command, if at all possible.

In addition to the above rules, if it is necessary to store any information with individual Gets, store it as a dictionary entry in the Get:cargo instance variable.

If you follow these rules, your Read Layer has a much better chance of working with extensions you intend to use with the default Read Layer. It also lets you revert to the default Read Layer if you have problems with your code. Most importantly, your code will also be much more maintainable by other programmers (or by you at a later date) because it will be obvious where you have made modifications to the system.

Implementation Steps

Creating a Read Layer that follows the above rules is relatively painless:

1. Create a copy of the file UserRead.tem and rename it appropriately.

2. Determine where any initialization code should be located. It is important that your Read Layer be *reentrant* (i.e., your Read may be called from within itself via a VALID or WHEN clause). Therefore, take care that you save and restore any filewide statics as needed. Your initialization code will therefore be split between your Read function and the static function ClearGetSysVars().

3. Rename the function UserReadFunc() and make any appropriate modifications. At this point, you have defined the interface.

4. Add any necessary supporting functions, declaring as public any functions that will be called by the user (either directly or through a command), and declaring as static all other functions.

5. Create a header file to define the commands that implement your new Read Layer.

Example: READ VALID

The first example of a custom Read Layer lets you control exit from the Read with the VALID clause.

A copy of UserRead.tem was made and UserReadFunc() was renamed ReadValid() (below). ReadValid() accepts two parameters, the GetList and a code block that executes when the user attempts to leave the Read. If the code block returns false (.F.), the user is left on the same Get, otherwise the Read is exited. Note that the code block is passed both the GetList and a logical value that indicates whether or not any of the Gets in the Read were changed.

Because the addition of this capability does not require any initialization and adds no new "states" to the Read Layer, you can completely ignore the re-entrancy issue. This greatly simplifies the implementation:

Creating a New Read Layer

```
// Restricting a Read Based on a VALID Clause
//
// ReadValid()
// Read until condition is true

FUNCTION ReadValid(aGetList, bValid)
   LOCAL oGet
   LOCAL nPos, nLastPos
   LOCAL aSavedGetSysVars

   IF (VALTYPE(ReadFormat()) == "B")
      EVAL(ReadFormat())
   ENDIF

   IF (EMPTY(aGetList))
      // S87 compatibility
      SETPOS(MAXROW() - 1, 0)
      RETURN (.F.) // NOTE
   ENDIF

   // Preserve state vars
   aSavedGetSysVars := ClearGetSysVars()

   // Set these for use in SET KEYs
   ReadProcName := PROCNAME(1)
   ReadProcLine := PROCLINE(1)

   // Set initial Get to be read
   nPos := Settle(aGetList, 0)

   DO WHILE (nPos <> 0)
      // Get next Get from list and post it as the
      // active Get
      oGet := aGetList[nPos]
      PostActiveGet(oGet)

      // Read the Get
      IF (VALTYPE(oGet:reader) == "B")
         EVAL(oGet:reader, oGet)    // Use custom reader
                                    // block
      ELSE
         GetReader(oGet)            // Use standard reader
      ENDIF

      // Save current position
      nLastPos := nPos

      // Move to next Get based on exit condition
      nPos := Settle(aGetList, nPos)

      // If I am about to exit, check to see if VALID
      // condition is true
      IF (nPos == 0)
         // VALID clause is passed current GetList,
         // whether or not the Read has been changed
         //
         IF (! EVAL(bValid, aGetList, UPDATED()))
            // If it isn't valid, we ain't going nowhere
            nPos := nLastPos
         ENDIF
      ENDIF
   ENDDO

   // Restore state vars
   RestoreGetSysVars(aSavedGetSysVars)
```

Creating a New Read Layer

```
   // S87 compatibility
   SETPOS(MAXROW() - 1, 0)

   RETURN (UPDATED())
```

A new command, READ VALID, makes it easy to use this new Read. Notice that the blockify result marker is used, allowing the code block to be either an expression or a block.

```
//    Translation Rules for READ VALID
//
//    Validrd.ch
//    READ VALID


#command READ VALID <valid> SAVE => ;
   ReadValid(GetList, <{valid}>)

#command READ SAVE VALID <valid> => ;
   ReadValid(GetList, <{valid}>)

#command READ VALID <valid> => ;
   ReadValid(GetList, <{valid}>) ;GetList := {}
```

You only need to #include Valid.ch and add the VALID clause where desired, to use the new Read Layer. A simple example follows:

```
//   Using the New Read Layer
//

#include "ValidRd.ch"

MEMVAR GetList

PROCEDURE Main()
   LOCAL cVar1, cVar2, cVar3
   cVar1 := cVar2 := cVar3 := SPACE(20)

   CLS
   @ 10, 10 SAY "Variable 1" GET cVar1
   @ 11, 10 SAY "Variable 2" GET cVar2
   @ 12, 10 SAY "Variable 3" GET cVar3

   // The following ensures that none of the Gets are left
   // empty, and is equivalent to:
   //
   //   READ VALID (! EMPTY(cVar1)).AND. (! EMPTY(cVar2)) ;
   //      .AND. (! EMPTY(cVar3))
   //
   READ VALID {|aGets, lUpdated| NoneEmpty(aGets, lUpdated)}
   RETURN


//   NoneEmpty()

//   Return .T. if none of the Gets is empty

STATIC FUNCTION NoneEmpty(aGets, lUpdated)
   LOCAL nGet
```

Creating a New Read Layer

```
// See if there are any empty Gets
nGet := ASCAN(aGets, {|oGet| EMPTY(oGet:varGet())})

// nGet will equal 0 if there were no empty Gets
RETURN (nGet == 0)
```

Example: Gets With Messages

A slightly more complex Read Layer example displays a message on the screen for every Get. This requires that you define a method for specifying and storing the message as well as for actually displaying it.

As outlined earlier, the message is stored as a dictionary entry in the Get:cargo instance variable through the following command:

```
// Translation Rule for @...GET...MESSAGE

// @...GET...MESSAGE.

#command @ <row>, <col> GET <var> [<clauses, ...>] ;
   MESSAGE <msg> [<moreClauses, ...>];

=> @ <row>, <col> GET <var> [<clauses>] [<moreClauses>];
      ;IF ATAIL(GetList):cargo == NIL;
      ;ATAIL(GetList):cargo := DictNew();
      ;END;
   ;DictPut(ATAIL(GetList):cargo, "cMessage", <msg>)
```

In the following example, the Read Layer retrieves the message from the dictionary and displays it as specified by the SET MESSAGE TO command, saving and restoring that section of the screen:

```
//  Retrieving and Displaying a Message
//

#include "Set.ch"

// ReadMessage()
//
FUNCTION ReadMessage(aGetList)
   LOCAL oGet
   LOCAL nPos
   LOCAL aSavedGetSysVars
   LOCAL cMsg, lCenter, nRow
   LOCAL nSaveRow, nSaveCol
   LOCAL cSaveScr

   IF (VALTYPE(ReadFormat()) == "B")
      EVAL(ReadFormat())
   ENDIF

   IF (EMPTY(aGetList))
      // S87 compatibility
      SETPOS(MAXROW() - 1, 0)
      RETURN (.F.)
   ENDIF

   // Retrieve settings from SET MESSAGE TO
```

Creating a New Read Layer

```
nRow := SET(_SET_MESSAGE)
IF nRow == NIL
   nRow := 0
ENDIF

lCenter := SET(_SET_MCENTER)
IF lCenter == NIL
   lCenter :=.F.
ENDIF

// Preserve state vars
aSavedGetSysVars := ClearGetSysVars()

// Set these for use in SET KEYs
ReadProcName := PROCNAME(1)
ReadProcLine := PROCLINE(1)

// Set initial Get to be read
nPos := Settle(aGetList, 0)

DO WHILE (nPos <> 0)
   // Get next Get from list and post it as the
   // active Get
   oGet := aGetList[nPos]
   PostActiveGet(oGet)

   // If a message is supplied and messages are enabled,
   // display it
   cSaveScr := NIL
   IF (nRow > 0).AND..(oGet:cargo != NIL)

      // Retrieve message from dictionary
      cMsg := DictAt(oGet:cargo, "cMessage")

      IF (cMsg != NIL)
         // Save message row and cursor settings
         cSaveScr := SAVESCREEN(nRow, 0, nRow, ;
            MAXCOL()-1)
         nSaveRow := ROW()
         nSaveCol := COL()

         IF lCenter
            SETPOS(nRow, ((MAXCOL()/2) - (LEN(cMsg)/2)))
            DISPOUT(cMsg)
         ELSE
            SETPOS(nRow, 0)
            DISPOUT(cMsg)
         ENDIF

         // Restore cursor
         SETPOS(nSaveRow, nSaveCol)
      ENDIF

   ENDIF

   // Read the Get

   IF (VALTYPE(oGet:reader) == "B")
      EVAL(oGet:reader, oGet)   // Use custom reader
                                // block
   ELSE
      GetReader(oGet)           // Use standard reader
   ENDIF

   // If screen line was saved, restore it
```

Programming and Utilities Guide 6–21

Creating a New Read Layer

```
      IF (cSaveScr != NIL)
         RESTSCREEN(nRow, 0, nRow, MAXCOL() - 1, cSaveScr)
      ENDIF

      // Move to next Get based on exit condition
      nPos := Settle(aGetList, nPos)
   ENDDO

   // Restore state vars
   RestoreGetSysVars(aSavedGetSysVars)

   // S87 compatibility
   SETPOS(MAXROW() - 1, 0)

   RETURN (UPDATED())
```

Finally, create commands to access this new Read Layer, in addition to the command that was already created for the Gets.

```
// Creating Commands to Access the New Read Layer
//
// GetRead.ch


// @..GET...MESSAGE.

#command @ <row>, <col> GET <var> [<clauses, ...>] MESSAGE ;
   <msg> [<moreClauses, ...>] ;

   => @ <row>, <col> GET <var> [<clauses>] ;
      [<moreClauses>] ;
      ;IF ATAIL(GetList):cargo == NIL;
      ;ATAIL(GetList):cargo := DictNew();
      ;END;
      ;DictPut(ATAIL(GetList):cargo, "cMessage", <msg>)


// READ MESSAGES

#command READ MESSAGES SAVE => ReadMessage(GetList)

#command READ SAVE MESSAGES => ReadMessage(GetList)

#command READ MESSAGES => ;
   ReadMessage(GetList); GetList := {}
```

To use this new Read Layer, add the MESSAGE clause to any Gets that should have a message. SET MESSAGE TO sets the row for the messages and whether or not they should be centered. Finally, use the READ MESSAGES command instead of READ. This example demonstrates this:

```
//  Using the New Read Layer

PROCEDURE Main()
   LOCAL GetList := {}
   LOCAL cVar1, cVar2, cVar3

   cVar1 := cVar2 := cVar3 := SPACE(20)

   SET MESSAGE TO (MAXROW() - 1) CENTER
```

```
CLS
@ 10, 10 SAY "Var 1" GET cVar1 MESSAGE ;
   "Enter Variable 1"
@ 11, 10 SAY "Var 2" GET cVar2
@ 12, 10 SAY "Var 3" GET cVar3 MESSAGE ;
   "Enter Variable 3" PICTURE "@!" ;
   VALID (! EMPTY(cVar3))

READ MESSAGES

RETURN
```

The Reader Layer

As flexible as the ability to create new Read Layers is, it allows no control over individual Gets. You gain control over an individual Get through that Get's *reader*. A reader intercepts the keystrokes and applies them to the Get; it is completely responsible for implementing the behavior of that particular Get. You can see this most clearly by examining the default reader, GetReader().

GetReader() communicates with the Read Layer through the current Get's Get:exitState instance variable. This tells the Read Layer how the Get was exited so it can determine which action it should take next. Getexit.ch defines the possible values for Get:exitState.

GetReader() then passes through three phases: prevalidation, assignment, and postvalidation. The Get function, GetPreValidate(), performs the prevalidation step. If this step fails, the Get is not entered and Get:exitState is set to GE_WHEN.

In the assignment phase, INKEY(0) intercepts keystrokes, then GetApplyKey() acts upon the ASCII value of the keystroke. When a user strikes an exit key, Get:exitState is set to other than GE_NOEXIT and control passes to the postvalidation phase.

The postvalidation phase evaluates whether the Get fulfills the requirements set in the VALID clause. If the requirements are met, execution passes back to the Read Layer. If they are not met, the assignment phase is re-entered.

```
//   CA-Clipper's Default Reader
//
// GetReader()
// Standard modal read of a single Get

PROCEDURE GetReader(oGet)

   // Read the Get if the WHEN condition is satisfied
   IF (GetPreValidate(oGet))
```

```
    // Activate the Get for reading
    oGet:setFocus()

    DO WHILE (oGet:exitState == GE_NOEXIT)

        // Check for initial typeout (no editable
        // positions)
        IF (oGet:typeOut)
            oGet:exitState := GE_ENTER
        ENDIF

        // Apply keystrokes until exit
        DO WHILE (oGet:exitState == GE_NOEXIT)
            GetApplyKey(oGet, INKEY(0))
        ENDDO

        // Disallow exit if the VALID condition
        // is not satisfied
        IF (! GetPostValidate(oGet))
            oGet:exitState := GE_NOEXIT
        ENDIF
    ENDDO

    // Deactivate the Get
    oGet:killFocus()
ENDIF

RETURN
```

Creating New Reader Layers

Though GetReader() implements all of the standard *CA-Clipper* Get behaviors, you may add new behaviors that fulfill special needs.

Basic Guidelines

As with the decision to extend the default Read Layer or to create a new one, it is not always clear when you should implement a feature through a new Read Layer and when you should implement it through a new reader. This indecision is especially strong when the feature involves taking some action before and/or after every Get.

The message Read Layer created earlier in this chapter is an example of this; you could have implemented it as a reader rather than as a Read Layer. Unfortunately, it would then have been incompatible with other readers. Since it is not uncommon to use several readers at a time, this is a larger drawback than incompatibility with other Read Layers.

Adding some features requires creating new behavior at all layers. In most cases, the change is so drastic that it counts as a new form system and is no longer part of the Get System. In these cases, you should consider creating a completely new form tool rather than drastically modifying the existing system. The creation of new form systems is outside the scope of this chapter.

Important Implementation Rules

Once you decide that creating a new reader is the correct thing to do, there are a couple of rules you should follow. As in creation of a Read Layer, the default behavior of the system should never be modified. When creating a new reader this means:

- Do not modify Getsys.prg or redefine any of the functions contained within it, including the GetReader() and GetApplyKey() functions.

- Do not redefine the basic @...GET command; again, you may add clauses through a new command definition.

- If a clause is added to the @...GET command, the new command should map to the supplied @...GET command, if at all possible.

To function well, your reader must adhere to the protocol laid out by the default reader, GetReader(). It must also store data in a way that won't harm other extensions that may be storing data. This means that the reader will:

- Use Get:exitState as its means of communicating with the controlling Read Layer.

- Use the function GetPreValidate() from Getsys.prg to ensure that the Get can be entered.

- Use the function GetPostValidate() from Getsys.prg to ensure that the Get can be exited.

- Use the Get function, GetApplyKey() from Getsys.prg for *all* keystrokes, or define a unique "apply key" function that handles all keystrokes. This enables you to use the new reader in non-modal form systems.

- Conform to all of *CA-Clipper*'s defined editing keystrokes unless there is a compelling reason to do otherwise.

- Handle all possible valid picture clauses for the data type being handled.

- Store any data that must accompany the Get as a dictionary entry in the Get:cargo instance variable. This requires use of the functions found in the Dict.prg sample program shipped with *CA-Clipper*.

Creating New Reader Layers

- Check the Get:cargo instance variable, when storing such data, to see if it is NIL data type. If it is type NIL, create and assign a dictionary to the instance variable. If it is non-NIL, assume that Get:cargo is already a dictionary and add the entry.

By following these rules you ensure that your reader is compatible with the broadest possible range of Read Layers, new form tools, and other readers.

Example: Incremental Date Get

Our first example of a custom reader is one in which a date is changed by incrementing it one day with the plus key (+) and decrementing it a day with the minus key (-). All normal editing keys are still allowed.

Because this reader only requires special handling for two keys, GetReader() can handle most of the keystrokes. This example, following our guidelines, creates two functions, DateGetReader() and DateGetApplyKey(), to implement this reader. DateGetReader() is a virtual duplicate of GetReader() except that it calls DateGetApplyKey() instead of GetApplyKey().

DateGetApplyKey() intercepts the plus and minus keys and implements special handling for them, passing control to GetApplyKey() for all other keystrokes. This is the least painful way to add correct support for all the default editing keystrokes.

```
// New Get Reader
#include "Getexit.ch"

  New Get Reader

// DateGetReader()
//
// Replacement for GetReader(). Logic is identical, except
// DateGetApplyKey() is  process keystrokes.

PROCEDURE DateGetReader(oGet)

   // Read the Get if the WHEN condition is satisfied
   IF (GetPreValidate(oGet))

      // Activate the Get for reading
      oGet:setFocus()

      DO WHILE (oGet:exitState == GE_NOEXIT)

         // Check for initial typeout (no editable
         // positions)
         IF (oGet:typeOut)
            oGet:exitState := GE_ENTER
         ENDIF
```

6-26 CA-Clipper

Creating New Reader Layers

```
      // Apply keystrokes until exit
      DO WHILE (oGet:exitState == GE_NOEXIT)
         DateGetApplyKey(oGet, INKEY(0))
      ENDDO

      // Disallow exit if the VALID condition
      // is not satisfied
      IF (! GetPostValidate(oGet))
         oGet:exitState := GE_NOEXIT
      ENDIF
   ENDDO

   // Deactivate the Get
   oGet:killFocus()
ENDIF

RETURN


//   DateGetApplyKey()
//
//   Replacement for GetApplyKey().  Performs incr/decr
//   on a date get for '+' and '-' keys, standard behavior
//   otherwise.
PROCEDURE DateGetApplyKey( oGet, nKey )
   LOCAL cKey := CHR(nKey)

   DO CASE
   CASE (cKey == '+'.AND. oGet:type == 'D')
      oGet:buffer := TRANSFORM(oGet:unTransform() + 1, ;
         oGet:picture)
      oGet:changed :=.T.
      oGet:display()

   CASE (cKey == '-'.AND. oGet:type == 'D')
      oGet:buffer := TRANSFORM(oGet:unTransform() - 1, ;
         oGet:picture)
      oGet:changed :=.T.
      oGet:display()

   OTHERWISE
      GetApplyKey(oGet, nKey)

   ENDCASE

   RETURN
```

To make this new reader painless to use, you must create a new command that includes support for it. Notice that the following command is implemented in a way that adheres to the guidelines:

```
//   Translation Rule for @...GET...DATEINC
//
//   DateGet.ch

// @...GET DATE

#command @ <row>, <col> GET <var> [<clauses>, ...] ;
   DATEINC [<moreClauses>, ...] ;

   => @ <row>, <col> GET <var> [<clauses>] ;
```

```
SEND reader := {|get| DateGetReader(get)} ;
[<moreClauses>]
```

This example demonstrates the ease of implementing the new reader. It includes the header file and uses the @...GET...DATEINC command to create the Gets.

```
//   Using The New Reader
//

#include "Dateget.ch"

PROCEDURE Main
    LOCAL dOne, dTwo

    dOne := CTOD("3/17/65")
    dTwo := CTOD("5/20/91")

    CLS
    @ 10, 10 GET dOne DATEINC
    @ 11, 10 GET dTwo DATEINC

    READ

    RETURN
```

The Get Function Layer

While the previous example uses GetApplyKey() to implement the majority of its keystroke behavior, this isn't always the case. Sometimes it is necessary to create a new apply key function from scratch. To do that, it is important to know exactly what GetApplyKey() does.

GetApplyKey() goes through several steps to implement the keystroke behavior associated with Gets:

- Checks and executes SET KEYs, if appropriate.

- Checks movement keys and sets Get:exitState to GE_UP, GE_DOWN, GE_TOP, or GET_BOTTOM, if appropriate.

- Checks other keys that exit Gets and sets Get:exitState to GE_ENTER, GE_ESCAPE, or GE_WRITE, if appropriate.

- Checks and acts upon special purpose keystrokes like *Insert* and *Ctrl-U* (undo), if appropriate.

- Discards all invalid keystrokes, or inserts/overstrikes them into the Get, based on the state of the insert flag.

The Get Function Layer

```
//   Implementing the Keystroke Behavior
//   Associated with Gets

PROCEDURE GetApplyKey(oGet, nKey)
   LOCAL cKey
   LOCAL bKeyBlock

   // Check for SET KEY first
   IF ((bKeyBlock := SETKEY(nKey)) <> NIL)
      GetDoSetKey(bKeyBlock, oGet)
      RETURN // NOTE
   ENDIF

   DO CASE
   CASE (nKey == K_UP)
      oGet:exitState := GE_UP
   CASE (nKey == K_SH_TAB)
      oGet:exitState := GE_UP
   CASE (nKey == K_DOWN)
      oGet:exitState := GE_DOWN
   CASE (nKey == K_TAB)
      oGet:exitState := GE_DOWN
   CASE (nKey == K_ENTER)
      oGet:exitState := GE_ENTER
   CASE (nKey == K_ESC)
      IF (SET(_SET_ESCAPE))
         oGet:undo()
         oGet:exitState := GE_ESCAPE
      ENDIF
   CASE (nKey == K_PGUP)
      oGet:exitState := GE_WRITE
   CASE (nKey == K_PGDN)
      oGet:exitState := GE_WRITE
   CASE (nKey == K_CTRL_HOME)
      oGet:exitState := GE_TOP

#ifdef   CTRL_END_SPECIAL
      // Both ^W and ^End go to the last Get
   CASE (nKey == K_CTRL_END)
      oGet:exitState := GE_BOTTOM
#else
      // Both ^W and ^End terminate the Read
      // (the default)
   CASE (nKey == K_CTRL_W)
      oGet:exitState := GE_WRITE
#endif

   CASE (nKey == K_INS)
      SET(_SET_INSERT, ! SET(_SET_INSERT))
      ShowScoreboard()
   CASE (nKey == K_UNDO)
      oGet:undo()
   CASE (nKey == K_HOME)
      oGet:home()
   CASE (nKey == K_END)
      oGet:end()
   CASE (nKey == K_RIGHT)
      oGet:right()
   CASE (nKey == K_LEFT)
      oGet:left()
   CASE (nKey == K_CTRL_RIGHT)
      oGet:wordRight()
   CASE (nKey == K_CTRL_LEFT)
```

```
         oGet:wordLeft()
CASE (nKey == K_BS)
   oGet:backspace()
CASE (nKey == K_DEL)
   oGet:delete()
CASE (nKey == K_CTRL_T)
   oGet:delWordRight()
CASE (nKey == K_CTRL_Y)
   oGet:delEnd()
CASE (nKey == K_CTRL_BS)
   oGet:delWordLeft()
OTHERWISE
   IF (nKey >= 32 .AND. nKey <= 255)
      cKey := CHR(nKey)

      IF oGet:type == "N".AND.;
         (cKey == ".".OR.cKey == ",")
         oGet:toDecPos()
      ELSE
         IF (SET(_SET_INSERT))
            oGet:insert(cKey)
         ELSE
            oGet:overStrike(cKey)
         ENDIF

         IF (oGet:typeOut.AND.! SET(_SET_CONFIRM))
            IF (SET(_SET_BELL))
               ?? Chr(7)
            ENDIF

            oGet:exitState := GE_ENTER
         ENDIF
      ENDIF
   ENDIF
ENDCASE
RETURN
```

Creating New Get Function Layers

Creating a completely new apply-key routine that doesn't use GetApplyKey() is a complicated task and should not be undertaken lightly. You should only do this when you must define entirely new behaviors, that are not "naturally" supported by the Get class. For example, when you want to redefine cursor movement, create new picture template characters or have characters insert from the right instead of the left.

To create a reader that requires this type of support, start by creating a new module that contains a renamed copy of GetReader() and GetApplyKey(), as well as the static function ShowScoreBoard() and its associated defines. Then throw out most of the code, and rewrite from scratch. Working from the copies of these functions, ensures that you have covered all of the default keystrokes.

There are many pitfalls in creating this type of reader. They include:

- Failure to accommodate all possible picture clauses, including those produced by the picture function "@R."

- Failure to correctly handle numeric Gets with no picture specified. In most cases it is easiest to handle this by calculating the picture and, if there isn't one defined, assigning it upon entry to the reader.

- Use of cursor movement and editing methods provided by the Get class. These methods are specific to the defined pictures and behaviors associated with the default readers. If you can use these methods, you probably should not be creating a Get Function Layer.

- Failure to obey Get:clear and clear the Get before accepting more keystrokes.

- Failure to correctly handle floating point numbers. This includes failure to truncate insignificant digits and comparison of floating point numbers exactly without rounding to a specific precision.

Summary

The open architecture of *CA-Clipper's* Get System gives you uncommon power over the user interface of an application. At heart it is the familiar @...GET command set, but in its object orientation, and in the unprecedented access to the language fundamentals as exemplified by the functions of the Read Layer and the user-configurable Reader, it gives you programming flexibility, and data and system control.

This chapter discussed creation of Get objects, basic and advanced use of the Get:block, the layers of the Get System and their primary functions. The principles that generated the guidelines expressed here will serve you as you use the other functions and methods of the Get System. Remember, though, that the greater the modifications you make to the system, the greater is your responsibility.

Accessed at the highest level, the Read Layer, the Get system is both familiar and predictable. Accessed at its lower levels, the Reader Layer and the Get Function Layer, it provides enhanced programming power, but it also requires attention to detail and careful adherence to correct practices. Properly used, the Get System creates robust, reusable, extensible code.

Summary

Newly Documented Methods and Instance Variables

Exported Instance Variables

Get:clear   (Assignable)

> Contains a logical value indicating whether the editing buffer should be cleared on the next valid character. Get:clear is set true (.T.) by Get:setFocus() and Get:undo() when the Get object is a numeric type, or Get:picture contains the "@K" picture function. At all other times it contains false (.F.).

Get:minus   (Assignable)

> Contains a logical value indicating whether the minus (-) has been added to the editing buffer. Get:minus is set to true (.T.) only when the Get object is a numeric type, the current value of the editing buffer is zero, and the last change to the editing buffer was the addition of the minus sign. It is cleared when any change is made to the buffer.

Exported Methods

Get:unTransform() → *xValue*

> Converts the character value in the editing buffer back to the data type of the original variable. Get:assign() is equivalent to Get:varPut(Get:unTransform()).

Get:delete() → *self*

> Deletes the character under the cursor.

Get:delEnd() → *self*

> Deletes from the current character position to the end of the Get, inclusive.

Get System Functions

> Some of the functions in the Getsys.prg have been made public so that they can be used when implementing customized Get readers. These functions are listed on the following pages.

GETACTIVE() function

Return the currently active Get object

Syntax

GETACTIVE() → *objGet*

Returns

GETACTIVE() returns the current active Get object within the current READ. If there is no READ active when GETACTIVE() is called, it returns NIL.

Description

GETACTIVE() is an environment function that provides access to the active GET object during a READ. The current active Get object is the one with input focus at the time GETACTIVE() is called.

Examples

- This code uses a WHEN clause to force control to branch to a special reader function. Within this function, GETACTIVE() retrieves the active Get object:

```
@ 10, 10 GET x
@ 11, 10 GET y WHEN MyReader()
@ 12, 10 GET z
READ

// Called just before second get (above)
// becomes current
FUNCTION MyReader
   LOCAL objGet              // Active Get holder
   objGet := GETACTIVE()     // Retrieve current
                             // active Get
   BarCodeRead( objGet )
   RETURN (.F.)              // Causes Get to be
                             // skipped in READ
```

Files: Library is CLIPPER.LIB, source file is Getsys.prg.

See also: @...GET, READ, READMODAL()

GETAPPLYKEY() function

Apply a key to a Get object from within a Get reader

Syntax

    GETAPPLYKEY(<oGet>, <nKey>) → NIL

Arguments

    <oGet> is a reference to a Get object.

    <nKey> is the INKEY() value to apply to <oGet>.

Returns

    GETAPPLYKEY() always returns NIL.

Description

    GETAPPLYKEY() is a Get system function that applies an INKEY() value to a Get object. Keys are applied in the default way. That is, cursor movement keys change the cursor position within the Get, data keys are entered into the Get, etc.

    If the key supplied to GETAPPLYKEY() is a SET KEY, GETAPPLYKEY() will execute the set key and return; the key is not applied to the Get object.

Notes

- **Focus:** The Get object must be in focus before keys are applied. Refer to Get:setFocus and Get:killFocus for more information.

- **CLEAR GETS:** The Get object must be in focus before keys are applied. Refer to Get:setFocus and Get:killFocus for more information.

Files: Library is CLIPPER.LIB, source file is Getsys.prg.

See also: GETDOSETKEY(), GETPOSTVALIDATE(), GETPREVALIDATE(), GETREADER(), READMODAL()

GETDOSETKEY() function

Process SET KEY during Get editing

Syntax

GETDOSETKEY(<oGet>) → NIL

Arguments

<oGet> is a reference to the current Get object.

Returns

GETDOSETKEY() always returns NIL.

Description

GETDOSETKEY() is a Get system function that executes a SET KEY code block, preserving the context of the passed Get object.

Note that the procedure name and line number passed to the SET KEY block are based on the most recent call to READMODAL().

Notes

- If a CLEAR GETS occurs in the SET KEY code, Get:exitState is set to GE_ESCAPE. In the standard system this cancels the current Get object processing and terminates READMODAL().

Files: Library is CLIPPER.LIB, source file is Getsys.prg.

See also: GETAPPLYKEY(), GETPOSTVALIDATE(), GETPREVALIDATE(), GETREADER(), READMODAL()

GETPOSTVALIDATE() function

Postvalidate the current Get object

Syntax

GETPOSTVALIDATE(<oGet>) → lSuccess

Arguments

<oGet> is a reference the current Get object.

Returns

GETPOSTVALIDATE() returns a logical value indicating whether the Get object has been postvalidated successfully.

Description

GETPOSTVALIDATE() is a Get system function that validates a Get object after editing, including evaluating Get:postBlock (the VALID clause) if present.

The return value indicates whether the Get has been postvalidated successfully. If a CLEAR GETS is issued during postvalidation, Get:exitState is set to GE_ESCAPE and GETPOSTVALIDATE() returns true (.T.).

Notes

- In the default system, a Get:exitState of GE_ESCAPE cancels the current Get and terminates READMODAL().

Files: Library is CLIPPER.LIB, source file is Getsys.prg.

See also: GETAPPLYKEY(), GETDOSETKEY(), GETPREVALIDATE(), GETREADER(), READMODAL()

GETPREVALIDATE() function

Prevalidate a Get object

Syntax

GETPREVALIDATE(<oGet>) → lSuccess

Arguments

<oGet> is a reference the current Get object.

Returns

GETPREVALIDATE() returns a logical value indicating whether the Get object has been prevalidated successfully.

Description

GETPREVALIDATE() is a Get system function that validates the Get object for editing, including evaluating Get:preBlock (the WHEN clause) if it is present. The logical return value indicates whether the Get has been prevalidated successfully.

Get:exitState is also set to reflect the outcome of the prevalidation:

Get:exitState Values

Getexit.ch	Meaning
GE_NOEXIT	Indicates prevalidation success, okay to edit
GE_WHEN	Indicates prevalidation failure
GE_ESCAPE	Indicates that a CLEAR GETS was issued

Note that in the default system, a Get:exitState of GE_ESCAPE cancels the current Get and terminates READMODAL().

Files: Library is CLIPPER.LIB, source file is Getsys.prg.

See also: GETAPPLYKEY(), GETDOSETKEY(), GETPOSTVALIDATE(), GETREADER(), READMODAL()

GETREADER() function

Execute standard READ behavior for a Get object

Syntax

    GETREADER(<oGet>) → NIL

Arguments

<oGet> is a reference to a Get object.

Returns

GETREADER() always returns NIL.

Description

GETREADER() is a Get system function that implements the standard READ behavior for Gets. By default, READMODAL() uses the GETREADER() function to read Get objects. GETREADER() in turn uses other functions in Getsys.prg to do the work of reading the Get object.

Notes

- If a Get object's Get:reader instance variable contains a code block, READMODAL() will evaluate that block in lieu of the call to GETREADER(). For more information refer to the Get:reader reference.

Files: Library is CLIPPER.LIB, source file is Getsys.prg.

See also: GETAPPLYKEY(), GETDOSETKEY(), GETPOSTVALIDATE(), GETPREVALIDATE(), READMODAL()

READFORMAT() function

Return, and optionally set, the code block that implements a format (.fmt) file

Syntax

READFORMAT([<bFormat>]) → bCurrentFormat

Arguments

<bFormat> is the name of the code block, if any, to use for implementing a format file. If no argument is specified, the function simply returns the current code block without setting a new one.

Returns

READFORMAT() returns the current format file as a code block. If no format file has been set, READFORMAT() returns NIL.

Description

READFORMAT() is a Get system function that accesses the current format file in its internal code block representation. It lets you manipulate the format file code block from outside of the Get system source code.

To set a format file, use SET FORMAT (see the SET FORMAT entry in the *Reference* guide) or READFORMAT().

READFORMAT() is intended primarily for creating new READ Layers. The code block that READFORMAT() returns, when evaluated, executes the code that is the format file from which it was created.

Files: Library is CLIPPER.LIB, source file is Getsys.prg.

See also: READKILL(), READUPDATED(), SET FORMAT

READKILL() function

Return, and optionally set, whether the current READ should be exited

Syntax

READKILL([<lKillRead>]) → lCurrentSetting

Arguments

<lKillRead> sets the READKILL() flag. A value of true (.T.) indicates that the current read should be terminated, and a value of false (.F.) indicates that it should not.

Returns

READKILL() returns the current setting as a logical value.

Description

READKILL() is a Get system function that lets you control whether or not to terminate the current READ.

Unless directly manipulated, READKILL() returns true (.T.) after you issue a CLEAR GETS (see the CLEAR GETS entry in the *Reference* guide) for the current READ and otherwise returns false (.F.).

By accessing the function directly, however, you can control the READKILL() flag with its function argument and use it to create new READ Layers.

Files: Library is CLIPPER.LIB, source file is Getsys.prg.

See also: CLEAR GETS, READFORMAT(), READUPDATED()

READUPDATED() function

Return, and optionally set, whether any Get changed during a READ

Syntax

READUPDATED([<lChanged>]) → lCurrentSetting

Arguments

<lChanged> sets the READUPDATED() flag...A value of true (.T.) indicates that data has changed, and a value of false (.F.) indicates that no change has occurred.

Returns

READUPDATED() returns the current setting as a logical value.

Description

READUPDATED() is a Get system function intended primarily for creating new READ Layers. It is identical in functionality to UPDATED() (see the UPDATED() entry in the *Reference* guide), except that it allows the UPDATED() flag to be set.

READUPDATED() enables you to manipulate the UPDATED() flag from outside of the Get system source code.

Files:
Library is CLIPPER.LIB, source file is Getsys.prg.

See also:
READFORMAT(), READKILL(), UPDATED()

Chapter 7
Error Handling Strategies

Earlier versions of *CA-Clipper* responded to runtime errors by passing control to one of six routines you could write in *CA-Clipper*. Error creation, however, was still done by code deep in the heart of *CA-Clipper* so that you could only respond to a predefined set of errors. Your encounter with errors was purely reactive and while you could customize the reaction, you could only do it globally. You could not install and remove error handling code *on the fly*.

In *CA-Clipper* applications, the line between internal and your own error handling code has blurred to the vanishing point. Not only can you customize the response to an error, you can customize the error. The mechanisms for custom errors and *CA-Clipper* errors are the same, and you are free to create and handle your own errors in emulation of *system* errors.

The first part of this chapter is intended to give you some insight into theoretical error handling strategies without giving you any specifics on how you would implement those strategies in *CA-Clipper*. The idea is to give you an understanding of the basic design of the error system before the implementation details are discussed.

Next, you are introduced to the specific *CA-Clipper* language elements designed to implement the described error handling strategies. Each element is presented followed by an example that shows how best to use it. Examples are also provided that tie various error handling elements together.

This chapter is theoretical in nature and is not intended as a tutorial on how to use the *CA-Clipper* error handling system nor is it intended to compare the error handling systems of previous versions to the new system. It does, however, provide an in-depth analysis of error handling strategies as well as several examples which, if studied carefully, can be used as models for your own error handling implementations.

In This Chapter

This chapter introduces you to the *CA-Clipper* error system, starting with an overview of general exception handling theory then progressing through the role of exceptions in modular applications. The following major topics are discussed:

- Overview of exception handling concepts
- *CA-Clipper* exception mechanisms
- Error objects
- A baseline strategy
- Network processing
- Source code for standard exception handling

Overview of Exception Handling Concepts

Error processing works well with modular architecture, where applications are broken into component modules. Self contained, independent modules avoid interapplication dependencies and unforeseen side effects (bugs). *CA-Clipper* architecture promotes use of the modular model.

A module may be a simple function. More often, in *CA-Clipper*, a module is a multi-function subsystem. Such a system is defined in a single source file. In order to scope variables and functions to that file, a module utilizes *CA-Clipper*'s new STATIC declaration. There is, typically, a logical coherency among the functions you include in a subsystem. They all serve some common purpose, which is your rationale for assigning them to a module.

Other modular languages use different terminology, but the concept is the same: portions of an application with common functionality are combined in units. Among these units there is specific division of labor and allocation of responsibility. A complete application is a joining of these subassemblies. In a modular application, interaction between modules is formal, almost contractual (see Figure 1). There is an implied contract between a client module and a contractor module. The client assigns the contractor both responsibility and authority for performing the specialized job for which it was designed. Each module has specific rights and obligations. The client must supply the contractor with the necessary input parameters, or raw materials, in proper form and number. In return, the contractor owes the client the results of operation. The contractor may expect proper input, and the client may expect proper output.

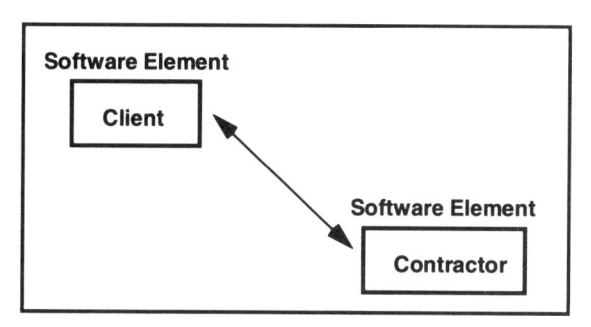

Figure 7-1—Programming by Contract

Error Scoping

Error processing is an essential element of modular programming, where localization is the key.

A module with responsibility for some well-defined part of the application's functionality should also take responsibility for possible malfunctions. The contractor should process departures from normal functionality. A "departure from normal" is called an *exception*, because it is exceptional to expected behavior. We use error and exception synonymously here, though distinctions can be made.

Processing encompasses both sides of a runtime error: generating (detecting, raising, asserting) and responding (reporting, handling). Runtime errors do not merely occur in *CA-Clipper*. They are explicitly created. The process of detection of an error condition and display of an alert message is called '*raising* an error.'

Exception handling should include exception handling as part of its primary functionality. It should not be patched in as an afterthought. The right place for code that handles errors pertaining to a certain module of an application is normally within that same module. This is called *localization* of error processing.

In-module coding is not the only way to localize error-handling, even though it is the most common method. You may also localize to sub-modular or supra-modular structure. Such localization may, with proper coding, span a function, a statement or statement sequence, or

Overview of Exception Handling Concepts

the error-handling may be invoked over several modules through calling routines or other activations.

Error processing, like other aspects of *CA-Clipper*, conforms to lexical scoping. That is, error handling procedures are part of the context of the software elements in which they are written, so they have similar scoping. Figure 2 shows three software elements, layered like an onion. Outer ones contain or call inner ones. Some error-processing code lies within each. An error raised by the innermost layer falls in that element's context and receives immediate processing by the inner element's own error handling code. It also lies within the context of the surrounding element and, if not resolved within the inner element, may be passed, for subsequent processing, to the error handling code in outer layers. Note that the contexts which surround an error depend on the calling sequence on any particular activation. Later, if a different set of callers invokes the same inner element, the same error would have a different overall context.

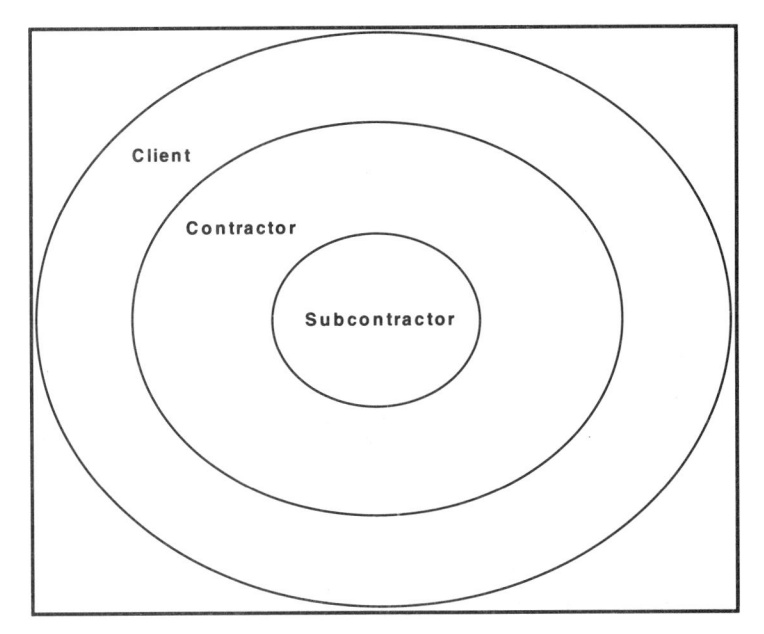

Figure 7-2—Context

What this does is localize the error handling to the place closest to its raising. It frees clients from involvement with a contractor's difficulties. Should the contractor successfully recover, the client proceeds unaware there was any incident. The author of the handler is likely to intelligently interpret the error and respond appropriately. Applications with this behavior are termed robust.

Raising Errors

Errors have their roots in *assertions*. An assertion is an inline test of runtime adherence to specification. It can be nothing more than an IF statement whose control condition returns true (.T.). Any failure of the IF test constitutes an exception, which is, by definition, an error. Assertions have the following general pseudocode form:

```
IF .NOT. AsItShouldBe
   REACT
ENDIF
```

Whether all exceptions are errors, or only some, is a subjective opinion. If under some circumstances it is OK for the condition to be false, but you code the assertion just because you still want to know about it, you probably would not label the exception erroneous.

Upon error detection, the reaction is to launch into error processing. This usually means transfer of control to application code designed for that purpose. Executing that code is called *handling* the error.

You can see that errors are not self-generating or self-defining. They don't just happen. Errors are reports of possible, though *exceptional*, behavior. Error-handling is premeditated.

The following pseudocode illustrates use of assertions in the contractual context. The assertions verify conformity to contractual responsibilities. In this simplified case, the module is a function that derives square roots. This function has the right to insist on only non-negative numbers, and has the obligation to return a correct value, one that returns the square when squared. An assertion at the top ensures that the function complains if it receives invalid input. Another at the bottom enforces valid output, testing it independently of the derivation algorithm. If that algorithm involves generating a mathematical series whose terms should converge, another assertion could sound an alarm if they began to diverge instead.

Overview of Exception Handling Concepts

```
FUNCTION TakeSquareRoot(nSquare)
   // Precondition: at this point nSquare had
   // better be non-negative */

   IF !(nSquare >= 0)
      REACT
   ENDIF
   .
   . <statements>
   .
   // Postcondition: at this point nRoot * nRoot
   // had better yield nSquare */

   IF !(nRoot * nRoot == nSquare)
      REACT
   ENDIF

   RETURN (nRoot)
```

It is natural to expect these errors. You should anticipate them, then insert appropriate assertions in your code. Otherwise you may encounter strange results without knowing why. Unanticipated errors can cause unpredictable program failures. An error is far preferable.

Handling Errors

As contingency, the architecture of an application should generally provide exception handling code. Such code should be strategically set aside in error handlers so it will be available when needed. On one hand, it is separate from main code; on the other, it is integral to the application. It is as important to an application's design as fire stairs are to a building's design. It may get less use than the main hallway but it is no less essential.

The location of this code within the source is less important than its availability to the application at runtime. When error-handling code is encountered during runtime, the main code is abandoned (see Figure 3). At the discretion of the handler, control may or may not return to main code. If it does, the main code must be written to retake control gracefully (retry, default handling, substitution options precoded).

Overview of Exception Handling Concepts

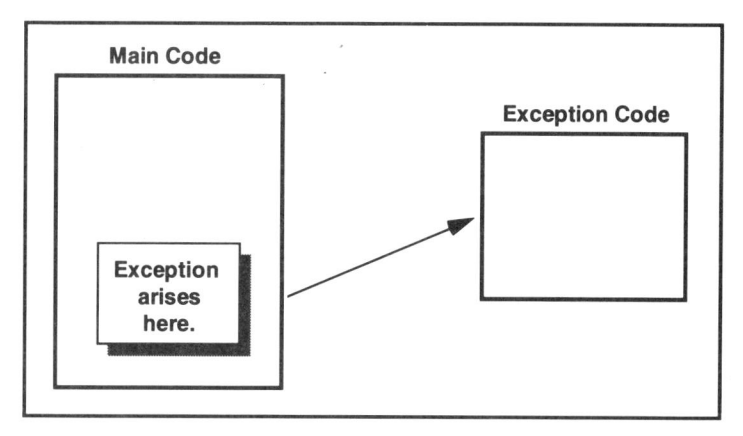

Figure 7-3—Exception Handling Code

When the handling code gains control there are only two theoretical options: *retry* or *terminate*. Generally, retrying means finding some way to proceed. That may amount merely to a simple, brute force retry. But usually it is more elaborate, including attempts to identify and remedy the cause of which the error condition is an effect.

The retry need not be performed by the handler. In case the error-originating code is written to retry on its own (equipped with a loop, for example), the handler can signal the original code to give it a try. This necessitates a protocol whereby the originating code indicates its readiness to retry if asked (in *CA-Clipper*: using the Error object's canRetry instance variable), and the handler asks by returning some predetermined value (in *CA-Clipper*, true (.T.)).

Variations on the theme are *default handling* and *substitution*. Suppose that error-generating code is written so it can proceed even if the error condition is not removed. If it communicates that fact when calling a handler (in *CA-Clipper*, e:canDefault), the handler has the option of returning control to the original code for continuation (in *CA-Clipper*: returning false (.F.)). Similarly, if the original code's purpose is to derive a value but it is written to accept a substitute value if derivation is impossible, a substitutability-aware handler has the option of returning a substitute value to the main code (e.g., zero to a routine that tried to divide by zero).

Termination also has a broader meaning than just halting and returning to the operating system. More generally, it means terminating the attempt to perform the current operation and signaling failure to the original invoker (client), with or without prior clean-up (e.g., closing files). The invoking process then decides whether it can tolerate the failure and what, accordingly, to do next. This is sometimes called propagating the error to the client or outer context (in *CA-Clipper*, BREAK).

The cardinal sin of error processing is to leave the client unaware of the contractor's failure. It is irresponsible for a contractor that fails in its specified mission to "pretend," by returning a result indistinguishable, by the client, from success.

CA-Clipper Exception Mechanisms

CA-Clipper imposes no error-processing protocol on applications, but gives you the tools to do so in your own way. These tools, and their practical application, are the focus of this section.

In *CA-Clipper*, there are two mechanisms for processing exceptions, the SEQUENCE construct, and a posted code block (the *error block*). With these mechanisms as tools, you can build different approaches to error handling. You can, if necessary, use these tools to emulate the fixed, more rigid behaviors of other programming languages.

The two mechanisms differ in the location of error-handling code and how it is called. In the SEQUENCE construct, error-handling code is inline (part of the RAM image of the loaded application) and is called by issuing a BREAK statement. In the error block approach it is stored in memory (in the code block) and is triggered by executing the EVAL() function.

In general, use SEQUENCE for application-specific exceptions (debits did not equal credits, inventory is negative, my boomerang won't come back), and error block for generic, low-level exceptions that usually deal with the computer and could arise across applications (device not ready, disk full, stack underflow).

The SEQUENCE Construct

The basic form of a SEQUENCE construct is shown in Figure 4. Between the BEGIN SEQUENCE and RECOVER statements is main code; between RECOVER and END is error-handling code. The error-handling code is called a RECOVER statement block.

```
BEGIN SEQUENCE

   .
   . <statements>
   .

RECOVER USING ---
                    ◀――――――――――  RECOVER statement block
   .
   . <statements>
   .

END
   .
   . <statements>
   .
```

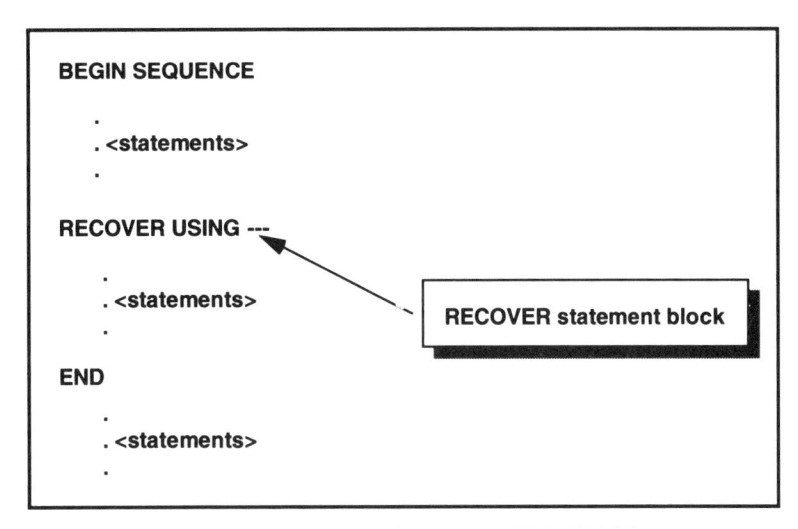

Figure 7-4—Locus of Exception Code with SEQUENCE

A BREAK statement permits the RECOVER statement block to gain control. Without one, the block never executes; instead, control leapfrogs past the END statement when the RECOVER statement is encountered.

BREAK, in a SEQUENCE construct, represents the concrete REACTion, as pseudocoded in earlier in this chapter. Any assertion in the main code should, when violated, provide a BREAK, abandoning the main code and passing control to the RECOVERy code. (Though this may be done indirectly; see the Baseline Strategy discussion below.)

BREAKs can be nested, as illustrated in Figure 5. Here an assertion could equivalently pass control to the RECOVER code whether appearing in the main code itself, one activation removed as in Udf1(), or two activations removed, as in Udf2().

CA-Clipper Exception Mechanisms

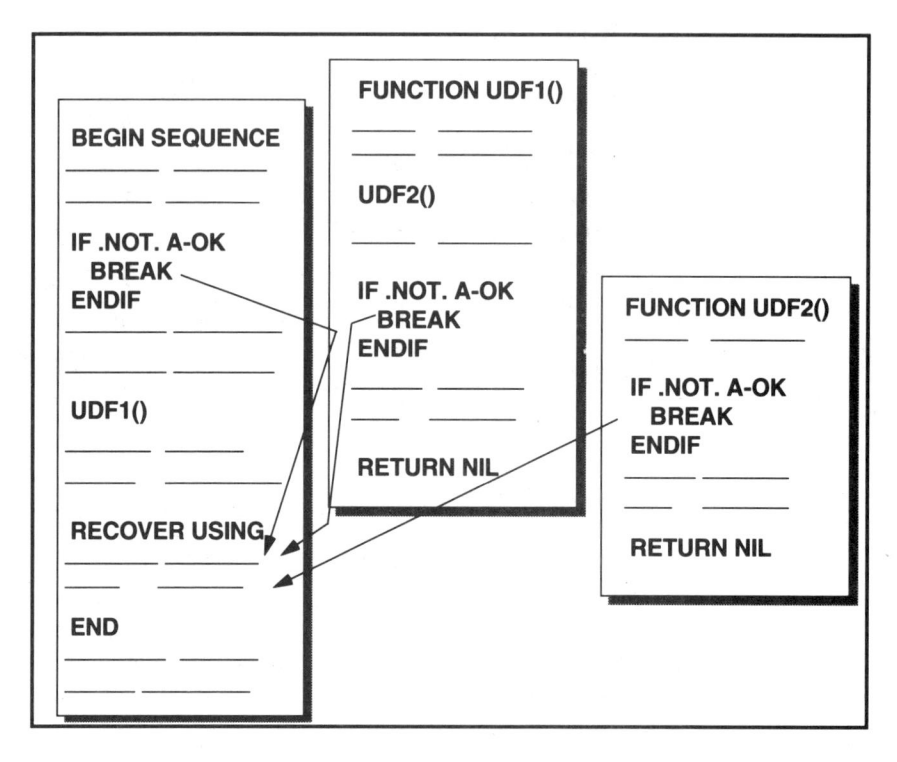

```
BEGIN SEQUENCE
  ___  ___
  ___  ___
  IF .NOT. A-OK
    BREAK
  ENDIF
  ___  ___
  ___  ___
  UDF1()
  ___  ___
  ___  ___
RECOVER USING
  ___  ___
  ___  ___
END
  ___  ___
  ___  ___

FUNCTION UDF1()
  ___  ___
  UDF2()
  ___  ___
  IF .NOT. A-OK
    BREAK
  ENDIF
  ___  ___
  ___  ___
  RETURN NIL

FUNCTION UDF2()
  ___  ___
  IF .NOT. A-OK
    BREAK
  ENDIF
  ___  ___
  ___  ___
  RETURN NIL
```

Figure 7-5—Nested BREAKs

You can nest SEQUENCE constructs, too (see Figure 6). Here, Udf1() contains its own SEQUENCE construct, so the assertions in Udf1() or Udf2() now call the RECOVERy code local to the context of Udf1(), not that of the original outer sequence. To return the error to the outer sequence would require a BREAK statement issued within Udf1()'s RECOVER statement block.

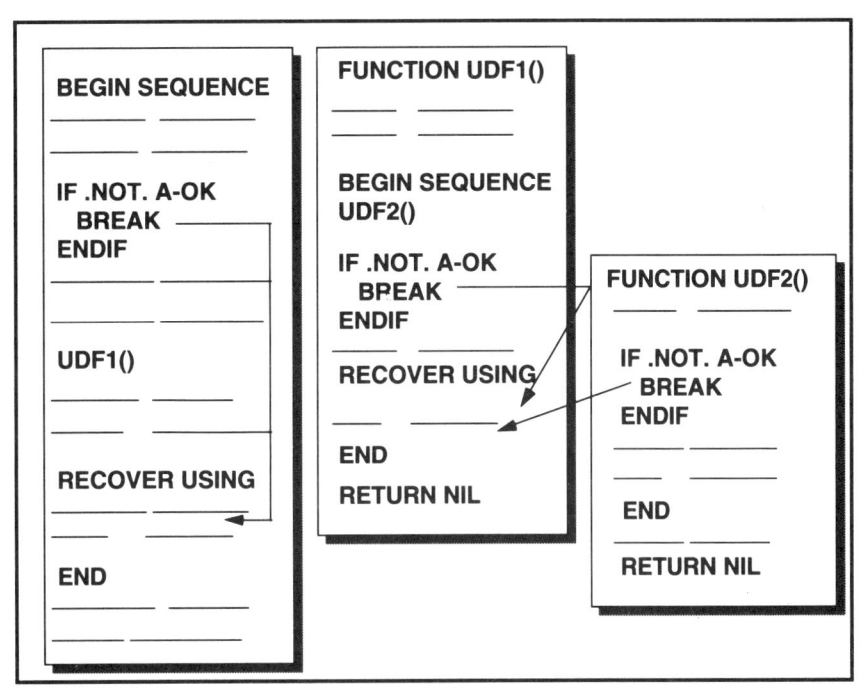

```
┌─────────────────────────────────────────────────────────────┐
│  ┌──────────────────┐  FUNCTION UDF1()                      │
│  │ BEGIN SEQUENCE   │  ___ ___                              │
│  │ ___ ___          │  ___ ___                              │
│  │ ___ ___          │                                       │
│  │                  │  BEGIN SEQUENCE                       │
│  │ IF .NOT. A-OK    │  UDF2()                               │
│  │   BREAK ────┐    │                                       │
│  │ ENDIF       │    │  IF .NOT. A-OK    ┌──────────────────┐│
│  │ ___ ___     │    │    BPEAK ────┐    │ FUNCTION UDF2()  ││
│  │ ___ ___     │    │  ENDIF       │    │ ___ ___          ││
│  │             │    │  ___ ___     │    │ ___ ___          ││
│  │ UDF1()      │    │  RECOVER USING    │ IF .NOT. A-OK    ││
│  │ ___ ___     │    │  ___ ___◄────┘    │   BREAK          ││
│  │ ___ ___     │    │                   │ ENDIF            ││
│  │             │    │  END              │ ___ ___          ││
│  │ RECOVER USING    │  RETURN NIL       │ ___ ___          ││
│  │ ___ ___◄────┘    │                   │ END              ││
│  │ ___ ___          │                   │ RETURN NIL       ││
│  │                  │                   └──────────────────┘│
│  │ END              │                                       │
│  │ ___ ___          │                                       │
│  │ ___ ___          │                                       │
│  └──────────────────┘                                       │
└─────────────────────────────────────────────────────────────┘
```

Figure 7-6—Nested SEQUENCE Constructs

This illustrates the practical side of error contexts in *CA-Clipper*. BEGIN SEQUENCE statements delimit context boundaries. Each time you encounter a new BEGIN SEQUENCE statement, you cross into a new context, with its own local (though optional) RECOVER statement block. You exit that context when you leave the main code sequence (whether entering or leapfrogging the RECOVER statement block).

Activation boundaries do not, necessarily, define contexts. For example, the same function can be part of different contexts at different times. If Udf2() is invoked by another caller, it is part of that caller's context. Since you can bracket any code you like with a SEQUENCE construct, *CA-Clipper* lets you define contexts quite flexibly. Contexts lie wherever you put your SEQUENCE constructs.

The following example shows the suggested general form of such a statement block. A CASE statement responds to whatever errors the programmer anticipated when he wrote the program. Where control passes next is governed by whether a QUIT, BREAK, or RETURN appears. If you supply no such code, control passes through the END SEQUENCE statement to whatever code follows.

This general description is adequate, though using BREAK, through EVAL(ERRORBLOCK(), oError), as described later, is the preferred form of error handling.

```
BEGIN SEQUENCE

//  General Form of SEQUENCE Construct and RECOVER Clause//

IF lProblem1
   oError := ERRORNEW()
   .
   . <stuff oError>
   .
   BREAK oError
ENDIF
.
.
.
IF lProblem2
   oError := ERRORNEW()
   .
   <stuff oError>
   .
   BREAK oError
ENDIF
.
.
.
RECOVER USING WhatItsAllAbout

   DO CASE

   CASE <obj-related condition>
      <handling>

   CASE--
      <handling>
      BREAK WhatItsAllAbout        // Optional propagation

   CASE__
      <handling>

   OTHERWISE
      BREAK WhatItsAllAbout        // Optional propagation

   ENDCASE

END SEQUENCE
```

There are several special cases you should consider. An application that does not use SEQUENCE constructs has DOS as its implicit outer context. A BREAK anywhere in the application leads to the DOS prompt. One with a single SEQUENCE construct surrounding the initial main menu module has that module as outer context, with any BREAK returning there like RETURN TO MASTER. Dot.prg in \CLIPPER5\SOURCE\SAMPLE\ subdirectory illustrates this case.

Ultimately, you can bracket single statements with a SEQUENCE construct, with RECOVERy code for that statement alone. If that statement were an invocation of DBEDIT(), ACHOICE(), or

MEMOEDIT(), strategically distributed BREAKs in the called user function could exploit the RECOVERy code. Otherwise, were it a USE command, there would be no apparent opportunity to issue a BREAK from "within" the USE. You can effect this in cases of system errors raised within USE, with the posted code block and indirect BREAKing mechanism discussed below.

The Posted Error Block

The *error block* is a code block deliberately written to handle errors. It is distinguished by being *posted* or *installed* in a designated memory location. Installation and later retrieval is done exclusively through the ERRORBLOCK() function. Those are ERRORBLOCK()'s two purposes.

ERRORBLOCK() views the block as data, not code. (The hallmark of code blocks is that they are hybrid: data that happens to contain code, or code that can otherwise be handled like data.) ERRORBLOCK() contains, in effect, a static variable, or slot for storing data accessible to itself alone, but ERRORBLOCK() itself is globally accessible. Because ERRORBLOCK() can install any code block, any part of the application can dynamically install its own new code.

The posted code block is strategically important because *CA-Clipper's* unconditional behavior when raising an error is to retrieve it with ERRORBLOCK(), then execute it with EVAL(). When you post an error block, you are prepositioning code for *CA-Clipper* to execute next time there is an error condition. You get to write *CA-Clipper's* response. Also, ERRORBLOCK() and EVAL() let you make your application's response the same as *CA-Clipper's* when one of your assertions is violated.

To illustrate, consider a deliberately erroneous call to a system function, say VAL(99). We know VAL()'s job is to return a numeric from a string of numerals, so it requires a character-type argument. We pass it a numeric data type instead. A one line Main.prg containing only the expression VAL(99) produces the following screen output:

```
Error BASE/1098  Argument error: VAL
Called from MAIN(1)
```

Studying the Errorsys.prg (the standard error handling routine provided as part of *CA-Clipper*) source code. You see a procedure named ERRORSYS. This special procedure runs automatically at the outset of every application. It contains a single line:

```
ERRORBLOCK({|e|DefError(e)}).
```

It causes the compiler to produce a code block that calls a UDF named DefError(), causing ERRORBLOCK() to install that block at runtime. Then, when *CA-Clipper* raises any error, a call is issued to DefError(). If you look at DefError(), also in Errorsys.prg, you can easily identify the code responsible for the error message.

The VAL() function is not tied to calling DefError(). It must, however, fetch and execute the currently posted error block. It neither knows nor cares about the block's code contents, who posted it, or how long ago. If the default posted block is still in place, DefError() will get a call. If, however, you replace the block, you replace the error behavior. If you wrote and installed a function that rings the bell or clears the screen:

```
ERRORBLOCK({|e|MyBellRinger(e)})
```

VAL(99) would ring the bell, or clear the screen as would all other *CA-Clipper* errors. (We will revisit VAL() to investigate the e parameter when we discuss error objects below.)

This example uses the basic error block form. The ERRORBLOCK() function installs the code block passed to it. However, it returns rather than discards the block that had been in place. You thus have the option of retention for possible later execution, or reinstallation. Here, it is retained by assignment to the variable oOldBlock.

```
// ERRORBLOCK()  --> {|e| <SomeExp1>}
bOldBlock := ERRORBLOCK({|e| MyErrHan(e)})
// ERRORBLOCK()  --> {|e| MyErrHan(e)}
// bOldBlock --> {|e| <SomeExp1>}
```

This error block retains, and later executes its predecessor:

```
oOldBlock := ERRORBLOCK({|e| MyFunc(e),EVAL(oOldBlock,e)})
```

This technique is the basis for code block *chaining*. The installed block's execution of its predecessor as its "final act" is equivalent to terminating and passing control to the predecessor, (i.e., chaining).

Notice the implications. The code block is built at compile time. Its installation and predecessor's assignment to oOldBlock, and its execution, take place during runtime. While the block contains a reference to a variable called oOldBlock, nonexistent at build time, that is not a problem as long as oOldBlock exists at evaluation. Depending on oOldBlock's storage class, it may or may not exist at any given moment during runtime. Declaring oOldBlock static ensures its permanent availability.

This code chains an error block to its predecessor.

```
STATIC bPreviousHandler
   .
   . <statements>
   .
   bPreviousHandler ;
      := ERRORBLOCK({|e| NewHandler(e,bPreviousHandler)})
   .
   . <statements>
   .

FUNCTION NewHandler(oErrObj,bMyAncestor)
   .
   . <statements>
   .
   RETURN (EVAL(bMyAncestor,oErrObj))
```

SEQUENCE vs. Error Block

With SEQUENCE and with error block, you can create two parallel chains of command. A group of code blocks that EVAL() one another in a chain, and a series of nested SEQUENCEs that BREAK from inner to outermost. It is natural to ask about the interrelationship. See Baseline Strategy.

Error code, embodied in an error block, can transfer control to error code in the current RECOVER statement block. Like any code, it does so by issuing a BREAK statement. (There is an equivalent BREAK() function specifically for encoding in expressions.)

The following example shows how a system-level runtime error can lead to RECOVER code. The error originates in code where a surrounding RECOVER statement block is pending. Like all system errors, it executes the installed error block. Because the block BREAKs, RECOVER code can receive control. The Baseline Strategy adopts this practice.

Error Objects

```
// ERRORBLOCK() --> {|e| <exp1>}
oOldBlock := ;
   ERRORBLOCK({|e| IF(!<lAssert>,BREAK(e),<exp>),<exp>})
.
. <statements>                      // cOldBlock --> {|e| <exp1>}
.
FUNCTION SomeSubsequentFunc
   BEGIN SEQUENCE
      .
      . <statements>
      .
      VAL(99)                       // Error arises here
      .
      RECOVER USING e
      .
      . <recovery code>
      .
      END
      .
      .
      RETURN (NIL)
```

Calling procedures differ between error blocks and RECOVER clauses. The difference is that you can RETURN from the former but not the latter. An error block executes through the EVAL() function, RECOVER executes after a BREAK from within a BEGIN SEQUENCE statement. Therefore, EVAL() lets you RETURN as from any other function call. BREAK does not. The BREAK's location is not saved as a return point for future reference. A RETURN from RECOVER goes to the most recent calling function, not the BREAK location. This means that in order to return to a chosen point from a RECOVER, you must write code loops. It also means that the discussion of error protocols below applies only to code block handlers, not RECOVER clause handlers.

There are four directions for the error block to go when called. As suggested, it can RETURN to the offending code. You must do this carefully to avoid producing undefined results. Secondly, it can go to its predecessor error block, by EVAL(). Third, it can branch to the SEQUENCEs in main application code, by BREAKing. Finally, it can QUIT.

Error Objects

Once you understand the mechanics of *invoking* error code, you must decide how to use it. That decision depends on the error's characteristics, and the vehicle for characterizing it is the *Error object*.

An Error object is a package of information items. (In that sense, it is akin to an array). The error-generating code supplies that information to the error-handling code. There are fourteen items, called instance variables, in each package. Most describe the error, but others tell the handler about the error-generating code itself. Error objects are passive. They do nothing (they have no methods). They are merely useful structures for holding information during error-processing.

The error object's role is that of go-between in communication between the error-generating and error-handling code. Communication is initiated by the former, and may become two-way dialog if the latter returns. All this implies a third step in the error-generating process. As mentioned earlier, errors are generated by assertions, which have a two-stage detect-react form:

```
IF .NOT. AsItShouldBe          // detect exception
   CALL HANDLER                // BREAK or EVAL(ERRORBLOCK())
ENDIF
```

The third step is preparation of an error object:

```
IF .NOT. AsItShouldBe
   PREPARE ERROR OBJECT
   CALL HANDLER WITH ERROR OBJECT
ENDIF
```

Preparing the object amounts to calling the ERRORNEW() function to generate an empty object, then assigning values to its instance variables.

When your error handler gains control, it has the error object on which to base a particular response. Typically, it uses a CASE statement. The CASE conditions are based on information about the error conveyed in the instance variables. This is called *specialization* of error handling.

The typical CASE statement provides localized handling for the module that contains it. It probably is most frequently invoked by errors that originate within the module. The CASEs are written accordingly, with those errors in mind. For example, in an accounting module, you are likely to find both assertions and handler code that address possibilities like "inventory fell below zero," or "debits and credits unequal," reflecting the specialized conventions of that module's world, and the errors they define. But the handler may receive other errors in the form of incoming error objects for errors that fit none of the CASEs. These objects probably originated in a called module that propagated them to this handler. They fall into the handler's OTHERWISE category, where the handler can further process them. In a RECOVER clause, processing amounts to issuing a BREAK <obj> where <obj> is the same error object the handler originally received.

In this way, an error raised but not handled to containment in an inner module is passed back up through higher contexts, each of which screens it and has the opportunity to respond. Ultimately it reaches the outermost context, namely the application's initial procedure. There, it experiences a hard landing to the DOS prompt if not bracketed by an outermost SEQUENCE construct. Or, if it finds a considerate containing sequence in the main program (a good practice), a graceful and informative message to soften the blow.

Specialization of an error object is the means by which an error object is passed among error handlers.

A Model Example

Let us reconsider VAL(99) for an illustration of the error object's role in raising an error, and the build it/fill it/pass it behavior followed by all *CA-Clipper* errors. We suggested that VAL(99) passes a parameter to the posted error block when calling it (see DefError() in Errorsys.prg listed at the end of this chapter). The parameter is an error object. You can intercept and study its contents by writing a diagnostic UDF that prints out instance variable, plus a two line main program that first installs a block to call your UDF then issues VAL(99). When VAL(99) balks, it hands control to the block, which hands it to the UDF, which displays the information on the screen.

This example, a pseudocode of the VAL() function of *CA-Clipper*, demonstrates that you can reproduce system error behavior in your own code. You can see that VAL(99) must contain code close to that shown here. Note that the values given to instance variables are arbitrary. Although VAL() may not be written in *CA-Clipper*, this pseudocode probably accurately portrays what VAL() is doing. (Note that significant portions of *CA-Clipper* have indeed been written in *CA-Clipper*, whether or not this is one of them).

```
//  Pseudocoded CA-Clipper VAL() Function
//FUNCTION VAL(p)

      LOCAL VErrObj     // Assertion to verify  type consistency

      IF VALTYPE(p) != "C"
         VErrObj              := ERRORNEW()       // Make object

         VErrObj:args         := {p}              // Stuff object
         VErrObj:canDefault   := .F.
         VErrObj:canRetry     := .F.
         VErrObj:canSubstitute := .T.
         VErrObj:description  := "Argument error"
         VErrObj:filename     := ''
         VErrObj:genCode      := 1
         VErrObj:operation    := "VAL"
         VErrObj:osCode       := 0
         VErrObj:subCode      := 1098
         VErrObj:subsystem    := "BASE"
         VErrObj:tries        := 0
         VErrObj:severity     := 2

         // Transfer control and ship object
         RETURN (EVAL(ERRORBLOCK(),VErrObj))
```

```
ENDIF
.
. <the string-to-numeric conversion code>
.
RETURN (<numeric result>)
```

The "System Error" Line-up: Error.ch

Any function may receive bad parameters from its caller, so "Argument Error" is a commonplace, predictable error. And it is not the only one in this "easily foreseeable" category. Argument error plus thirty others are so reliably expected that the *CA-Clipper* runtime support system raises them, explicitly. The header file, Error.ch, assigns each a distinguishing (version-dependent) code, and a corresponding (version-independent) name. This header file is located in the \CLIPPER5\INCLUDE directory.

The thirty-one anticipated errors in Error.ch have fairly obvious names like EG_ARG, EG_NOVAR, and EG_OPEN (refer to a non-existent variable to raise an EG_NOVAR error; USE a made-up filename to raise EG_OPEN). The names are shorthand for thirty-one generic codes, one of which appears in the genCode instance variable of any *CA-Clipper* runtime Error object. Error handlers can query the error object they have received to identify which type of error was raised.

As indirectly suggested in the example code, if you raise your own errors you must be thorough. Certain application-level errors you specifically foresee will be the ones for which you place assertions in your code, and which you depict in their Error objects.

Carrying Error-Processing Protocols

Error object instance variables supply information, not only about the error, but also the code that raised it. After dealing with the error, the handler may return control to the failed code for further purposes. But it generally has no internal knowledge of this code. Therefore, it is up to the failed code to tell the handler whether it is prepared to regain control, and, if so, to what return values it will respond.

That is the purpose of the canDefault, canRetry, and canSubstitute instance variables. The failed code might or might not be equipped to proceed despite the error condition, or perform a retry, or accept a substitute for a value it failed to derive. Respectively, the three instance variables signify whether the code is so equipped. If it is, it performs the default, retry, or substitution, contingent on regaining control from

Error Objects

an error handler, and upon the value the handler returns to it. In this sense, the error object is a protocol carrier.

A look at the pseudo-VAL() shows that it is unequipped to retry or default, but VAL() can substitute. First, the canSubstitute instance variable returned from a violated VAL() is true (.T.). The VAL() code communicates with us, telling us something about itself. It says that, if we return control and a value, it will return that value to its caller. If, for example, you post a code block that returns "Wacky Wacky," then write a program to print VAL(99), it prints "Wacky Wacky." Try it.

The pseudo-VAL() shows how to write code that can substitute: it embeds its call to the error block in a RETURN statement. Should the error block ever return to VAL() (which is optional, witness DefError()), the value returned by the block becomes the value returned by the EVAL().

By setting VErrObj:canSubstitute true (.T.), VAL() tells the error block about this arrangement. The error block, upon seeing canSubstitute, returns a substitute value to the failed function. Expressed in pseudocode:

```
ERRORBLOCK( {|e| SomeHandler(e)} )

FUNCTION SomeHandler(HisErrObj)
    .
    . <statements>
    .
    IF HisErrObj:canSubstitute == .T.
       RETURN "Wacky Wacky"
    ENDIF
    .
    . <statements>
    .
```

It returns the substitute value only upon ascertaining that the calling code is prepared to substitute it.

Similarly, there are conventions for holding a retry dialog, and for plowing ahead (defaulting) despite the error condition. (These two together are mutually exclusive.) Retry is possible if the operation that fails is surrounded with a looping structure to let it go back and repeat.

This example demonstrates a basic retry structure.

```
// Error-Handling Protocol Conventions
//
lFirstTry := .T.

DO WHILE lRetry
    .
    . <body of operation>
    .
    lRetry := .T.
```

```
IF .NOT. AsItShouldBe        // Your assertion
   IF lFirstTry
      obj                 := ERRORNEW()
      obj:canRetry        := .T.
      obj:canDefault      := .T.
      obj:canSubstitute   := .F.
      obj:tries           := 0
      lFirstTry           := .F.
   ENDIF
   obj:tries++
   lRetry := EVAL(ERRORBLOCK(), obj)    // Raise the error
   IF lRetry
      LOOP                      // Causes retry
   ELSE
      <default code>            // Otherwise execute your Plan B
   ENDIF
ELSE
   lRetry := .F.                // No violation, no retry
ENDIF
.
.<continuation of operation>
.
ENDDO
```

The error block, if canRetry is true, may seek to retry the original code.

On the other hand, the error code may go through a few motions, perhaps altering the original error condition, then return its request for another try. When it tries the second time, the code encounters the error condition once again. This time, instead of creating a new Error object, it increments the tries instance variable of the existing one and raises the error again. This may happen any number of specified times. At the specified limit, the error code may not return true (.T.), instead, giving up and branching elsewhere.

If default response is indicated, *<default code>* has several options. It may drop out of its IF, execute *<continuation code>* and represent its supplement. It may exit, short-circuiting *<continuation code>*, becoming its alternative, allowing the code beyond this operation to execute. Or it could disrupt continued flow with a QUIT, BREAK, or RETURN. The possibilities are numerous and entirely user-definable.

A Baseline Strategy

A hallmark of *CA-Clipper*'s error mechanisms is versatility. The mechanisms are generalized, so you can adapt them to emulate various error-handling approaches. Typically, *CA-Clipper* does not provide a particular approach. Instead, it provides tools to implement an approach of your choosing. Therefore, for code integrity, you should adopt certain conventions. Here are some suggestions:

A Baseline Strategy

- To provide more than 'say so and quit' response to errors, replace DefError() by posting a generic error block that BREAKs rather than quits.

  Retain the initial, special case code for zero division and network processing that appears in *CA-Clipper's* DefError(). During programming, the standard DefError() may be desirable, but a user, in the middle of a completed application, is less appreciative of cryptic error messages on a DOS screen.

- Place substantive processing of application-specific errors you raise in RECOVER clauses. Deal with them in the application, not in the error block.

- Instead of directly invoking RECOVER using BREAK in your assertions, do so indirectly through EVAL(ERRORBLOCK(), object). That works because the generic error block we posted BREAKs for us. This is the behavior of the *CA-Clipper*-supplied functions.

Allow subsystems to install their own error blocks freely, while assuring that such a subsystem does three things:

- Preserves a copy of the predecessor block upon entry.

- Restores (reinstall) that block upon exit.

- Provides a call to the predecessor when done, in case there is an error along the way (i.e., to chain).

The block should customize low-level handling, not substantively handle application-level exceptions. For example, it may keep track of more information before BREAKing, or it may intercept additional errors for 'say so and quit' treatment instead of BREAKing. For application-level processing, you should use the RECOVER clause local to the error, thereby heading back up the chain of error blocks to reach the top one, which BREAKs.

Provide OTHERWISE BREAK *<object>* for those situations in a RECOVER clause you cannot handle through a CASE option. That way, you guarantee that an unrecognized error eventually reaches the outermost context.

Provide an outermost RECOVER that gracefully QUITs. It takes over the job of DefError(), now that we have decided to replace it with a block that BREAKs. *CA-Clipper's* ALERT() function is designed for user interface under error conditions. It is sensitive to the presence or absence of the *CA-Clipper* full-screen I/O system. If the full-screen system is not present and normal I/O is compromised, ALERT()

produces a TTY-style display. Do not use ALERT() in other settings, where it may not produce expected screen output.

Finally, keep the Error object near at hand throughout. All assertions should build error objects in the first place. And subsequent error code invocations, both EVAL()s and BREAKs, should pass it along.

These conventions make the application's error behavior uniform. Wherever an error, whether user-defined or the system's, just call the current error block and you are finished.

A Comprehensive Example

The following application is a bare-bones demonstration of the error handling principles we have described. The example is not all-inclusive, but shows how control flows among the error handlers. The handlers are, essentially, empty (e.g., the outermost RECOVER clause does not contain "graceful" QUIT code).

VAL(99), in subsystem 2, functions as a practice error. DefError() is supplanted, in subsystem 1, by a replacement that BREAKs. Each subsystem installs then deinstalls its own handler. Upon installation, the predecessor block is always saved. These handlers chain. Following execution, each calls its predecessor.

You can discover some useful behaviors by compiling and executing variations of this example. For example, you can move VAL(99) among the subsystems, and change the mix of subsystems that do and do not install their own handlers in the chain.

Notice that the block triggered by VAL(99) is always the most recently installed, not necessarily the one that appears in the same subsystem as the VAL() call. Installing a code block globalizes it, even though, as here, it may call static functions. Similarly, when the error block chain BREAKs, notice which RECOVER clause is called. It is the one that belongs to the innermost SEQUENCE surrounding the BREAK, not necessarily the one that appears in the same subsystem as BREAK. When the VAL(99) in subsystem 2 leads to the BREAK that appears in subsystem 1, it is the RECOVER clause of subsystem 2, not of subsystem 1, that takes over. Also, remove the BREAK and watch DefError() kick back in.

```
// Error Subsystem Architecture

// First Subsystem (.prg)

FUNCTION Main
      STATIC oOldHandler
      // First Handler
      oOldHandler := ERRORBLOCK({|e| Hndlr1(e,oOldHandler)})
      CLEAR SCREEN

      BEGIN SEQUENCE

         SecondEntryPoint()

      RECOVER
          ? "RECOVERy in Main"

      END SEQUENCE

      ERRORBLOCK(oOldHandler)
      RETURN (NIL)


STATIC FUNCTION SUB1
    RETURN (NIL)

STATIC FUNCTION SUB2
    RETURN (NIL)

STATIC FUNCTION Hndlr1(obj,handler)
    ? "UDF called by First Handler"
    BREAK
    RETURN (EVAL(handler,obj))



//   Second Subsystem (.prg)

FUNCTION SecondEntryPoint
    STATIC oOldHandler
// Second Handler
    oOldHandler := ERRORBLOCK({|e| Hndlr2(e,oOldHandler)})

    BEGIN SEQUENCE

    ThirdEntryPoint()
    ? VAL(99)
    RECOVER
    ? "RECOVERy in SecondEntryPoint"

    END SEQUENCE

    ERRORBLOCK(oOldHandler)
    RETURN (NIL)



STATIC FUNCTION SUB2
    RETURN (NIL)

STATIC FUNCTION Hndlr2(obj,handler)
    ? "UDF called by Second Handler"
```

```
       RETURN ( EVAL(handler,obj) )

// Third Subsystem (.prg)
FUNCTION ThirdEntryPoint
   STATIC oOldHandler
// Third Handler
   oOldHandler := ERRORBLOCK({|e| Hndlr3(e,oOldHandler)})

   BEGIN SEQUENCE



   RECOVER
   ? "RECOVERy in ThirdEntryPoint"

   END SEQUENCE

   ERRORBLOCK(oOldHandler)
   RETURN  (NIL)

STATIC FUNCTION SUB1
   RETURN  (NIL)

STATIC FUNCTION SUB2
   RETURN  (NIL)

STATIC FUNCTION Hndlr3(obj,handler)
   ? "UDF called by Third Handler"

   RETURN (EVAL(handler,obj))
```

What Belongs in RECOVER?

Finally, consider application errors and responses to them, i.e., the conditions you put in your assertion IFs and your RECOVER CASEs. Here is a variety of observations:

Determining what is an error, and an appropriate response, is a subjective art. The error condition should be judged genuinely exceptional, and the response should not be overly complex. Ensure that the responses anticipate specific problems and are deliberately calculated to address those problems. One response may propagate the error, other responses may unwind a series of discrete events that could not be collectively completed. If your RECOVER code calls dedicated functions, consider making them static since they have no outside applicability.

There is a high degree of subjectivity in error processing. Suppose you have a loop to obtain and process numbers from a source (perhaps it takes an average). Some may be ill-formed (contain character values outside the legal range according to their representational format). Is

A Baseline Strategy

that an error? It depends why you need the numbers, and the derived result. Some results are good only if all prescribed inputs are good. Others can tolerate a percentage of throw-away inputs.

A bank, calculating next year's interest rate on a customer's variable mortgage, must factor in all 52 weeks of the contractually specified government index values. Omission or inaccuracy of a single value invalidates the whole result. On the other hand, tabulation of survey responses may require a result accurate within a certain margin of error. That may be statistically obtainable even if 10% of gathered responses are omitted. No need to agonize. As the programmer, assert an error or don't assert an error accordingly; use your subjective judgement.

The following pseudocode characterizes such typical subjectivity.

```
FUNCTION GatherValues

   BEGIN SEQUENCE
      DO WHILE
         <obtain next value>
         IF <ill-formed>
            BREAK indication-of-error
         ENDIF
         <process value>
      ENDDO
      RETURN (result)

   RECOVER USING indication-of-error
      CASE this-particular-error
         <clear up>
         QUIT

   END SEQUENCE
```

Beyond raising the error, even if that is necessary, QUIT may be overkill. That requires a subjective decision too. Your source might be an interactive user or a mainframe tape. If a user types "T" instead of "6," it is worth requesting the number again, but if a tape has an unreadable scratch and 100 retries have already failed, another request is futile. The 'correct' response is context-dependent.

Whatever your response, keep it in bounds. Do not let error code proliferate into a parallel application. Remember, localization is the key.

Also, do not use the error system to process conditions that are not errors, i.e., conditions not genuinely exceptional. By definition, errors are abnormalities.

The following example illustrates the difference between true errors and other conditions. It shows two operations, either of which could be accomplished using the error system or traditional control structures. The first seeks to advance a day-of-the-week array. It experiences a system error whenever the current day is Saturday. The second seeks to

A Baseline Strategy

divide two numbers. It experiences an error whenever the divisor is zero.

Either function could raise the error, address it reactively through RECOVER, or avoid the error by anticipating the boundary. While RECOVERy is preferred for division by zero, it is discouraged for day of the week. The former is a mathematical aberration; trying it is a mistake. The latter is a normal boundary condition, a built-in, a matter of course.

```
// Abnormalities Only, Please!

STATIC TheDays := {"Sunday","Monday","Tuesday","Wednesday",;
   "Thursday","Friday","Saturday"};
   i := 0

FUNCTION NextDay
   LOCAL retval
   BEGIN SEQUENCE
      retval := TheDays[++i]
   RECOVER USING --
      retval := TheDays[i:=1]
   END
   RETURN (retval)
```

or...

```
FUNCTION NextDay
   LOCAL retval
   IF i != 7
      retval := TheDays[++i]
   ELSE
      retval := TheDays[i := 1]
   ENDIF
   RETURN (retval)

   ** PREFERRED **
```

```
FUNCTION Divide(x, divisor)
   LOCAL retval
   BEGIN SEQUENCE
      retval := x/divisor
   RECOVER USING --
      retval := 0
   END
   RETURN (retval)

   ** PREFERRED **
```

or...

```
   LOCAL retval
   IF divisor != 0
      retval := x/divisor
   ELSE
      retval := 0
   ENDIF
   RETURN (retval)
```

The responses in RECOVER represent conscious anticipation of a specific set of error possibilities, as discussed above and illustrated in the first example. They are not haphazard or catch-all.

Propagation is a common option for errors that fall into the OTHERWISE category. It is also a possibility for those that do not, once desired processing is complete.

Error response can unwind; this is the basis for transaction processing.

The following example outlines unwinding. It also shows the technique for re-executing the SEQUENCE statement block and demonstrates the SEQUENCE construct as the enclosure for coherent, logical units of activity, not random statement groups. The activity is foreseen to have certain, predictable outcomes that call for remedy, and the RECOVER clause addresses those.

As a typical example, you may wish to commit either five fields or none to certain records. Assume the current error block eventually BREAKs. Then, bracketed in a SEQUENCE as shown in this example, you are guaranteed your five REPLACEs worked if you dropped out the END SEQUENCE. You would not need to code an assertion with explicit BREAK, as here. An assertion internal to the replace operation triggers the error block, thus BREAK, if any of the REPLACEs failed. In that event, back the successful REPLACEs out of the database (the error block helps you identify them) and try again.

```
//  Unwinding and Restarting
//
#command RESTART SEQUENCE => LOOP

DO WHILE .T.
   BEGIN SEQUENCE
      .
      . <do 3 out of 5 things>
      .
      IF <can't do 4th>
         BREAK--
      ENDIF
      .
      . <do 5th>
      .
      RECOVER USING--
      .
      . <undo those 3 things>
      .
      IF HavaNotherGo
         // Restarts SEQUENCE block as a whole
         RESTART SEQUENCE
      ENDIF
      .
      .
      .
   END SEQUENCE
   EXIT
ENDDO
```

Network Processing

The *CA-Clipper* function NETERR() functions differently from earlier versions of *CA-Clipper*. Two events that in version Summer '87 automatically set the NETERR() status true (.T.) no longer do so. They are failure of a USE command due to an "access denied" from the underlying DOS file-open call; and failure of APPEND BLANK due to the "phantom" record being locked by another workstation, perhaps by simultaneous APPEND BLANK attempts. (The phantom record is just beyond the last record.)

While USE and APPEND BLANK do preset the NETERR() state to false (.F.), upon failure, they merely call the current error block without further attention to NETERR(). Setting NETERR() to true is left to the error block; the commands themselves no longer contain code to do so. Formerly, you could not set NETERR(), but in *CA-Clipper* you do it by passing a true (.T.) or false (.F.) argument to the function.

The default error block, DefError() takes care of this:

```
// For network open error, set NETERR() and subsystem default
IF (e:genCode == EG_OPEN .AND. e:osCode == 32 .AND. ;
    e:canDefault)
   NETERR(.T.)
   RETURN (.F.)
ENDIF

// For lock error during APPEND BLANK, set NETERR() and
// subsystem default
IF (e:genCode == EG_APPENDLOCK .AND. e:canDefault)
   NETERR(.T.)
   RETURN (.F.)
ENDIF
```

One implication: if you supplant DefError(), include similar code in its replacement or network activity that hinges on NETERR() will change behavior. Also note the second action taken in DefError(), namely, returning false. USE has already indicated its readiness to regain control (e:canDefault was true). When it does, it executes its "default processing," which does nothing visible, and certainly does not open the file. Probably it does little but return. But that allows mainline code to retain control and continue. DefError()'s returning false (.F.) merely allows USE, and by extension, the ensuing code, to run to completion.

On another network issue, *CA-Clipper* now distinguishes between the two causes (rooted in the DOS function that opens files) of "access denied", that may occur when opening a file. One failure is the result of other workstations' use of the file; the other, of attributes of the file itself. The DOS open-a-file function returns the same error code in both cases, namely 5, obscuring any distinction between causes. But a secondary error code is available from DOS on request, and this does

Network Processing

allow the distinction. *CA-Clipper* obtains this secondary code and returns it in the Error object's osCode instance variable. In a clash with other workstations, the result is code number 32, entitled "sharing violation," as opposed to number 5, "access denied." In a network, 32 indicates inter-workstation contention while 5 reflects rights and attributes issues.

Equipped with tools and understanding, you can now enhance your applications with error processing. You will not seek to make them error-proof, but to make them error-systematic. You can raise abnormal conditions as errors, as supplied *CA-Clipper* functions do. You can provide localized, specialized handling in response. The approach is user-definable and the tools belong to you.

Transact.prg

The example below illustrates the use of *CA-Clipper* error features, the implementation of the transaction/rollback concept and the restart/retry of a block of statements.

The illustrated transaction supposes a one-to-many relationship between two data files. The transaction is defined to consist of applying DELETEs to all of the child records that correspond to a given parent record. (In this case, all the invoices associated with customer Jones.) Rollback is defined to be the reversal of all DELETions applied in case any other records were not successfully deleted. If the DELETion of any child record fails, rollback is the reversal (i.e., RECALL) of those deletions that succeeded.

The installed error block, embodies the recommendation that error block code eventually lead to a BREAK so that local code can be accompanied by local recovery specific to local activity, and all errors will reliably transfer control to that recovery.

(The error block shown here accomplishes this artificially. Moreover, it is the highest-level block in a possible chain that would perform the BREAK, upon which this code would rely.

Note that there is RECOVERy with the start-up code (GetReady) for the start-up code, and separately, RECOVERy with the multiple-delete code for the multiple-delete code. If an error arises in either context, execution of the localized RECOVERy code is thanks in both cases to one and the same error block. (Once posted, the error block does not "belong" to the routine that posted it, but to any and all routines that executes it while it remains installed.)

Network Processing

```
// TRANSACT.PRG
//
// Copyright (c) 1991 Computer Associates.
//                    All Rights Reserved.

// CLIPPER Transact /N/W/A
// RTLINK FILE Transact PLL base52

#include "Error.ch"

PROCEDURE Main

   LOCAL aDeleteds, cName, aNotRecalled, oWhatHappened
   FIELD LastName
   ERRORBLOCK({|e| BREAK(e)})       // Artificial but prototypical
                                    // Open files, Set relation
   IF .NOT. GetReady()
      ? "No I don't feel like it tonight"
      QUIT
   ENDIF

   IF .NOT. Invoice->(DBSEEK(cName := "Jones"))
      ? 'Oh come on. This example is hard-wired for the ;
        "Joneses."'
      QUIT
   ENDIF

   // parent record lock, gateway to its children
   IF Customer->(RecLock(5))

      // allow for possible retry
      DO WHILE .T.
         BEGIN SEQUENCE
            // Transaction Attempt (multiple DELETEs)
            aDeleteds := {}
            DO WHILE TRIM(Invoice->(LastName)) == "Jones"
               // may not succeed
               Invoice->(RLOCK())
               // in that case, this calls error block
               Invoice->(DBDELETE())
               // records "damage" already done
               AADD(aDeleteds, Invoice->(RECNO()))
               Invoice->(DBSKIP())
            ENDDO

         RECOVER USING oWhatHappened

            IF (oWhatHappened:genCode() == EG_UNLOCKED)
               //  Rollback (UnTransact) attempt
               // stay till they're squashed
               DO WHILE .NOT. EMPTY(aDeleteds)
                  aNotRecalled := {}
                  AEVAL(aDeleteds, {|RecNum| ;
                     UnDeleteIt(RecNum, aNotRecalled)})
                  aDeleteds := ACLONE(aNotRecalled)
               ENDDO
               IF (ErrBox("Rollback successful, Transaction ;
                  NOT Performed " + "Restart transaction ;
                  again?",    {"Yes", "No"}) == 1)
                  // reposition pointer
                  Invoice->(DBSEEK(cName := "Jones"))
                  // ...and start over
                  LOOP
               ENDIF
            ENDIF
```

Network Processing

```
            END       // SEQUENCE
            EXIT

      ENDDO
      Customer->(DBUNLOCK())
   ENDIF

   RETURNSTATIC PROCEDURE UnDeleteIt( RecNum, aNotRecalled )
   Invoice->(DBGOTO(RecNum))
   // either...
   IF Invoice->(RLOCK())
      Invoice->(DBRECALL())              //   uneventful RECALL
      Invoice->(DBUNLOCK())
   ELSE
      AADD(aNotRecalled, RecNum)         //.. or mark for revisit
   ENDIF
   RETURN



   STATIC FUNCTION GetReady
      /*
      Illustrates use of error system to treat network file
      "acquisition" as a transaction.  Dispenses with Summer
      '87 NETERR() check after USE attempt as a result.
      Streamlines code and makes it more readable.

      Need to accomplish 3-steps-in-1, or nothing: open 2 files
      and indexes, and set a relation.  That defines the
      transaction.  Relies on knowledge that error block
      will predictably BREAK.  Therefore if any part of the
      main sequence of statements fails the RECOVER statement
      block that gets control can undo the parts that did not.
      RETURN value reflects overall result.

      */

      LOCAL lReturn := .T.
      CLEAR SCREEN
      BEGIN SEQUENCE
         USE Invoice   INDEX Invoice   SHARED NEW
         USE Customer  INDEX Customer  SHARED NEW
         SET RELATION TO Customer->LastName INTO Invoice
      RECOVER // for ANY error; don't grab object, don't check why
         CLOSE SELECT("Invoice")
         CLOSE SELECT("Customer")
         lReturn := .F.
      END       // SEQUENCE
      RETURN (lReturn)


   FUNCTION ErrBox( cErrMsg, aOptions )
      LOCAL nChoice, i, nStrgLen, cOldColor, cOldScreen
      cOldColor := SETCOLOR("W/R+, W+/B")
      cOldScreen := SAVESCREEN(10, 0, 14, 79)

      @ 10, 0 TO 14, 79 DOUBLE
      @ 11, 1, 13, 78 BOX SPACE( 9 )
      @ 11, 2 SAY cErrMsg

      nStrgLen := LEN(aOptions) * 3
      AEVAL(aOptions, {|arr| nStrgLen += LEN(arr)})
      nStrgLen := (78 - nStrgLen) / 2

      FOR i := 1 TO LEN(aOptions)
```

```
        @ 13, nStrgLen PROMPT aOptions[i]
        nStrgLen += LEN(aOptions[i]) + 3
NEXT

nChoice := 1
MENU TO nChoice
SETCOLOR(cOldColor)
RESTSCREEN(10, 0, 14, 79, cOldScreen)

RETURN (nChoice)

// EOF - TRANSACT.PRG //
```

Chapter 8
CA-Clipper Compiler
CLIPPER.EXE

This chapter describes the basic operations of the *CA-Clipper* compiler (CLIPPER.EXE). CLIPPER.EXE takes as input a source file (.prg) containing one or more procedures and user-defined functions, and creates an object file (.OBJ) that can be linked with other *CA-Clipper*-compiled and foreign object files to form an executable file (.EXE).

In This Chapter

The following topics are discussed in this chapter:

- Invoking the *CA-Clipper* compiler
- Specifying options with CLIPPERCMD
- The compiler script file
- The compiler return code
- How the *CA-Clipper* preprocessor works
- How *CA-Clipper* compiles
- The compile and link batch file
- Header files
- Output files
- Changing the size of the environment
- Compiler options

Invoking the CA-Clipper Compiler

The *CA-Clipper* compiler can be executed from the DOS prompt with the following general syntax:

```
CLIPPER [<sourceFile> | @<scriptFile> [<option list>]]
```

<sourceFile> is the name of the program file to compile to an object file. If no extension is specified, a (.prg) extension is assumed. The filename may optionally include a drive designator and a path reference.

<scriptFile> is the name of a script file containing a list of source files to compile into a single object file. If an extension is not specified, a (.clp) extension is assumed as a default.

<option list> is a list of one or more options to control the course of the compilation. Options may be specified in either upper or lowercase and must be prefaced with a slash (/) or a dash (-) character. Each compiler option is discussed in detail in the Compiler Options section later in this chapter.

If the compiler command line is specified with no arguments, a list of the compiler options and descriptions displays to the console.

Note: The compiler requires a minimum 25 file handles. This requires DOS 3.3 or greater and "Files=25" in CONFIG.SYS.

Specifying Options with CLIPPERCMD

The *CA-Clipper* compiler uses the environment variable CLIPPERCMD to allow the specification of options without supplying them on the compiler command line. To define CLIPPERCMD, use the DOS SET command as follows:

```
SET CLIPPERCMD=[<option list>]
```

<option list> is a list of compiler options that will be read and processed each time the compiler is invoked.

When defining CLIPPERCMD, options are specified just as they would be on the compiler command line (i.e., they can be either upper or lowercase and must be separated by a space and prefaced with a slash (/) or a dash (-) character). Options defined in CLIPPERCMD are processed before command line options. If an option has already been defined in CLIPPERCMD, it is overridden with the command line definition.

To save yourself from having to enter this SET command repeatedly, it can be placed in your AUTOEXEC.BAT file where it will be processed automatically each time you reset your computer.

Examples

- The following example causes *CA-Clipper* to suppress line numbers and use a new standard header file each time the compiler is invoked:

    ```
    SET CLIPPERCMD=/L /UNewstd.ch
    ```

- The next example specifies a directory in which to place output object files:

    ```
    SET CLIPPERCMD=/OC:\CLIPPER5\OBJ\
    ```

The Compiler Script File

A script file (sometimes referred to as a *clip list*) is a text file containing a list of source files to compile into a single object file. The resulting object file has the same name as the script file unless the /O option is specified.

The list may include program, procedure, and format files. Referenced files (including procedure files) are not automatically compiled as they would be if the application was compiled on the *CA-Clipper* command line. This is true regardless of whether the /M option is specified.

Use the following rules to create a script file:

- Source files are separated by a carriage return/linefeed.

- Source files may be specified with or without an extension. If not specified, a (.prg) extension is assumed.

- Drive designators and path references may also be specified as part of any source filename.

The Compiler Return Code

If a fatal error is encountered or the user presses *Ctrl-C* or *Ctrl-Break*, the compiler terminates with a DOS return code of 1. If the compilation ends normally, the return code is set to 0. A compile session ends normally even if there are warnings. The DOS return code can be tested in a batch file using the DOS ERRORLEVEL keyword.

If an error occurs, an error message is displayed and no output file is generated. For a list of error messages and their meanings, see the *Compiler Error Messages* chapter in the *Error Messages and Appendices* guide. To direct these messages to a file, use DOS redirection. For example, the following compiler command line:

```
C>CLIPPER <sourceFile> > ERRFILE.TXT
```

directs the compiler output to a file called ERRFILE.TXT.

Refer to The Compile and Link Batch File section later in this chapter for an example that uses DOS ERRORLEVEL to check for a failed compile.

How the CA-Clipper Preprocessor Works

The *CA-Clipper* compiler has two distinct phases: the preprocessor phase and the compilation phase. The portion of the compiler engine that performs the preprocessor phase is called the preprocessor, and the portion that performs the compilation phase is called the compiler.

Before the compilation phase takes place, the preprocessor scans the source file from top to bottom for certain directives and translates them into regular *CA-Clipper* source code that can be compiled. The output of the preprocessor is then used as input to the compiler.

The /P compiler option can be used to write this preprocessor output file (.ppo) to disk so you can see the source code that was used as input to the compilation phase. This option is especially useful if you have used the #command and #translate directives to create user-defined commands. You can look at the contents of the (.ppo) file to see whether or not commands translated as you expected.

The following table summarizes the preprocessor directives. For more information on their usage, see the Directive items in the *Reference* guide.

Summary of Preprocessor Directives

Directive	Meaning
#command	Specify a user-defined command or translation directive
#define	Define a manifest constant or pseudofunction
#error	Generate a compiler error and display a message
#ifdef	Compile a section of code if an identifier is defined
#ifndef	Compile a section of code if an identifier is undefined
#include	Include a file into the current source file
#stdout	Casues the compiler to output the literal text to the standard output device
#undef	Remove a #define definition
#xcommand	Specify a user-defined command or translation directive without abbreviations

How CA-Clipper Compiles

The *CA-Clipper* compiler converts source code written in *CA-Clipper* into Intel object code files. The source code programs are identified by the DOS filename extension (.prg). Object code files are identified by the DOS filename extension (.OBJ). These object files can be linked with a linker to produce an executable file, identified by the DOS filename extension (.EXE).

Unless a script file or the /M option is specified, *CA-Clipper* compiles the specified program file and all program files referenced with DO, SET PROCEDURE, and SET FORMAT. Within each referenced file, any other files referenced with these commands and statements are also compiled until the entire system of source files is processed. This proceeds according to the following rules:

- The program specified on the compiler command line is compiled first.

- Procedure files, procedures referenced with DO, and format files referenced in the current program are compiled in the order specified.

- Procedure references made with DO and not already compiled are compiled if found as (.prg) files in the current directory. If not found, *CA-Clipper* returns a warning message.

- Format references made with SET FORMAT and not already known are assumed to have a (.fmt) extension unless otherwise specified. If a procedure with the same name as a format reference has already been compiled, the format will refer to the compiled procedure and not to the (.fmt) file on disk.

If a script file is used, each file is compiled in the order listed according to these rules with one exception—no referenced files are compiled.

Note that a DO requires that the procedure be known or available in the current directory as a file of the same name with a (.prg) extension. If the procedure is called with function-calling syntax and not known, *CA-Clipper* does not search the disk for a program file of the same name, but instead assumes the procedure is external and proceeds without any warning message.

The Compile and Link Batch File

The default installation of the *CA-Clipper* development system places a sample compile and link batch file, CL.BAT in the \CLIPPER5\BIN directory. This batch file can be used to compile and link single program files (.prg) to executable files (.EXE) only if there are no errors during compilation. If there is a compiler error, CLIPPER.EXE terminates with a return code of 1, and the batch file in turn terminates without attempting to link.

The compiler batch file is invoked as follows:

```
CL <sourceFile>
```

<sourceFile> is the name of the program file to compile and link. The filename can be specified including both a drive designator and/or a path reference. However, specifying <sourceFile> with an extension will cause a linker error.

Header Files

Header files, also referred to as *include files*, contain preprocessor directives and command definitions. Header files have no default extension and are specified using the #include preprocessor directive or the /U compiler option. The INCLUDE environment variable is used to locate header files and is defined using the DOS SET command as follows:

```
SET INCLUDE=<pathSpec>
```

The INCLUDE path is searched if a requested header file cannot be found in the current directory.

In the default installation configuration, INCLUDE is set to \CLIPPER5\INCLUDE. To change this default, alter the SET INCLUDE command in your AUTOEXEC.BAT file. To search for header files in a particular directory on the D: drive, you might change it to the following:

```
SET INCLUDE=D:\INCLUDE
```

Note: When searching for header files, *CA-Clipper* searches the directory specified with the /I compiler option after the current directory and before the INCLUDE path. In a sense, this option adds the specified directory to the front of the INCLUDE path specification.

Output Files

There are three types of output files produced by the *CA-Clipper* compiler. Any specified output filename may include a full or partial path name to explicitly indicate where *CA-Clipper* will write the file on disk. Unless otherwise specified, files are created in the current directory.

Default extensions for output files, with the exception of preprocessor output files, may be overridden with explicit extensions specified as part of the filename.

Object Files

A single object file is created as a result of a successful compile. By default, the name assigned to the file is the same as the source or script filename specified on the compiler command line, but with an .OBJ extension. You may, however, specify the object filename and destination with the /O compiler option described in the Compiler Options section of this chapter.

Temporary Files

Normally, *CA-Clipper* generates one or more temporary files during the compilation process. You can control where these files are created by setting up an environment variable called TMP with the DOS SET command as follows:

Changing the Size of the Environment

```
SET TMP=<pathName>
```

The compiler normally deletes temporary files after they have served their purpose. Therefore, you may never see them on your disk. The main purpose of the TMP variable is to write the temporary files to another disk drive if you are running short of space on the drive that you normally use to compile. This way, the temporary files will not take up disk space that is better used by permanent files.

For example, if you are working on the C: drive, you could put temporary files in a directory on another disk drive as follows:

```
SET TMP=D:\TEMP
```

Note: The /T compiler option may be used to specify a temporary file directory. If this option is specified, it overrides the TMP directory setting.

Preprocessed Output Listing

The /P compiler option allows you to write a preprocessed output listing to a file. The name assigned to the file is the same as the program or script filename specified on the compiler command line with a (.ppo) extension.

Changing the Size of the Environment

When the environment variables, CLIPPERCMD, INCLUDE, or TMP are specified, it is possible to run out of environment space. In DOS version 3.2 and above, you can change the size of the environment by loading COMMAND.COM with the SHELL directive in CONFIG.SYS.

For example, the following CONFIG.SYS command line sets the environment size to 2048 bytes:

```
SHELL=C:\COMMAND.COM C:\ /P /E:2048
```

Note that in addition to specifying the /E option to define the environment space, you must also give the name and location of COMMAND.COM as well as force COMMAND.COM to remain resident with the /P option.

Compiler Options

Compiler options are switches that control compilation behavior. Besides the preprocessor directives and the standard header file, compiler options are the main mechanism for controlling the compiler and preprocessor. Options can be specified on the *CA-Clipper* command line and in the CLIPPERCMD environment variable. Command line options take precedence if there is a conflict.

Compiler options fall into the following three basic categories:

- Variable configuration
- Preprocessor switches
- Compiler switches

When specifying options, they can be either upper or lowercase, and must be prefaced with either a slash (/) or a dash (-) character. Options may be specified in any order and must be separated by a space.

Some compiler options have arguments. If an option has arguments they are specified after the option, and no space is allowed between the option and any of its arguments.

/A Automatic Declaration of PUBLIC and PRIVATEs

Any variable included in a PRIVATE, PUBLIC, or PARAMETERS statement is automatically declared as MEMVAR.

/B Include Debugging Information

Includes line numbers, local and static variable names, source filenames, and other debugging information in the object file. If /L is also specified, line numbers are not included as part of the debugging information.

Using the /B option causes the size of the resulting object file to increase. The amount of the size increase depends on the number of lines in the source code file as well as the number of local and static symbols defined. The number of additional bytes is approximately (4 + 3 per line + size of each static and local symbol + size of source filename symbol). Note that each symbol size includes a null terminator byte (e.g., the symbol Cust_Name requires 10 bytes) and that the 3 bytes per line figured into this size increase is eliminated by the use of the /L option.

Compiler Options

/CREDIT   **Display Credits to the Console**

This option displays the *CA-Clipper* credits to the console.

/D   **Define an Identifier**

```
/D<identifier>[=<text>]
```

Defines an identifier to the preprocessor. If *<text>* is not specified, *<identifier>* is given an empty value. Note that if double quote marks are used in the *<text>* as literal characters, they must be preceded by the backslash (\) character.

This option is designed for use with the conditional compilation directives to allow you to define a manifest constant on the compiler command line and control whether a section of the source file is compiled or not. Refer to the Directive items in the *Reference* guide for more information on conditional compilation.

/ES   **Exit Severity level 0**

Default exit severity level. If warnings are encountered during compilation, the compiler does not set the DOS errorlevel upon exit. This maintains compatiblity with *CA-Clipper* 5.0x.

/ES0   **Exit Severity level 0**

Same as /ES.

/ES1   **Exit Severity level 1**

Specifies an exit severity level of 1. If warnings are encountered during compilation, the compiler sets the DOS errorlevel upon exit.

/ES2   **Exit Severity level 2**

Specifies an exit severity level of 2. If warnings are encountered during compilation, the compiler does not generate an object file (.OBJ) and sets the DOS errorlevel upon exit. This effectively promotes warnings to error status at the DOS level.

/I   **Expand Include Directory Search List**

```
/I<pathName>
```

Adds the directory specified by *<pathName>* to the front of the list of directories to be searched for header files (as specified by the INCLUDE environment variable). Multiple /I options may be specified in the same compiler session; each one causes an additional path to be added to the front of the header file directory search list.

/L **Suppress Line Numbers**

Excludes program source code line numbers from the object file. The effect of this option is to reduce the object file size by three bytes for each line containing a program statement.

It should be used only on completed programs since its use prevents the reporting of source code line numbers when there are runtime errors. Additionally, a program compiled using the /L option cannot be debugged with *The CA-Clipper Debugger*.

If both /L and /B are specified, line numbers are not included in the debug information written to the object file.

/M **Compile Only Current (.prg)**

Compiles only the current (.prg) file, suppressing the automatic search for (.prg) files to satisfy unresolved external references. If this option is specified, procedures referenced with a DO statement, SET FORMAT, or SET PROCEDURE commands and not found in the current (.prg) are assumed external.

In addition, procedure files specified by SET PROCEDURE commands are not compiled. Header files specified with the #include directive, however, are compiled.

/M has no effect when compiling with a script file.

/N **Suppress Automatic Main Procedure**

Suppresses the automatic definition of a procedure with the same name as the (.prg) file. This option must be used if you have filewide variable declarations in a program (.prg) file.

If the /N option is not specified, each (.prg) file is compiled with an implicit procedure consisting of all code from the top of the program file to the first PROCEDURE or FUNCTION declaration statement.

The first procedure to execute when you invoke an .EXE file, also referred to as the *starting procedure*, is the first procedure or function encountered at link time. When you compile without the /N option, the implicit procedure serves as the startup procedure. When you compile with the /N option, the first explicitly declared procedure or function in the first object file listed at link time becomes the startup procedure.

Compiler Options

/O — Compile to a Specified Object Filename

```
/O<objFile>
```

Writes the output object file to <objFile>. If <objFile> contains only a path specification, it must end with a backslash (\) character. The file is written to the specified directory with a base name the same as the first source filename and an .OBJ extension.

/P — Produce a Preprocessed Output Listing

Produces a preprocessed output listing. The listing is written to a file with the same name as the first source filename, but with a (.ppo) extension. The location and name of this file is unaffected by the /O option.

/Q — Suppress line number display

Suppresses line numbers from displaying on the screen when the compilation is in progress. Note that line numbers are still written to the object file unless /L is also specified.

/R — Tell the Linker Where to Search for Unresolved Externals

```
/R[<libFile>]
```

Embeds in the object file a request to the linker to search <libFile> for unresolved external references. Multiple /R options for the same compiler session are additive, causing the name of each library referenced to be embedded. Specifying /R without a <libFile> causes no libraries to be embedded.

By default, *CA-Clipper* embeds a request for CLIPPER.LIB, EXTEND.LIB, and DBFNTX.LIB in the object file. Note that /R overrides this default request.

/S — Check Syntax

If specified, the syntax of the current (.prg) file and all referenced source code files is checked, but no object file is generated. To check the syntax of the current program file only, add the /M option to the compiler command line.

/T — Specify Location of Temporary Files

```
/T<pathName>
```

Specifies a directory for temporary files generated during compilation. If this option is specified, it overrides the directory setting specified in the TMP environment variable.

/U   Preprocess with User Standard Header File

/U[<userStandardHeaderFile>]

Identifies an alternate standard header file to preprocess in place of the supplied STD.CH which is used automatically. If the /U option is specified without a <userStandardHeaderFile>, no standard header is used.

The user-defined standard header file is a replacement for STD.CH and therefore cannot contain anything other than preprocessor directives. *CA-Clipper* uses the same searching order for this header file as it does for all other header files: the current directory, the /I directory, and the INCLUDE path.

/V   Treat All Ambiguous Variable References as Dynamic Variables

Forces the compiler to assume that all references to undeclared or unaliased variable names are public or private variables. The default is to treat ambiguous references as fields. The /V option has the same effect as using the dynamic variable alias (MEMVAR->).

/W   Generate Warning Messages for Ambiguous Variable References

This option generates warning messages for undeclared or unaliased variable references. It is useful when converting private and public variables to locals and statics.

/Z   Suppress Shortcutting for Logical Operators

This option suppresses shortcutting optimizations on the logical operators .AND. and .OR. It is provided as an aid to isolating code that depends on the behavior of older versions of *CA-Clipper*. Note that code executed using the runtime macro operator (&) is always optimized with shortcutting—optimization cannot be disabled in the macro system.

Examples

- The following compiler command line checks the syntax of Main.prg only, without displaying line numbers:

  ```
  CLIPPER Main /S /Q /M
  ```

- The following compiler command line compiles all files in the Account.clp script file. The options in this case cause debugging information to be included in the object file which will be placed in the C:\CLIPPER5\OBJ directory and given the name ACCTPAY.OBJ:

  ```
  CLIPPER @Account /B /OC:\CLIPPER5\OBJ\ACCTPAY
  ```

  The script file Account.clp may look something like the following:

  ```
  AccMain
  AccPay
  AccRec
  AccRep
  ```

Summary

In this chapter, you learned how to compile your *CA-Clipper* source code programs. The *CA-Clipper* compiler is full-featured and versatile. It can be used to compile a single program file or to compile several files at once by taking its input from a script file. To get a list of options and command line syntax, enter CLIPPER at the DOS prompt without any arguments. In the next chapter, you will learn how to link the .OBJ files produced by the compiler to create executable files (.EXE).

Chapter 9
The CA-Clipper Linker
RTLINK.EXE

CA-Clipper is supplied with *.RTLink*™, the *CA-Clipper* linker. It is a full-featured linker that supports both a positional syntax similar to Microsoft LINK and a freeformat syntax similar to Plink86-Plus. This chapter is a reference guide to *.RTLink* that discusses both syntax forms as well as the various input modes for the linker.

Also included in this chapter are explanations of the most commonly used linker options with examples. Note that the version of *.RTLink* provided with *CA-Clipper* is a special version designed exclusively for use with the *CA-Clipper* compiler.

In This Chapter

The following topics are discussed in this chapter:

- Overview of *.RTLink*
- The *.RTLink* files
- Invoking *.RTLink*
- Configuring the *.RTLink* defaults
- The *.RTLink* return code
- The compile and link batch file
- Output files
- How *.RTLink* searches for files
- Linker options
- Dynamic overlaying of *CA-Clipper* code
- Prelinking
- Incremental linking
- Static overlaying

Overview of .RTLink

The *CA-Clipper* linker, *.RTLink*, is a dynamic overlay linker designed specifically for *CA-Clipper*-compiled code. It combines object files (.OBJ) created by the *CA-Clipper* compiler and other compilers and assemblers with standard libraries to form a stand-alone executable file (.EXE) that can be run from the operating system.

.RTLink dynamically overlays all *CA-Clipper*-compiled code allowing programs larger than available memory to run. It does this by paging *CA-Clipper* code from disk on a least-used basis. More frequently executed code remains in memory while less frequently executed code is swapped out of memory to make room for called routines not currently in memory. If performance becomes a question, a *CA-Clipper*-compiled module can be made to always reside in memory by specifying the RESIDENT option at link time.

Dynamic overlays are automatically placed within a single monolithic executable file (.EXE), but can optionally be placed in external overlay files (.OVL) using the DYNAMIC INTO option. This allows you to break up an application program into files that fit onto floppy disks for delivery to users. There is no performance advantage to either configuration, simply the option to specify the organization of a program's executable file structure.

.RTLink also supports prelinking and incremental linking. Prelinking allows you to link commonly used code into a prelinked library (.PLL) and then use the prelinked library when creating executable files (.EXE) to dramatically reduce link times. Prelinked libraries can also reduce disk space since more than one executable file (.EXE) can use the same prelinked library (.PLL).

Incremental linking also speeds up linking of *CA-Clipper*-compiled code by linking only those modules that have changed since the last link.

With *.RTLink*, there is no support for dynamic overlaying of C and Assembler code but there is the option to create static overlays as one would with a standard overlay linker. Refer to the Static Overlay discussion later in this chapter.

The .RTLink Files

Several files are needed to run the *.RTLink* linker. Each of the filenames is listed below along with an explanation of when the file is needed:

- RTLINK.EXE

    This is the program file you will be using to create executable files (.EXE) and prelinked library files (.PLL). The standard DOS file searching rules are used to locate RTLINK.EXE.

- RTLINKST.COM

    This file contains the startup code needed each time you generate a prelinked library file (.PLL) with *.RTLink*. The code from this file is written to the .PLL file so that it is available when dependent .EXE files are executed. RTLINKST.COM is therefore needed only when you generate a .PLL file with the PRELINK option and not when executing an .EXE file that depends on a .PLL file. RTLINKST.COM must reside in the same directory as RTLINK.EXE.

- RTLUTILS.LIB

    This library file contains the static overlay manager. RTLUTILS.LIB is needed when you use *.RTLink* to link programs that use any of the static overlay linker options.

- RTLINK.DAT

    This file contains all of the *.RTLink* error messages as well as the data needed to parse FREEFORMAT input. RTLINK.DAT must be present whenever you run RTLINK.EXE and must reside in the same directory as RTLINK.EXE.

- RTLINK.HLP

    This file contains the help text displayed when you invoke *.RTLink* with the /HELP option. This file must reside in the same directory as RTLINK.EXE.

In the default installation of the *CA-Clipper* development system, RTLUTILS.LIB is located in the \CLIPPER5\LIB directory and all other *.RTLink* files are located in the \CLIPPER5\BIN directory.

Invoking .RTLink

There are two interfaces provided by .*RTLink*: POSITIONAL patterned after Microsoft LINK and FREEFORMAT, patterned after Plink86-Plus. The two interfaces differ by the manner input files are specified to the linker.

Two interfaces are provided to allow transition to .*RTLink* from whatever class of linker you are currently using. If you are not already familiar with the POSITIONAL interface through the use of Microsoft LINK, it is strongly suggested that you use the FREEFORMAT interface which is both more readable and flexible. No matter which interface you choose, .*RTLink* may be invoked using any of the input modes explained in the table below.

RTLink Input Modes

Mode	Input Specified
Command line	On the RTLINK command line
Prompt	In response to onscreen prompts
Script File	Via an ASCII text file

This section discusses the syntax for invoking both of the .*RTLink* interfaces in each of the input modes.

The FREEFORMAT Interface

Unless configured otherwise, the default .*RTLink* interface is FREEFORMAT. If you prefer POSITIONAL, .*RTLink* must be configured as explained in the next section. The syntax for invoking .*RTLink* using the FREEFORMAT interface is as follows:

```
RTLINK [FILE <objFile list>
   [OUTPUT <outputFile>]
   [LIBRARY [<libFile list>]
   [<linkOption list>]]
```

Unlike the POSITIONAL interface, FREEFORMAT requires each linker argument prefaced with a keyword allowing arguments to be specified in any order. The order suggested above corresponds closely to the POSITIONAL order but is presented this way only for consistency. You can specify these arguments in any order you like.

FILE

FILE <*objFile list*> specifies a comma-separated list of object or library files to link. If a library is specified as an argument of the FILE option, all object files within the library are linked. If a file is specified without an extension, an .OBJ extension is assumed.

If more than one FILE option is encountered in the same linker session, the <*objFile lists*> are added together causing the files in all FILE lists to be linked. FILE may be abbreviated to FI.

OUTPUT

OUTPUT <*outputFile*> specifies the name of the output file to generate. If not specified, the name of the first file on the *.RTLink* command line is used.

If PRELINK is *not* specified, the <*outputFile*> generated is an executable file with an .EXE extension. If PRELINK is specified, *.RTLink* creates both a prelinked library and a transfer file with extensions of .PLL and .PLT, respectively.

If more than one OUTPUT argument is specified, the last one encountered is used. OUTPUT may be abbreviated to OU.

LIBRARY

LIBRARY <*libFile list*> specifies a comma-separated list of libraries and/or object files to search in order to resolve any undefined symbols. If an extension is not specified, a .LIB extension is assumed. LIBRARY may be abbreviated to LI.

Specifying a <*libFile list*> when creating a prelinked library (.PLL) does not automatically include all symbols from the specified library in the resulting .PLL file. To include a specific symbol in a .PLL file, use the REFER option. To include the entire contents of a library, specify the library file in the FILE <*objFile list*>.

Note: Do not specify the name of a .PLL in the <*libFile list*> list if you are linking with a prelinked library (.PLL) . Specify the PLL option with the prelinked library (.PLL) instead.

Linker Options

<*linkOption list*> specifies a list of linker options described later in this chapter. Unlike the POSITIONAL syntax, FREEFORMAT does not require that options be preceded by a forward slash (/) character.

Invoking .RTLink

Unless the forward slash is used, however, options must be separated by a space.

Although the *<linkOption list>* appears at the end of the syntax representation, linker options may be placed anywhere within the command line.

Command line Mode

If *.RTLink* is executed using the FREEFORMAT interface with FILE, OUTPUT, and/or LIBRARY command line arguments, the linker session begins immediately after you press *Return* to execute the command. This method of usage is called the command line mode because all filenames and options necessary to perform the link session are assumed to be specified on the command line. Note that the FILE argument is required in the command line mode, and if not included the link results in an error.

If a command line exceeds the DOS command line limit of 128 characters, use a script file as discussed below.

Prompt Mode

If *.RTLink* is specified with no command line arguments or options, the linker enters an interactive mode referred to as the *prompt mode*. In this mode, you can enter any options in any order. When specifying input in the prompt mode, there are a number of considerations:

- What you enter in response to a prompt must be one of the *.RTLink* command line arguments (i.e., FILE, OUTPUT, or LIBRARY), a linker option, or a continuation of the previous prompt response. The syntax is the same as if you were entering the argument or option on the command line.

- To go to a new prompt, press *Return*.

- Long commands can be continued on a new prompt line by pressing *Return* and continuing your response at the next prompt.

- To specify more than one filename in response to a prompt, separate each file with a comma as you would on the linker command line.

- To end a prompt session, type a semicolon or *Ctrl-Z*, and then press *Return* in response to the current prompt.

Script File Mode

An alternative syntax for invoking *.RTLink* is as follows:

```
RTLINK @<scriptFile> <linkOption list>
```

<scriptFile> is the name of an ASCII text file used as the linker script file. If an extension is not specified, .LNK is assumed. In this mode, *.RTLink* input is taken from the script file instead of from the command line or prompt responses, although linker options may still be specified on the command line.

Script files are used to automate linker sessions that you tend to repeat, in addition to sessions that are too long to enter from the command line. As indicated in the FREEFORMAT syntax specification, you may enter linker options on the command line in this mode. Remember that options specified on the command line are overridden by the same options specified in the script file.

In the FREEFORMAT interface, options can be specified as on the command line. The only rule governing the format of the file is that it contains valid command line arguments and linker options.

Commands need not be placed one to a line. The new line character, like a space or a *Tab*, is considered to be white space. Therefore, if a command will not fit on one line, you can continue it on a new line simply by pressing *Return*—no continuation character is necessary.

The # character designates a comment. When encountered, the rest of the line is ignored.

Script files can be nested by prefacing a filename with the (@) symbol. The named script file is read and executed by *.RTLink* in its entirety. When it terminates, control returns to the current script file.

When you execute *.RTLink* using a script file, all responses are echoed to the screen so that you can inspect the linker session unless you specify the SILENT option.

The POSITIONAL Interface

The alternate interface to FREEFORMAT is POSITIONAL. If you want to use POSITIONAL as the default interface, specify it with the RTLINKCMD environment variable, like this:

```
SET RTLINKCMD=/POSITIONAL
```

Invoking .RTLink

There may, however, be other linker options that you want to specify in RTLINKCMD. See the Configuring The *.RTLink* Defaults section for more information on the *.RTLink* configuration variable.

Once POSITIONAL has been specified as the default, the syntax for invoking *.RTLink* becomes:

```
RTLINK [<objFile list>,
    [<outputFile>],
    [<mapFile>],
    [<libFile list>]
    [<linkOption list>]] [;]
```

As indicated in the syntax representation, optional items are omitted by specifying a comma instead of the option (i.e., using two consecutive commas). For this reason, list items in the syntax must be separated using a space, a *Tab*, or a plus character—not a comma.

Object Files

<objFile list> is the same as the object file list in the FREEFORMAT syntax except that each file specified is separated by a space instead of a comma.

The Output File

<outputFile> is the same as the output file in the FREEFORMAT syntax.

The Map File

<mapFile> specifies the name of an ASCII text file containing one or more reports about the link session. This argument names the map file but does not cause the file to be generated. To generate a map file and define what reports it will contain, use the /MAP option as described in the Linker Options section in this chapter.

Library Files

<libFile list> is the same as the library file list in the FREEFORMAT syntax except that each file specified is separated by a space instead of a comma.

Linker Options

<linkOption list> is a list of linker options to control the linker session. In the POSITIONAL interface, all options must be preceded by a

forward slash (/) character with no intervening space between the slash and the option name.

Although the <linkOption list> appears at the end of the syntax representation, linker options may be placed anywhere within the command line, including in between other command line arguments and within a single argument's file list. This is important when specifying dynamic overlays in the POSITIONAL syntax.

Command line Mode

If the semicolon is included at the end of the .RTLink command line, the linker session begins immediately after you press *Return* to execute the command. This method of usage is referred as the *command line mode* because all filenames and options necessary to perform the link session are included on the command line. If you specify files up to and including at least one library file, .RTLink assumes the command line mode even if you do not include the semicolon.

Prompt Mode

Like the FREEFORMAT interface, the POSITIONAL interface has an interactive prompt mode. You enter the prompt mode if you specify .RTLink without any arguments, omit the terminating semicolon at the end of the command line, or leave unanswered prompts in the command line specification.

In the prompt mode, you are queried for the filenames to include in the link session. The prompts are always displayed in the same order as specified on the command line.

Depending on what files have already been specified on the command line when you invoke .RTLink, the first prompt displayed will be different. For example, if you specify the object and executable files on the command line, the prompt session will begin by asking for a map filename.

The following guidelines should be followed when responding to the prompts:

- To specify more than one filename in response to a prompt, separate each file specified with a space, *Tab*, or plus (+) character as you would on the command line.

- To continue a response to a prompt from one line to the next, terminate the current response line with a plus (+) character. The same prompt is issued again and you can continue responding to the current prompt. This process may be continued until you have completed the response to the current prompt.

- To specify a linker option, enter the option anywhere in your response to the current prompt.

- To terminate a response and proceed to the next prompt, press *Return*. If you do not enter any response before pressing *Return*, the default response displayed in square brackets next to the prompt is assumed.

- To terminate the prompt session, press *Return* from the last prompt. To end a session prematurely, type a semicolon (;) or *Ctrl-Z*, and then press *Return* in response to the current prompt. This saves some time if you will be accepting the default responses for the remaining prompts.

Script File Mode

Like the FREEFORMAT interface, the POSITIONAL interface supports a script file. To specify a script file, you use the following command line syntax:

```
RTLINK @<scriptFile> <linkOption list>
```

When *.RTLink* is invoked like this, it takes input from the *<scriptFile>* instead of from the command line or prompt responses. As with FREEFORMAT, the script file is an ASCII text file with a default extension of .LNK. This, however, is where the similarities end.

Note: This explanation for using script files with the POSITIONAL interface is included only for completeness. If you want to use script files, it is recommended that you do so with the FREEFORMAT interface where script files are much easier and more straightforward.

The entries in a POSITIONAL script file are processed on a line-by-line basis with each new line treated as the next expected linker response. Lines in the script file, therefore, must be in order according to the POSITIONAL command line syntax specification. This is also the same order in which prompts are presented in prompt mode. Each POSITIONAL argument must be placed on a new line in the file, and a blank line must be left for missing responses. The format of a POSITIONAL interface script file is as follows:

```
<objFile list>[<linkOption list>]
[<outputFile>]
[<mapFile>]
[<libFile list>][;]
```

As with entering responses in prompt mode, lines may be continued with a plus character. As in the command line syntax, linker options may be placed anywhere in a script file. In the script file specification above, they are placed at the end of the object file list since this is the only required line in the file.

Unless all linker prompts are answered within the script file, *.RTLink* enters the prompt mode to obtain the remaining responses required to link. To prevent this from happening, terminate the last line of the script file with a semicolon.

When you execute *.RTLink* using a script file, all responses are echoed to the screen so that you can inspect the linker session. Use the SILENT option to suppress this display.

Configuring the .RTLink Defaults

Most of the linker options described later in this chapter allow you to change some default behavior of *.RTLink*. If there are one or more defaults you want to change on a more permanent basis, you can do so by defining an environment variable, creating a configuration file, or a combination of the two.

The RTLINKCMD Environment Variable

The preferred method for changing the *.RTLink* defaults is by defining the environment variable called RTLINKCMD. Each time *.RTLink* is invoked, it reads the options specified in the RTLINKCMD environment variable before processing the command line arguments and options. RTLINKCMD is defined using the DOS SET command like this:

```
SET RTLINKCMD=<linkOption list>
```

<linkOption list> is a list of linker options to read and process each time you invoke *.RTLink*. In FREEFORMAT mode, the list may also include command line arguments such as FILE and LIBRARY.

When defining RTLINKCMD, options are specified just as they would be on the *.RTLink* command line. The options defined in RTLINKCMD are processed before those on the command line and those in the script file. This is to allow the newly specified defaults to be overridden.

The following example makes the default input mode positional and sets incremental linking on:

Configuring the .RTLink Defaults

```
SET RTLINKCMD=/POSITIONAL /INCR:50
```

RTLINK.CFG

In addition to the RTLINKCMD environment variable, *.RTLink* also supports the configuration file, RTLINK.CFG. Like RTLINKCMD, its purpose is to save commonly used linker options in order to override the defaults.

Each time you invoke *.RTLink*, the linker searches for the RTLINK.CFG file as follows:

- The current directory
- The directory where RTLINK.EXE resides

If found, the contents of the file are read and processed. If not found, no error occurs.

The options in RTLINK.CFG are processed before RTLINKCMD, the *.RTLink* command line, or the script file. This gives options defined in RTLINK.CFG the lowest precedence, allowing them to be overridden at any point later in the link process.

Configuration Options

The configuration options are special linker options that may only be specified as part of the RTLINKCMD environment variable, within RTLINK.CFG, or on the *.RTLink* command line if you are linking with a script file.

These options are not allowed under any other circumstances. Furthermore, the forward slash is required regardless of whether the current input mode is FREEFORMAT or POSITIONAL. The syntax is as follows:

```
/FREEFORMAT | /POSITIONAL
```

These options configure the linker interface with /FREEFORMAT as the default. This means that if you do not specify either option, *.RTLink* will be configured to use the FREEFORMAT interface. The abbreviation for these two options are /FREE and /POSI.

The change in the interface takes effect when *.RTLink* switches to a new input stream. For example, if you specify /POSITIONAL on the command line, the remainder of the command line itself is still processed as FREEFORMAT. Any script files would then, however, be processed as POSITIONAL. Also, the entire command line is processed before the script file is processed, so both of the following would cause MY.LNK to be processed in POSITIONAL:

```
RTLINK /POSITIONAL @MY

RTLINK @MY /POSITIONAL
```

/POSITIONAL and /FREEFORMAT options in RTLINK.CFG and RTLINKCMD, therefore, take effect when processing the command line. /POSITIONAL and /FREEFORMAT options on the command line take effect when processing script files.

The .RTLink Return Code

If *.RTLink* encounters an error or is interrupted by the user pressing *Ctrl-C* or *Ctrl-Break*, it terminates with a return code (DOS ERRORLEVEL) of 1. The DOS return code is also set to 1 if there are any unresolved symbol warnings during link (but not prelink) mode. If the linker session ends normally, the return code is set to 0.

Some examples of typical errors that are encountered during a linker session are the inability of *.RTLink* to locate a file or to interpret an option because of an improper syntax specification. If an error occurs, an error message is displayed and no output file is generated. For a list of error messages and their meanings, see the *.RTLink Error Messages* chapter in the *Error Messages and Appendices* guide.

The Compile and Link Batch File

The default installation of the *CA-Clipper* development system places a compile and link batch file, CL.BAT in the \CLIPPER5\BIN directory. This batch file can be used to compile and link single program (.prg) files to executable files (.EXE) only if there are no errors during compilation. If there is a compiler error, CLIPPER.EXE terminates with a return code of 1, and the batch file in turn terminates without attempting to link.

The batch file is invoked as follows:

```
CL <sourceFile>
```

The *<sourceFile>* is the name of the program file to compile and link. The filename can be specified with a drive designator and/or a path reference. However, specifying the *<sourceFile>* with an extension will cause a linker error.

Warning! The compile and link batch file is written using the FREEFORMAT syntax for .RTLink. Changing the linker interface to POSITIONAL in RTLINK.CFG or RTLINKCMD will hinder the operation of this batch file.

Output Files

There are several types of files produced as output by *.RTLink*. This section describes each type separately.

Any output filename that is specified may include a full or partial path name to explicitly indicate where *.RTLink* places the file on disk. Unless otherwise specified, files are created in the current directory.

Except in certain cases pointed out in the ensuing discussion, default extensions for output files may be overridden with explicit extensions specified as part of the filename. To indicate a file with no extension, include a period at the end of the filename.

Prelinked Library Files (.PLL)

A prelinked library file is created by *.RTLink* when you specify the PRELINK option. The name assigned to the file is the output filename with a .PLL extension. The extension cannot be overridden.

When you link object files that are dependent on a prelinked library with the PLL option, the resulting .EXE file depends on the prelinked library file (.PLL) that you name. The .PLL file contains the executable code loaded into memory whenever you execute an .EXE file that is dependent on that .PLL file. This means that each time the .EXE is executed, the corresponding .PLL file must be available. If this program architecture is distributed to the end user, the .PLL file must be included.

The order in which .PLL files are searched for at runtime is as follows:

- The directory in which the .EXE is located
- The path specified in the PLL environment variable
- The path specified in the LIB environment variable

If the file is not found, a runtime error is generated.

Prelinked Transfer Files (.PLT)

Each time you create a prelinked library file (.PLL), an associated transfer file is created. The name assigned to this file is the same as the prelinked library, but with a .PLT extension. As with prelinked library files, the file extension of the transfer file cannot be overridden. The file is always placed in the same directory as its associated .PLL file.

The .PLT file contains information about addresses and segments within the .PLL file that will be needed by *RTLink* to create dependent executable files. It is the .PLT file that is needed to link files with the PLL option.

A .PLT file is needed only at link time and is not required in order to run *RTLink*-generated executable files. There is, therefore, no need to distribute .PLT files with your system.

Warning! *If you distribute an application program including a .PLL file, do not distribute the .PLT file with the application if you are concerned about the proprietary nature of your library code. This file contains information that will allow your users to link with the .PLL file calling routines within the .PLL file.*

Executable Files (.EXE)

Unless you specify the PRELINK option, *RTLink* runs in link mode and produces an executable file (.EXE) as its output. The name assigned to the file is the output filename with an .EXE extension. The extension for executable files cannot be overridden.

The .EXE extension is used to designate a file that can be executed directly by the operating system shell, COMMAND.COM. The order in which .EXE files are searched for at runtime is dictated by DOS as follows:

- The current directory
- The path specified in the PATH environment variable

If the file is not found, a runtime error is generated.

An .EXE file produced by *RTLink* is identical to a conventional .EXE file as long as it does not depend on a .PLL file. Those .EXE files that depend on prelinked libraries are, however, internally different from conventional DOS .EXE files.

An *.RTLink* .EXE that depends on a .PLL does not contain the final memory images that are present in conventional .EXE files but instead contains segment images. It also contains additional information about the .PLL that you would not find in a conventional .EXE file. This information is needed by the *.RTLink* startup code in order to find the needed .PLL and form the memory image.

Warning! *When running under any DOS version prior to 3.0, do not rename an .EXE file that was created using .RTLink or you will not be able to execute the file.*

Map Files (.MAP)

A map file is generated by *.RTLink* using the MAP linker option. The default extension for this file type is .MAP.

A map file is simply an ASCII text file containing information about symbol and segment addresses within the memory image. You specify the exact content and order of the file contents with the MAP option. A map file produced in link mode has much more information than one produced in prelink mode since *.RTLink* knows the final memory layout in the former mode, and only symbol names and some relative addresses in the latter.

Overlay Files (.OVL)

Overlay files are created by *.RTLink* with the DYNAMIC and SECTION options. The default extension for this file type is .OVL.

When you execute an .EXE file that has associated overlay files, the .OVL files must be available when they are needed at runtime. This means that they must be distributed with your system.

When an .OVL file is needed during execution, only the directory where the .EXE was originally found is searched—no other path settings are used. If the file is not found, a runtime error is generated.

Temporary Files

On occasion, *.RTLink* generates temporary files during the linking process. You can control where these files are created by setting up an environment variable called TMP with the DOS SET command as follows:

```
SET TMP=<pathName>
```

Placing this command in your AUTOEXEC.BAT file automatically defines the TMP variable so that the temporary files generated by *.RTLink* are automatically placed in the desired location. If the TMP environment variable is not specified, temporary files are created in the current directory.

The linker deletes temporary files after they have served their purpose. Therefore, you may never see them on your disk. You can use the TMP variable to write the temporary files to another disk drive if you are running short of space on the drive that you normally use to link. To speed up the linking process you can also set TMP to write temporary files on a RAM disk. This way, the temporary files will not take up disk space that is better used by permanent files. For example:

```
SET TMP=D:\TEMP
```

Information Files (.INF)

If you specify the INCREMENTAL linker option to allow incremental linking, *.RTLink* generates and maintains an information file in order to keep up with what has changed since the last time you linked. The name of this file is the same as the *<outputFile>* but with an .INF extension.

How .RTLink Searches for Files

There are several types of files that serve as input to *.RTLink*. Each type of file is searched for according to a prescribed set of rules as follows:

- The current directory
- The path in a specific DOS environment variable

When searching for a file, *.RTLink* prompts you for a directory if it cannot locate the file in any of the search locations. To prevent this prompt, use the BATCH option.

Any filename may include a full or partial path name to explicitly indicate where *.RTLink* should find the file. If, however, a file is not specified with an explicit path, the prescribed rules are used to locate the file. If the file is not found after all of the possible locations have been searched, *.RTLink* terminates with an error message, the DOS return code is set to 1, and no output file is produced.

How .RTLink Searches for Files

This section describes how *.RTLink* attempts to locate each possible input file type. The rules stated above apply to all file types, so they will not be repeated. Instead, the specific search environment variable is defined for each file type and any deviations from the general rules are noted.

Library Files (.LIB)

Library files have a default extension of .LIB and are specified to *.RTLink* using either the LIBRARY command line argument in FREEFORMAT mode or by specifying files in the library file position in POSITIONAL mode. The environment variable used to locate library files is called LIB, and it is defined from DOS using the SET command as follows:

```
SET LIB=<pathSpec list>
```

In the default installation configuration, LIB is set to \CLIPPER5\LIB. To change this default, you would alter the SET LIB command in your AUTOEXEC.BAT file. For example, to include a directory containing third-party libraries, you might change it to the following:

```
SET LIB=C:\CLIPPER5\LIB;C:\CLIPPER5\THIRDPARTY\LIB
```

Note: When searching for RTLUTILS.LIB only, *.RTLink* looks in the directory where RTLINK.EXE resides before resorting to the LIB path list. RTLUTILS.LIB is not a library file that you specify, but rather one which *.RTLink* automatically looks for under certain circumstances.

Object Files (.OBJ)

Object files have an .OBJ extension as a default and are specified to *.RTLink* using either the FILE command line argument in FREEFORMAT mode or by specifying files in the object file position in POSITIONAL mode. The environment variable used to locate object files is called OBJ, and it is defined from DOS using the SET command as follows:

```
SET OBJ=<pathSpec list>
```

In the default installation configuration, OBJ is set to \CLIPPER5\OBJ. To change this default, you would alter the SET OBJ command in your AUTOEXEC.BAT file. For example, if your object files are located in a directory called OBJ, you might change it to the following:

```
SET OBJ=C:\OBJ
```

When searching for .OBJ files, *.RTLink* searches the LIB path before the OBJ path.

Script Files (.LNK)

Script files have a default extension of .LNK and are specified on the *.RTLink* command line by preceding the filename with an @ symbol. There is no associated environment variable to specify where to search for script files.

Prelinked Transfer Files (.PLT)

When you link object files that are dependent on a prelinked library with the PLL linker option, *.RTLink* actually searches for the associated transfer file (.PLT). The prelinked library file (.PLL) is needed when you execute the dependent .EXE file. Both files are created when you use *.RTLink* in prelink mode by specifying the PRELINK option.

The environment variable used to locate prelinked library and transfer files is called PLL, and it is defined using the DOS SET command as follows:

```
SET PLL=<pathSpec list>
```

For example:

```
SET PLL=C:\CLIPPER5\PLL;C:\CLIPPER5\MYPLL
```

Note: The LIB environment variable used to locate library files is also used when searching for prelinked library (.PLL) and transfer (.PLT) files. The LIB path list is searched after the PLL list. The LIB and PLL environment variables are used at link time and runtime to locate files

Linker Options

In both the POSITIONAL and FREEFORMAT interfaces for *.RTLink*, the <linkOption list> command line argument refers to the use of one or more of the linker options described in this section. The options are categorized and then organized alphabetically, and the representation of the syntax is the same as that used for *CA-Clipper* commands and functions. Unless otherwise noted, options are assumed to be global which means that they affect the entire linker session.

Linker Options

For many options, two syntax specifications are given. The POSITIONAL interface requires using the syntax that is preceded by a forward slash. FREEFORMAT allows you to use either syntax, and where there is only one syntax representation, allows you to omit both the slash and the colon.

Precedence of Options

Since linker options may be specified at several different levels, an established order of precedence is necessary to determine which option to use when conflicting options are specified. That order is as follows:

- Linker script file
- *.RTLink* command line
- RTLINKCMD environment variable
- RTLINK.CFG configuration file

These precedence rules come into effect only when there is a direct conflict between options specified at different levels (e.g., the specification of opposite linker options). For example, suppose that you are linking with a script file that contains /NOIGNORECASE, and you execute the link as follows:

```
RTLINK @LNKFILE /IGNORECASE
```

Since the linker script file takes precedence over the command line and there are conflicting options, the option in the script file (i.e., /NOIGNORECASE) is used and case sensitivity is enabled.

Similarly, suppose that you have set up an RTLINKCMD variable establishing INCREMENTAL linking. Command line options take precedence over environment variable settings, so to disable this feature for the current linker session you could invoke *.RTLink* as follows:

```
RTLINK NOINCREMENTAL FI Main
```

Abbreviations

As with the *.RTLink* command line arguments, all linker options may be abbreviated. The abbreviations are given on an individual basis in the text describing each option, but you may use additional letters above and beyond the abbreviated form to make options more readable. For example, the abbreviation for INCREMENTAL is INC, but INCREMENT is also an acceptable abbreviation.

Entering Numbers

When a linker option requires a numeric argument, the default base for the number differs depending on the syntax you use. With FREEFORMAT, numbers are interpreted as hexadecimal or, base sixteen. With POSITIONAL, they are interpreted as decimal or, base ten. If you wish to override these defaults, you may do so by specifying a base modifier after the number. The following table defines the modifier characters for all of the available number systems.

Base Modifiers

Modifier	Base
.	Decimal (10)
B	Binary (2)
H	Hexadecimal (16)
O	Octal (8)

There should be no space between the number and its modifier. Base modifiers may not be combined. For example, *11BH* is interpreted as 11B in hexadecimal with B interpreted as a hex digit rather than the binary base modifier.

The NIL Keyword

The word NIL has special meaning to *.RTLink* in the FREEFORMAT input mode. It is used to represent an object with a null name (i.e., one with no characters). For example, you could match a module with the name ("") using the linker option MODULE NIL.

Batch Mode Options

Toggle Prompting For Input

```
/BATCH  |  /NOBATCH
```

NOBATCH is the default of these two, mutually exclusive linker options. This means that if you do not specify either option, NOBATCH is assumed causing *.RTLink* to prompt you for a path name when library (.LIB), object (.OBJ), and prelinked transfer (.PLT) files cannot be found. Since it is the default, you would not need to specify NOBATCH except to override a previous BATCH setting (e.g., in RTLINKCMD). The abbreviation for NOBATCH is NOBAT.

BATCH prevents the linker from prompting you when it is unable to locate a file. This is called batch mode because the linker will not interrupt for any input. If files cannot be found, an error message is generated and *.RTLink* terminates with a DOS return code of 1. The abbreviation for BATCH is BAT.

Case Sensitivity Options

Toggle Case Sensitivity

/IGNORECASE | /NOIGNORECASE

IGNORECASE is the default of these two, mutually exclusive linker options. If you do not specify either option, IGNORECASE is assumed causing *.RTLink* to treat upper and lowercase letters as if they are identical. This means that *MyFunc* and *myFunc* would refer to the same symbol. Since it is the default, you would not need to specify IGNORECASE except to override a previous NOIGNORECASE setting (e.g., in RTLINKCMD). The abbreviation for IGNORECASE is IG.

NOIGNORECASE causes case to be significant in symbol and segment names. *MyFunc* and *myFunc*, therefore, would be treated as different names. Generally, you would not specify this option when linking *CA-Clipper* programs since the identifiers are not case-sensitive; however, there may be some conflicts when linking C or Assembler routines that necessitate case sensitivity and the use of NOIGNORECASE. The abbreviation for NOIGNORECASE is NOIG.

Configuration Options

Set the .RTLink Interface Mode

/FREEFORMAT | /POSITIONAL

These options configure the linker interface, and /FREEFORMAT is the default. This means that if you do not specify either option, *.RTLink* will be configured to use the FREEFORMAT interface. Since it is the default, you would not need to specify /FREEFORMAT except to override a previous /POSITIONAL setting (e.g., in RTLINKCMD). The abbreviation for these two options are /FREE and /POSI.

The configuration options are special linker options that may only be specified as part of the RTLINKCMD environment variable, within RTLINK.CFG, or on the *.RTLink* command line if you are linking with a script file. These options are not allowed under any other circumstances. Furthermore, the forward slash is required regardless of whether the current input mode is FREEFORMAT or POSITIONAL.

The change in the interface takes effect when .*RTLink* switches to a new input stream. Thus, configuration options in RTLINK.CFG and RTLINKCMD take effect when processing the command line. Configuration options on the command line take effect when processing script files.

Dynamic Overlaying Options

Place CA-Clipper Modules into Dynamic Overlays

```
/DYNAMIC[:<ovlFile>]
DYNAMIC [INTO <ovlFile>]
```

DYNAMIC is the opposite of RESIDENT and is the default of the two. This means that if you do not specify either option, DYNAMIC is assumed. DYNAMIC may be abbreviated to DYN.

DYNAMIC forces .*RTLink* to make any subsequent *CA-Clipper*-compiled modules within the <*objFile list*> or <*libFile list*> dynamic instead of resident. This feature is limited to programs that are compiled with *CA-Clipper*—modules written in another language such as C or Assembler are always linked as RESIDENT.

If the <*ovlFile*> is omitted, modules are placed in the executable file (.EXE). If the <*ovlFile*> is specified, any *CA-Clipper* modules loaded are written to the <*ovlFile*>. The default extension for this file is .OVL. If a subsequent DYNAMIC option names the same <*ovlFile*>, those modules are appended to the file.

Since DYNAMIC is the default, you do not need to specify DYNAMIC in order to have dynamic overlaying in your program unless you have a RESIDENT setting elsewhere (e.g., in RTLINKCMD). The default overlay, however, is internal to the .EXE file. To have an external dynamic overlay file, you must specify DYNAMIC with an <*ovlFile*>.

Place CA-Clipper Modules in the Root Section

```
/RESIDENT
```

RESIDENT may be abbreviated to RES, and it is the opposite of DYNAMIC. RESIDENT causes subsequent *CA-Clipper*-compiled modules to be loaded into the root section of the current program or the prelinked library. Use of this option overrides the automatic creation of dynamic overlays for all *CA-Clipper*-compiled code.

Examples

On a single *.RTLink* command line, you can specify selective modules to link as DYNAMIC or RESIDENT by interspersing these two commands within in the object file list. In the POSITIONAL syntax, for example, you could do the following:

```
RTLINK /DYNAMIC Test1 Test2 /RESIDENT Test3 /DYNAMIC:NewOvl Test4;
```

In FREEFORMAT using a script file, the equivalent would be:

```
RTLINK @MYSCRIPT
```

where MYSCRIPT.LNK is as follows:

```
DYNAMIC
FILE Test1, Test2
RESIDENT
FILE Test3
DYNAMIC INTO NewOvl
FILE Test4
```

See the Dynamic Overlaying of *CA-Clipper* Code section in this chapter for more information on dynamic overlays.

Help and Debugging Options

Display Static Overlay Loading at Runtime

```
/DEBUG
```

DEBUG provides a method to debug programs that use static overlays. When this option is specified, the overlay manager displays a message that identifies each overlay as it is loaded into memory during program execution. DEBUG may be abbreviated to DEB.

List .RTLink Options

```
/HELP
```

HELP lists the linker options to the console. When the listing is complete, *.RTLink* terminates with a return code of 0 and no output file is produced. The abbreviation for HELP is H.

List Public Symbols

```
/MAP[:<mapOption list>]
MAP [= <mapFile>] [<mapOption list>]
```

MAP causes *.RTLink* to generate a map file containing one or more reports about the link session. MAP may be abbreviated to MA.

<mapFile> is the name of the file to generate. In the POSITIONAL syntax, the map filename is specified on the command line, and therefore not included in the option syntax. If not specified, the map file is given the same name as the *<outputFile>*, but with a default extension of .MAP.

<mapOption list> is a list of map options that determine which reports are written to the specified map file. If not specified, all possible reports are written to the map file. If specified, any combination of the options shown in the following table may be used.

MAP Report Options

Option	Meaning
S	List segments with assigned addresses
N	List public symbols with addresses sorted by symbol
A	List public symbols with addresses sorted by address

In POSITIONAL syntax, you could do the following to generate a map file with the default name rather than explicitly specifying it. Even though no map filename is explicitly specified on the command line, the /MAP option causes it to be generated. In this example, it will be called TEST1.MAP:

```
RTLINK Test1 Test2 /MAP:sn;
```

On the other hand, you could specify the map filename and linker option. In this example, the file will be called MYMAP.MAP:

```
RTLINK Test1 Test2,, MYMAP /MAP:sn;
```

In this corresponding FREEFORMAT example, even though no map filename is explicitly specified with the MAP option, the map file is generated with a default name of TEST1.MAP:

```
RTLINK FILE Test1, Test2 MAP sn
```

On the other hand, you could specify the map filename as part of the MAP option. In this example, the file will be called MYMAP.MAP:

```
RTLINK FILE Test1, Test2 MAP=MYMAP sn
```

Suppress .RTLink Script File Input

/SILENT

SILENT suppresses the display of the linker prompts and responses when *.RTLink* is invoked in script file mode. Its abbreviated form is SI.

Display .RTLink Process Information

/VERBOSE[:<level>]

VERBOSE causes *.RTLink* to display brief status messages during the linker session which allow you to see what modules are being linked and in what order. During INCREMENTAL linking, this option displays the current percentage of wasted space in the .EXE file. The abbreviation for VERBOSE is VE.

<level> must be 0, 1, or 2. This argument specifies the amount of information to be displayed.

Incremental Linking Options

Toggle Incremental Linking

/INCREMENTAL[:<wastedSpace>] | /NOINCREMENTAL

NOINCREMENTAL is the default of these two, mutually exclusive linker options. This means that if you do not specify either one, NOINCREMENTAL is assumed. Since it is the default, you would not need to specify NOINCREMENTAL except to override an INCREMENTAL setting that you had specified elsewhere (e.g., in RTLINKCMD).

NOINCREMENTAL disables incremental linking of *CA-Clipper* modules causing the linker to link the entire program no matter what has changed. It may be abbreviated to NOIN.

INCREMENTAL may be abbreviated to INC. This option allows incremental linking of *CA-Clipper*-compiled modules. Simply put, when you incrementally link a program, *.RTLink* checks to see which modules have changed and only relinks those modules. The result is much faster linking.

See the Incremental Linking section in this chapter for more information on this subject.

Miscellaneous Options

Toggle Default Library Search

/DEFAULTLIBRARYSEARCH | /NODEFAULTLIBRARYSEARCH

DEFAULTLIBRARYSEARCH is the default of these two, mutually exclusive linker options, which means that if you do not specify either option DEFAULTLIBRARYSEARCH is assumed. Since it is the default, you would not need to specify DEFAULTLIBRARYSEARCH except to override a previous NODEFAULTLIBRARYSEARCH setting (e.g., in RTLINKCMD). The abbreviations for these two options are DEFA and NOD.

Compilers often embed default library names in object files to cause certain libraries to be searched automatically by the linker. *CA-Clipper*, for example, embeds CLIPPER.LIB, EXTEND.LIB, and DBFNTX.LIB, so that you do not have to specify any of these libraries when you link. DEFAULTLIBRARYSEARCH causes *.RTLink* to use these embedded names.

NODEFAULTLIBRARYSEARCH causes *.RTLink* to ignore the embedded names. If there are undefined symbols after loading all modules specified in the *<objFile list>* and you have not explicitly specified any libraries to search, *.RTLink* will list these symbols as undefined.

Toggle Extended Dictionary Search

/EXTDICTIONARY | /NOEXTDICTIONARY

/EXTDICTIONARY is the default of these two, mutually exclusive linker options which means that if you do not specify either option, /EXTDICTIONARY is assumed. Since it is the default, you would not need to specify /EXTDICTIONARY except to override a /NOEXTDICTIONARY setting (e.g., in RTLINKCMD). EX and NOE are the abbreviations for these options.

The extended dictionary is a list of symbol locations and dependencies that *.RTLink* uses to speed up library searches.

/EXTDICTIONARY causes *.RTLink* to use the extended dictionary.

/NOEXTDICTIONARY causes the linker not to search the extended dictionary. You may need to specify this option to prevent a warning when there are two possible sources for the same public symbol (e.g., if you redefine a library routine in your program).

Control the Stack Size

/STACK:<sizeBytes>

STACK specifies a program stack overriding the stack size specified in the object module. The abbreviation is ST.

<sizeBytes> is the number of bytes to allocate and can be any positive number up to 65,535 decimal.

STACK may be specified in either prelink or link modes. If used when prelinking, the stack size will be the same for all subsequent dependent executable files (.EXE) unless it is explicitly overridden (with the STACK option) when the .EXE is created.

Prelink Options

Exclude Modules from the Link

/EXCLUDE:<symbol>
EXCLUDE <symbol list>

EXCLUDE causes any module defining the specified symbol to be excluded from the link. Even if symbols in the module are referred to elsewhere, including in the REFER option, the module will not be linked. EXCLUDE may be abbreviated to EXC.

An example of a situation in which you will use EXCLUDE is if you redefine a routine that is defined in one of the language libraries that you use. If the .EXE file containing the new routine definition depends on a prelinked library, a potential problem exists.

If the routine that you redefine is general (i.e., widely used by other routines), chances are that it will be placed in the .PLL because of a call made to it by another routine even if you did not explicitly request it with REFER.

If this happens, there are two available definitions for the same routine, one in the .PLL library and one in the .EXE file. When the .EXE file is executed, references to the routine within the prelinked library are resolved to the library version and references to the routine within the .EXE are resolved to the .EXE version. If you wanted the routine to be globally redefined, this is a disastrous situation.

To solve the problem, EXCLUDE the routine when you create the .PLL so that the library version will not be included in the prelinked library. Later, when you link to create the .EXE file that depends on this .PLL, all references to the routine will be resolved to the new definition in the .EXE file.

As an alternative solution, you could also place the .OBJ file that redefines the routine into the .PLL file. Since the routine will be defined before the libraries are searched, this will also keep the library version of the routine from being pulled into the .PLL. Of the two solutions, this is the better one if all .EXE files that reference the .PLL use the replacement version of the routine.

You may specify as many symbols as you want using a single EXCLUDE option in the FREEFORMAT input mode by using a comma-separated list as the argument. In POSITIONAL mode, you may specify multiple /EXCLUDE options to indicate more than one symbol during a single run of .RTLink.

Specify a Prelinked Library (.PLL)

/PLL:<prelinkLib>

PLL is used in link mode to specify a prelinked library file that the resulting executable file depends on. This option may also be used in prelink mode in order to create a prelinked library that depends on another prelinked library; however, this nesting of prelinked libraries is allowed only one level deep. For example, you can create a prelinked library, B.PLL, that depends on A.PLL, but you cannot in turn create a library, C.PLL, that depends on B.PLL. The abbreviation for PLL is PL.

PLL needs only the prelinked transfer file (.PLT) to generate the dependent .EXE file—the actual prelinked library file (.PLL) need not be present when you link with the PLL option. The library file will be needed later on when you execute the dependent .EXE file.

Warning! *Because the PLL option always assumes a .PLT extension, <prelinkLib> should be specified without an extension. Similarly, the dependent .EXE file, when executed, will assume a .PLL extension. For these reasons, the default file extensions for the prelinked library and transfer files should never be changed.*

Enable Prelinking Mode

/PRELINK

PRELINK changes the mode of .RTLink from link to prelink. Use of this option causes the <outputFile> generated by .RTLink to be a prelinked library file instead of an executable file. The default extension of this file is .PLL. An additional file, called the transfer file, is also generated with the same name and a .PLT extension. All other information provided as input to the linker is similar to the default link mode. PR is the abbreviation for PRELINK.

Note: You can ignore any reports of undefined symbols during prelinking. Reports of undefined symbols when creating an .EXE, however, are more serious and should not be ignored. Unresolved symbols are expected during a prelink, but at link time all program code should have been seen by *.RTLink* and all symbols resolved. For this reason, *.RTLink* terminates with a DOS return code of 1 when undefined symbols are encountered in link mode.

Include Modules in the Link

```
/REFER:<symbol>
REFER <symbol list>
```

In prelink mode, REFER causes *.RTLink* to treat the specified symbols as if they have been referred to, but not yet defined. This forces the linker to search all specified libraries in order to load the necessary code to define the symbols. The abbreviation for REFER is REF.

In link mode, a symbol is marked as undefined by loading an object module and at the same time declaring the symbol external. In prelink mode, however, you would list the library modules you want to include with REFER. Mainly, this is done with commonly used library modules so that the code does not have to be included in each of the .EXE files that is dependent on the prelinked library.

You may specify as many symbols as you want using a single REFER option in the FREEFORMAT input mode by using a comma-separated list as the argument. In POSITIONAL mode, you may specify multiple /REFER options to indicate more than one symbol during a single run of *.RTLink*.

See the Prelinking section in this chapter for more information on this subject.

Static Overlaying Options

Note: The options in this section are available only under the FREEFORMAT interface. There are no corresponding POSITIONAL linker options. See the Static Overlaying section in this chapter for more information and examples on using static overlays.

Designate a Static Overlay Area

```
BEGINAREA
    .
    . <statements>
    .
ENDAREA
```

BEGINAREA is used in link mode to designate the beginning of a static overlay area. The end of the area is designated with ENDAREA. All sections created between these two options become overlay sections within the overlay area.

Static overlay areas may be nested by specifying one BEGINAREA...ENDAREA construct inside of another. The abbreviations for BEGINAREA and ENDAREA are BE and EN, respectively.

Move Segments into Current Static Overlay Section

```
MODULE <moduleName list>
```

MODULE moves the segments from the specified modules into the current static overlay section. MODULE overrides the FILE and LIBRARY options. The major utility of MODULE is to pull library code into an overlay section. MODULE may be abbreviated to MO.

The *<moduleName list>* is comma-separated. To specify a null name in the list, use the NIL keyword instead.

Preload a Static Overlay Section at Runtime

```
PRELOAD
```

PRELOAD causes the current static overlay section to be loaded into memory before the program begins execution. Normally, only the resident sections of the program are loaded into memory before execution begins. The abbreviation for PRELOAD is PRE.

Place PRELOAD after the SECTION you want to preload. For example:

```
SECTION = Area1
PRELOAD
FILE Test
SECTION Area2 INTO Extern
PRELOAD
FILE Test2
```

Create a Static Overlay Section

```
SECTION [= <sectionName>] [INTO <ovlFile>]
```

SECTION creates a static overlay section and causes segments within any non-*CA-Clipper* object modules in subsequent FILE or LIBRARY options to become part of that section. The abbreviation for SECTION is SEC.

If a *CA-Clipper* module is specified in an overlay SECTION, it is treated the same as with the DYNAMIC <*ovlFile*>. The structure of SECTIONS and static overlay areas is meaningless where *CA-Clipper* modules are concerned except that they go into the specified <*ovlFile*>.

An overlay section can be stored in a separate overlay file by specifying the optional INTO <*ovlFile*> clause. The default extension for this file is .OVL, and it is usually referred to as an *external* overlay. Unless this clause is specified, the section will be part of the final executable file (.EXE). In other words, it will be an *internal* overlay.

Note: INTO is a clause of SECTION and cannot be separated by any other options. Specify SECTION INTO first, followed by FILE and other options

Dynamic Overlaying of CA-Clipper Code

When *.RTLink* encounters a *CA-Clipper*-compiled module, it handles the code in one of two ways depending on whether you designate the module with the DYNAMIC or RESIDENT option. You can specify these options on a module-by-module basis allowing a combination of dynamic and resident code in your program.

By default, all *CA-Clipper*-compiled modules are assumed to be DYNAMIC which means that you can take advantage of dynamic overlaying without doing anything special. This feature is limited to programs that are compiled with *CA-Clipper*—modules written in another language such as C or Assembler are always linked as RESIDENT.

Dynamic Modules

An .EXE file created with *.RTLink* automatically creates dynamic overlays for *CA-Clipper*-compiled code. The following command:

```
RTLINK FILE Test1, Test2, Test3
```

is equivalent to:

```
RTLINK DYNAMIC FILE Test1, Test2, Test3
```

When you execute such a file, the code is paged from disk to memory on a *least recently used* basis. This means code that has not been used for a while is overwritten in memory to make room for the new code, leaving frequently used code in memory.

The obvious advantage in executing dynamic code is that programs much larger than 640K can be run without actually specifying overlays. A program in which most modules are dynamic can run even on machines with very limited available memory. The penalty is that the program will run more slowly than if more memory were available.

The reason that dynamic programs run more slowly in less memory is that if the program needs a routine that is not currently in memory, it has to go to the disk in order to bring the routine into memory. The less memory you have, the more often the program has to access the disk. However, the actual swapping from disk is minimal unless memory is extremely tight because frequently used modules stay in memory and the dynamic memory manager can very efficiently find pages that are already in memory.

Resident Modules

You can also cause part or all of your program to remain in memory at all times by using the RESIDENT option. Modules that are specified as resident get loaded into memory once when you originally execute the program. The advantage of using resident programs is that since the program remains in memory, there is never a need to load it from disk again. The disadvantage is that the amount of memory required to run the program is determined by the program size and, because of DOS limitations, cannot exceed 640K.

The next example shows how to make part of a program resident and part dynamic by placing certain modules in a dynamic overlay file:

```
RTLINK RESIDENT FILE Test1 DYNAMIC INTO Test1 FILE Test2, Test3
```

This example causes the code in Test1 to be linked as resident into TEST1.EXE and the code in Test2 and Test3 to be linked into a dynamic overlay file called TEST1.OVL. Using a dynamic overlay file in this strategy has the advantage that you can isolate code that you know is used infrequently in order to enhance the performance of your program, while keeping the code that is used most often in memory.

Prelinking

RTLink provides a feature allowing you to link portions of your program that are not subject to change in advance of creating an actual executable file. This is called prelinking and is accomplished using the PRELINK linker option to generate a prelinked library. Then, the library is identified with the PLL option when you link other .PLL and .EXE files that depend on it. This section describes when and how to use prelinked libraries.

Deciding to Use a Prelinked Library

There are two reasons why you would want to use a prelinked library. The first is to shorten link time during program development above and beyond the time saved using incremental linking (see the Incremental Linking section in this chapter).

For example, during development you could place all of the stable, unchanging code (e.g., portions of CLIPPER.LIB, EXTEND.LIB, and possibly some third-party libraries) that is used by your program into a prelinked library. Each time you link your program during development, *.RTLink* does not relink the modules defined in the specified prelinked library because, by definition, they do not need to be relinked. Only the object and library files specified on the *.RTLink* command line will be relinked, which can save considerable time.

After the program is developed and ready for shipping, you would probably want to link it without using the prelinked library. There is never any advantage to shipping a prelinked library unless your application consists of more than one .EXE file—you will not save disk space, and your program will not execute any faster.

The second use for prelinking is to save disk space when you have multiple .EXE files in your application that share much of the same code. To do this, you would design a prelinked library containing the code that was common to all of the .EXE files. By putting the common code in a single prelinked library file and having all of .EXE files access that library, you cut down on the number of times code is repeated and thereby save disk space.

Building a Prelinked Library

Once you have decided to use a prelinked library, you need to build it. To do so, you follow these steps that are explained individually in this section:

- Identify the object files that make up each program
- Identify the libraries that your program needs
- Identify external references in your object files
- Identify the modules that make up each executable
- Design the prelinked library contents
- Identify symbols that cause needed modules to be pulled into the prelinked library
- Build the prelinked library

Each of these points is discussed separately below.

Identify Object Files

The first step in building a prelinked library is to identify which object files are a part of each program. Each program would normally be made up of source files which you write. The source files are compiled or assembled into object files, and the object files are linked to form your executable program.

If you have been using a linker to link your program without prelinked libraries, you had to specify the object files that make up the program.

Identify Libraries

The second step is to identify the language libraries such as CLIPPER.LIB and EXTEND.LIB as well as any third-party libraries that your program needs.

If you have been using a linker to link your program without prelinked libraries, you either had to specify the library files used by the program or the compiler embedded them in the object file (e.g., *CA-Clipper* embeds CLIPPER.LIB, EXTEND.LIB, and DBFNTX.LIB).

Prelinking

Identify External References

The third step is to identify the symbols that your object files declare as external which are not declared as public by other object files in your program. These are the symbols that the linker would normally resolve from libraries.

You can accomplish this by forcing *.RTLink* to report symbols being resolved from libraries. Simply link your object files as you normally would but without naming any library files, and specify the NODEFAULTLIBRARYSEARCH linker option to prevent *.RTLink* from searching embedded library names. As a safeguard to make sure that no modules are being pulled in from libraries, you may also want to specify the VERBOSE option. When you link your program in this manner, *.RTLink* will not be able to resolve any references to library symbols and will report them as undefined.

Identify Modules in Executables

The next task is to determine the modules that make up each executable. Normally, a linker has only two sources for these modules: the object and library files.

Identifying your object files has already been done, so identifying modules which come from each object file is also already done—for most compilers there is one module per object file.

To identify library modules, run *.RTLink* on the object and library files as you normally would to link the program and specify the VERBOSE option. This option causes the linker to display the names of library modules as it pulls them into the executable.

Design Prelinked Library

The decision of what will go into a prelinked library varies depending on what you are trying to accomplish—disk space reduction or linking speed.

If you are seeking increased linking speed, put only stable modules that are not likely to change into the prelinked library and only link the modules which change during link mode when the .EXE is produced.

If you are attempting to minimize disk space, include all modules that are needed by every executable file that refers to the prelinked library. You may also want to include modules that are needed by many, but not all, of the executable files. The general idea is that if code is used by more than one executable, it is a candidate for placement in the prelinked library, and you should try various combinations to optimize disk usage.

One important point in using static overlays with *.RTLink* is that code and data which are to go in a static overlay section cannot be placed in a prelinked library. If you want a certain piece of code or data to go into a static overlay section, make sure that it is not included in a referenced prelinked library and is only specified and included at the final, .EXE producing link step.

Identify REFER Symbols

The next step is to identify the symbols that will cause the modules identified in the previous step to be pulled into the prelinked library. If a module comes from an object file, you will not need a symbol to cause it to be pulled in since you can explicitly specify object files when creating the prelinked library.

Build the Prelinked Library

To build the prelinked library, run *.RTLink* in prelink mode by specifying the PRELINK option. The POSITIONAL syntax is:

```
RTLINK [<objFile list>,
   [<outputFile>],
   [<mapFile>],
   [<libFile list>]
   /PRELINK
   [/REFER: <symbol> [/REFER:<symbol>...]]
   [/EXCLUDE:<symbol> [/EXCLUDE:<symbol>...]];
```

The FREEFORMAT syntax is:

```
RTLINK [FILE <objFile list>
   [OUTPUT <outputFile>]
   [LIBRARY [<libFile list>]
   PRELINK
   [REFER <symbol list>]
   [EXCLUDE <symbol list>]
```

Specify the object files that you want to include in the prelinked library file by putting them in the *<objFile list>*.

Specify the library files that you identified by putting them in the *<libFile list>*.

Use the REFER option to specify the symbols that you identified to pull in the needed modules from libraries. In order to resolve these symbols, *.RTLink* will pull the module which defines each symbol into the prelinked library along with any other modules that it references.

EXCLUDE causes any module defining the specified symbol to be excluded from the link. Even if symbols in the module are referred to elsewhere—including in the REFER option—the module will not be linked. An example of a situation in which you will use EXCLUDE is if you redefine a routine that is defined in one of the language libraries that you use.

The <outputFile> in prelink mode specifies the name of the prelinked library, and it cannot include a file extension. Actually, two files are generated. They are the prelinked transfer file (.PLT) which is used when you build an .EXE file that is dependent on a prelinked library, and the prelinked library file (.PLL) which is used when you execute a file that is dependent on a prelinked library.

Building a Dependent Executable

The final thing you need to know about prelinked libraries is how to build an executable file that depends on one. To do this, link the program with the PLL option to specify the name of the prelinked library.

Besides the PLL option, you need to specify the object files that make up your program (but not object files which are in the prelinked library) and, if there are modules to come in from libraries which are not in the prelinked library, you need to specify the libraries that need to be searched. You may also specify other linker options if you want to perform the final link.

The prelinked transfer file (.PLT) is needed to perform this step. Later on, when you execute the resulting .EXE file, the prelinked library file (.PLL) will be needed.

Note: BASE52.LNK is a library link file that creates a PLL out of the standard *CA-Clipper* Runtime libraries. BASE52.LNK can be found in \CLIPPER5\PLL.

Incremental Linking

Developers generally dislike waiting for their linker to run. Compilers are getting faster and faster, and a developer can divide a program into modules and separately compile them so that the entire program need not be recompiled after every change. Normally, however, this kind of control is not available with the linking process.

The linker must link the entire program after each change, even if only a small percentage of the program has changed. Thus, a larger percentage of the wasted time between program change and program test is being taken up by linking.

The idea behind incremental linking is to take advantage of the fact that most of the program does not change between links. As an incremental linker, *.RTLink* can detect which modules changed and only relink those modules, thus resulting in much faster linking.

By default, your programs are linked in the traditional manner with *.RTLink* with each module being relinked without regard to what has changed. To turn on the incremental linking feature, specify the INCREMENTAL option when you link.

This option causes an information file (.INF) to be generated that contains a list of each *CA-Clipper* module in the program along with other information. It is with the use of this file that *.RTLink* is able to tell what has changed in the program. The information file has the same base name as the *.RTLink <outputFile>*.

Incremental linking necessarily results in wasted space in the output file, and at a certain point when the amount of wasted space in the output file becomes too large, the entire program will be relinked. You can specify the amount of wasted space as part of the INCREMENTAL option. It is specified as a percentage of the program size, and the default is 25%. The VERBOSE option can be used to show the current amount of wasted space.

Static Overlaying

Static overlays are defined using the linker options described previously in the Static Overlaying Options section in this chapter. This section is designed to give you additional information on using these options to create static overlays.

Note first that static overlays cannot be used with *CA-Clipper*-compiled modules. If a *CA-Clipper* module is encountered within a static overlay section, it will be dynamically overlaid as described in the *Dynamic Overlaying* section in this chapter. For the most part, you will only use static overlays if you have a large system where much of it is written in C. Static overlays currently work only with large memory models.

Overlays are portions or segments of a program that do not reside in RAM until the execution of the segment is required. The main program code, or root portion of the program, is always resident in memory.

Static Overlaying

Overlays are stored on disk until the root calls them into memory. Thus, the entire program need not be in memory at any one time.

The portion of memory that overlays share is called the overlay area. An overlay is loaded whenever a routine in the overlay is called. When the root calls another overlay that is not currently in memory, the new overlay overwrites the first. Note that the largest overlay determines the amount of memory allocated for all of the overlays.

Deciding to Use Static Overlays

There are two main factors that affect the decision of whether or not to use static overlays: your memory requirements and the execution speed that you want.

The main reason why you need to use overlays is to reduce the amount of memory required to run your program. If you cannot run in a small amount of memory or if you want to use as little memory as possible, then overlays are probably what you need.

The trade off involved when using overlays is that execution speed is reduced. This is because any time you make a call to a routine in a static overlay section, the overlay manager must first check to see if that section is already in memory and must bring the section into memory from disk if necessary. The process of checking which section is currently in the overlay area is usually a very quick one. The process of bringing in the overlay from disk takes more time.

If you build overlays wisely so that this swapping happens infrequently, you should notice only a very slight degradation in execution speed. In fact, in a well designed system you may not notice any degradation at all.

Designing Static Overlays

When designing overlays you must identify the modules of your program that you want to place in each overlay section. A module cannot reside in more than one overlay section, so modules referenced by more than one section would normally go in the root section of your program.

The *.RTLink* overlay manager is included in the RTLUTILS.LIB library. This file must be present when linking a program that uses static overlays, but you do not need to specify it to *.RTLink*—the linker searches for this file automatically.

Static Overlaying

In using overlays with .*RTLink*, code and data that are to go in a static overlay section cannot be placed into a prelinked library. If you want a certain piece of code or data to go into a static overlay section, make sure that the code/data is not included in a referenced prelinked library and is only specified and included at the final, .EXE producing step.

Keep in mind that swapping overlays in and out slows execution, so you will want to try to organize overlay sections so that this swapping is minimized.

The size of a static overlay area is the size of the largest overlay section in that area. Remember that if you are using the *.RTLink* overlay manager, the DEBUG option displays messages when overlays are loaded and unloaded. This option can be used to debug executables that use static overlays and to minimize overlay swapping so that you can optimize your section layout.

The segment listing portion of the map file shows the sizes of each overlay section and area.

For example, if your program consists of a main menu which performs different functions depending on the selection, you may want to place the modules which implement each function into sections that share one overlay area since once you have chosen one option, you are assured that you will not choose other options until after the first one finishes. Further, the time necessary to swap in a static overlay section will be at least partially hidden in the user's response time to the menu prompt.

External Static Overlays

You can optionally have a static overlay section stored in a file that is distinct from the .EXE. This concept is called an external overlay. In order to use external overlays with *.RTLink*, you use the INTO portion of the SECTION option to specify the name of the external overlay file.

The speed of execution of a program using external overlays and a program using internal overlays should be the same if that is the only difference between the two programs. It should not take any more time to get an external overlay than it does to get an internal overlay.

The drawback to using external overlays is that there are more files for a user to deal with than would be the case if you had used internal overlays. External overlays give no performance or memory benefits over internal overlays, and they may not be shared between programs.

Examples

- The following generates a program called MYAPPL.EXE with one overlay area that is shared by three sections:

```
FILE MAINMENU, UTILS
VERBOSE MAP OUTPUT MYAPPL
BEGINAREA
    SECTION
        FILE CHOICE1
    SECTION
        FILE CHOICE2
    SECTION
        FILE CHOICE3
ENDAREA
```

- The next example creates a file called MYAPPL.EXE using a nested overlay structure:

```
FILE MAINMENU, UTILS VERBOSE MAP OUTPUT MYAPPL
BEGINAREA
    SECTION
        FILE CHOICE1
    SECTION
        FILE CHOICE2
    SECTION
        FILE CHOICE3
    BEGINAREA
        SECTION
            FILE SUB1
        SECTION
            FILE SUB2
        SECTION
            FILE SUB3
    ENDAREA
ENDAREA
```

Summary

In this chapter, you learned how to link .OBJ and .LIB files to create DOS executable files with *RTLink*. You were introduced to the concepts of incremental linking, dynamic overlays, and prelinked libraries as well as other *RTLink* features. To get a list of all the linker options, enter RTLINK /HELP from the DOS prompt.

Chapter 10
CA-Clipper Debugger CLD.EXE

Welcome to *The CA-Clipper Debugger*, sometimes referred to more simply as the Debugger. Although you may already be familiar with the process of debugging a compiled application, we recommend that you take the time to read this chapter.

Once you have finished writing a program, you must test it to make sure there are no errors. The compiler and linker will spot some errors that you can correct without having to actually execute the program, but it is not at all unusual for a program to contain errors which you can only detect while the program is running. Examples of this type of error are runtime and execution errors, cosmetic and other display errors, and logic errors. It is with this type of error that a debugger comes in handy.

As its name suggests, a debugger is a tool that helps you track down and remove errors (bugs) from your source code. *The CA-Clipper Debugger* is a source code debugger which means that it allows you to look at your source code while your program is running. You use the Debugger to execute your application so that you can switch back and forth between running the program and viewing the source code to spot potential problem areas.

The Debugger provides you with certain features that make it easier for you to isolate and identify errors. Among these features are the ability to:

- Trace source code line-by-line (Single Step Mode)
- Watch a variable as it is updated (Watchpoint)
- Monitor variables by storage class
- Inspect work areas and set values
- Change the value of a variable
- Create new variables and change the values of existing ones

- Execute procedures and user-defined functions linked into your application
- Stop program execution when a variable changes value (Tracepoint)
- Stop program execution at a line of code or when a function is called (Breakpoint)

When you execute your application using the Debugger, all of these features and more are available to you so that you can tightly control how your application runs and keep an eye on things as it does. In this manner, you can quickly and easily identify errors in your source code.

Whenever you notice an error, you simply edit the code, recompile, and relink, although you must perform these three steps outside the context of the Debugger. You may need to repeat this cycle (debug/edit/compile/link) several times if your program has a lot of errors.

In This Chapter

This chapter documents the Debugger command line syntax and its various modes of operation, describes the Debugger display including all windows and menus that you will encounter, explains how to debug a program using the Debugger, and offers a complete menu command reference. The following major topics are covered:

- Starting the Debugger
- The Debugger display
- Debugging a program
- Menu command reference

Starting the Debugger

In order to debug an application, you must first create an executable file (.EXE) using the compiler option for debugging. You can then invoke the Debugger and begin the debugging session. Some parts of your application may require source code changes in order to be properly debugged.

Preparing Your Programs for Debugging

To take full advantage of the features offered by *The CA-Clipper Debugger*, there are certain things that you should consider. Some programming practices inhibit the debugging process while others require the knowledge of special Debugger features. Also, there are a couple of compiler options that you need to know about before you can debug a program. This section discusses special programming, debugging, and compiling issues that you need to consider before attempting to debug your application.

Programming Considerations

Certain parts of your code present special problems when debugging. This section explains the techniques for debugging multistatement command lines, header files, code blocks, and macros.

Multistatement Command Lines
CA-Clipper allows you to place more than one program statement on a single line. For example:

```
nNewPage := (nLineNo > 55); ReportPage(nNewPage)
```

When you use *Uparrow* or *Dnarrow* in the Debugger Code Window, the cursor moves up and down one complete line at a time. This is true regardless of how many statements are on the line in question. In the above example, the Debugger does not allow you to step through the first and second statements independently nor does it allow you to set a Breakpoint at either statement. The entire line is treated as a single entity.

The code in the above example should be broken up into two lines, as follows:

```
nNewPage := (nLineNo > 55)
ReportPage(nNewPage)
```

This makes debugging easier and also makes the code more readable.

Header Files
Header files are files that are referenced using the #include preprocessor directive. They typically contain manifest constant and pseudofunction definitions. The Debugger does not automatically display header files in the Code Window, but you can view them with the File Open command.

Code Blocks
A code block contains executable program code that is treated by *CA-Clipper* as data and can be passed as a parameter to other programs. Inspecting a code block in the Debugger only reveals its name, not the actual code.

Starting the Debugger

However, code blocks can be executed from within the Debugger. The following command, when entered in the Command Window, causes the code block *bMyBlock* to be evaluated. Any code within *bMyBlock* is executed and the results are displayed accordingly.

```
> ? EVAL(bMyBlock)
```

In addition to allowing you to execute code blocks interactively, the Debugger also traces code blocks back to their definition as long as Options Codeblock is on. When using Single Step Mode to execute a piece of code containing a code block, the Debugger moves the Execution Bar to the line of code where the block was created each time the code block is evaluated. This allows you to see the contents of the block (which are unavailable during normal inspection) and occurs regardless of whether the block was declared in the current routine.

Macro Substitution

The Debugger treats macros as ordinary character strings. As such, they can be inspected or viewed using the ? | ?? command. For example:

```
> ? &macroVar
```

displays the contents of the expanded macro variable *macroVar*.

Compiling Your Source Code

You cannot debug an application with the Debugger unless you compile it using the /B option. This option tells the compiler to include debugging information in the object file (.OBJ). For example:

```
CLIPPER MainProg /B
```

In this example, all programs called by MainProg.prg are also compiled with the /B option.

If, however, you use a (.clp) file or compile each program in your application separately, you must specify the /B option every time you invoke the compiler. During the debugging session, any subsidiary programs which have been compiled without this option are ignored.

In addition to using the /B option, you may also want to compile with the /P option if you use any of the preprocessor directives and are interested in viewing the preprocessor output while debugging.

As their name suggests, preprocessor directives are instructions to the preprocessor which do not exist at runtime. This means, for example, that you cannot inspect #define constants and pseudofunctions in the same way as other expressions when you are debugging.

If you compile a program using the /P compiler option, a file with a (.ppo) extension is produced. This file shows the preprocessor output which can be viewed using the Options Preprocessed command. With

Starting the Debugger

this command, the Debugger displays each line of source code in the Code Window with the output from the preprocessor shown underneath.

Warning! *In order to use The CA-Clipper Debugger, your application must be compiled with CA-Clipper.*

Invoking the Debugger

There are several ways to invoke the Debugger, each of which is discussed below. All of the methods require that the application be compiled with the /B option—debugging is not possible otherwise.

From DOS

To invoke the Debugger from the DOS prompt, use the following syntax:

```
CLD [[/43 | /50 | /S] [@<scriptFile>]
    <exeFile> [<argument list>]]
```

CLD.EXE is *The CA-Clipper Debugger* executable file. If you installed the *CA-Clipper* development system using the default configuration, CLD.EXE is located in the \CLIPPER5\BIN directory and your DOS PATH should be altered accordingly.

/43 | /50 | /S specifies the screen mode used by the Debugger. These three options are mutually exclusive and, if more than one is specified, CLD.EXE uses the last one that occurs on the command line. The default screen mode is 25 lines, with the Debugger using the entire screen.

/43 specifies 43-line mode and is available on EGA monitors only.

/50 specifies 50-line mode and is available on EGA and VGA monitors only.

/S is available on EGA and VGA monitors only. This option splits the screen between your application and the Debugger, allowing you to view the application and the Debugger simultaneously. On a VGA monitor /S uses 50-line mode, and on an EGA monitor it uses 43-line mode. In split screen mode, the top 25 lines of the screen are used by your application, and the remaining lines are used for the Debugger display.

<scriptFile> is the name of a script file with a default extension of (.cld). CLD searches for the specified *<scriptFile>* in the current directory and then searches the DOS PATH. A script file is simply an ASCII text file

Programming and Utilities Guide 10-5

Starting the Debugger

containing one or more Debugger commands, with each command appearing on a separate line. When the Debugger is invoked with a script file, each command in the file is executed automatically after the <exeFile> file is loaded. See the Using a Script File section later on in this chapter for more information on this subject.

In addition to any script file called for on the CLD command line, the Debugger automatically searches for a script file with the name *Init.cld*. If a file by this name is located in the current directory or anywhere in the DOS PATH, the Debugger executes it as a script file. If both Init.cld and a command line script file are present, Init.cld is executed first followed by the command line script file.

<exeFile> is the name of the executable (.EXE) file you want to debug. CLD searches for the <exeFile> in the current directory only—the DOS PATH is not searched. If this file has not been compiled using the /B compiler option to embed debugging information, debugging is not possible.

<argument list> is the argument list for <exeFile>. There must be a space between <exeFile> and <argument list>.

Note that with the exception of the argument list, all other CLD arguments must come before the executable filename on the command line. If no command line arguments are specified, CLD displays a brief help screen.

When the Debugger is invoked in this manner, the source code associated with the <exeFile> appears in the Code Window, and you can begin running and debugging your application.

Linking the Debugger

Under certain circumstances, you may want to link the Debugger into your application. Doing this would allow you, for example, to debug a program at a customer's site in the event that an unforeseen problem occurred after the application was already in the field.

To include the Debugger as part of your application, the Debugger library, CLD.LIB, must be linked into your program as an object file (not as a library file) as in the following example:

```
RTLINK FI Myprog, \CLIPPER5\LIB\CLD.LIB
```

Again, your program must be compiled with the /B option. When the Debugger is linked into a program in this manner, it does not come up automatically when you run your application. Instead, it must be invoked with *Alt-D*.

Starting the Debugger

Note that if you ship your application with the Debugger linked as an object file, you should also ship the help file, CLD.HLP, located in \CLIPPER5\BIN in the default installation. When you ask for help, the Debugger will first search the current directory and then all directories specified in the DOS PATH. If CLD.HLP is not found in any of these locations, the Debugger will prompt you for a directory name. If CLD.HLP is still not found, you will not have access to help text within the Debugger.

Invoking with Alt-D

When an application is linked with CLD.LIB as one of the object files, the Debugger is enabled and can be invoked by pressing *Alt-D* while the application is running. Doing this invokes the Debugger with the Execution Bar positioned on the current line of executing code.

You can also use *Alt-D* when the Debugger is invoked using the CLD command line. When executing the application in Run Mode, simply press *Alt-D* at any time to terminate the application and return control to the Debugger. Run Mode is one of the many execution modes that is available to you when you debug an application using *The CA-Clipper Debugger*. For more information on modes of execution, see the section entitled Debugging a Program in this chapter.

Using ALTD()

The ALTD() function serves two purposes: it allows you to control whether or not the Debugger is enabled or disabled, and it allows you to invoke the Debugger if it is enabled. ALTD() can be used regardless of whether the Debugger is invoked using the CLD command line or using *Alt-D*.

When it appears in a program compiled with the /B option, ALTD() used with no arguments acts like a Breakpoint by stopping the application and giving control to the Debugger.

ALTD(0) temporarily disables the Debugger so that any code following the function call is executed as if the Debugger were not present. When the Debugger is disabled, all Breakpoints and Tracepoints are ignored and even calls to ALTD() and pressing *Alt-D* will not invoke the Debugger.

ALTD(1) is used to enable the Debugger once it has been disabled. Note that for programs compiled with /B, the default state for the Debugger is enabled.

For applications that are linked with CLD.LIB, you can use ALTD() to prevent the user from inadvertently invoking the Debugger. By

Starting the Debugger

including code similar to the following at the beginning of the main program, the ability to invoke the Debugger is controlled by the presence a command line parameter:

```
PARAMETERS dBug

IF dBug != NIL
    ALTD(1)         // Invoke Debugger with ALTD() or Alt-D
ELSE
    ALTD(0)         // Debugger cannot be invoked
ENDIF
```

Using the code shown above, the Debugger is disabled unless the application is invoked with an argument. Even though the program may contain calls to ALTD() and the user may press *Alt-D*, the Debugger cannot be invoked unless the application command line argument is specified. If your application accepts other arguments, use *dBug* as the last one in the list.

How the Debugger Searches for Files

When you make a file open request to the Debugger, you may specify the drive, directory and file extension explicitly as part of the filename. Otherwise, the Debugger makes certain assumptions about the filename and location.

Depending on the context of the file open request, the default file extension supplied by the Debugger varies. For example, when the request is for a script file the Debugger assumes a (.cld) extension and for a preprocessed output file the default extension is (.ppo). Default file extensions are supplied only when you do not specify an explicit file extension.

If no explicit file location (i.e., drive or directory) is specified or if the file location cannot be obtained from the application, the Debugger looks for the file in the current directory. Then, with the exception of the .EXE file, the Debugger searches the DOS PATH—the Debugger searches for the executable file in the current directory only. Only after these locations are exhausted will an error message be displayed indicating that the file could not be found.

Using a Script File

The Debugger allows you to record commands in a script file and execute the commands directly from the file. A script file is simply an ASCII text file that can be created and edited using any word processor.

Starting the Debugger

The file consists of one or more Debugger commands, with each command on a new line in the file, and has a default extension of (.cld).

Some programs may require several debug/edit/compile/link cycles in order to trace persistent errors. Recording repetitive commands in a script file eliminates the amount of typing necessary to reach the same position in your application each time. For example, the following script file sets a Breakpoint at the first call to the function ViewData(), designates the variables *lMadeChanges* and *lSaveChanges* as Watchpoints, and specifies *nFieldNum* and *nFileArea* as Tracepoints:

```
BP ViewData
Point Watchpoint lMadeChanges
Point Watchpoint lSaveChanges
Point Tracepoint nFieldNum
Point Tracepoint nFileArea
```

Another advantage to using script files is to record preferred Debugger settings so that you do not have to set them each time you debug. The easiest way to record option settings in a script file is to set them using the Debugger menus. Then, use the Options Save command as in the following example:

```
> Options Save MyScript ↵
```

The script file, MyScript.cld, is saved to disk and can be used or modified at any time. The settings that you can save in this way include most Options menu settings, the Monitor menu settings, and case sensitivity and Callstack status.

As an alternative to setting the options and saving them to a script file using the Options menu, you can type the appropriate menu commands directly into a script file. Menu commands are formed using the menu name followed by the first word of the menu option you want to set. Any necessary arguments are placed at the end of the command. An example of some menu command settings recorded in a script file follows:

```
View Callstack
Monitor Local
Options Codeblock
```

When the Debugger is active, you can execute a script file with the Options Restore command as in the following example:

```
> Options Restore MyScript ↵
```

You can also execute a script file from the Debugger command line as in the following example:

```
CLD @MyScript MainProg
```

Starting the Debugger

Each time the Debugger is invoked, it automatically searches for a script file with the name Init.cld. If a file by this name is located in the current directory or anywhere in the DOS PATH, the Debugger executes it as a script file. If both Init.cld and a command line script file are present, Init.cld is executed first followed by the command line script file.

Getting Help

The CA-Clipper Debugger offers online help in the form of the Help Window that is divided into two panes: the left pane contains a list of topics for which help is available, and the right pane contains the help text for the currently highlighted topic. You can activate the Help Window using the Help command or simply by pressing *F1*.

There are several main topics of discussion in the Help Window including About Help, Keys, Windows, Menus, and Commands. When the Help Window is first activated, one of these topics is highlighted on the left, and the text on the right discusses the topic.

To get help on a particular topic, highlight it using *Uparrow* or *Dnarrow*. As the highlight moves, the associated help text on the right changes to reflect the current topic.

If the indicator at the bottom right of the window shows more than one page of information, use *PgUp* and *PgDn* to scroll the help text.

Press *Esc* to remove the Help Window and continue debugging.

The text that the Debugger displays in the Help Window is contained in a separate file, CLD.HLP, which must be present on your disk. When you ask for help, the Debugger will first search the current directory and then all directories specified in the DOS PATH. If CLD.HLP is not found in any of these locations, the Debugger will prompt you for a directory name. If CLD.HLP is still not found, you will not have access to help text within the Debugger.

For more information on the Help Window, see the section entitled The Debugger Display in this chapter.

Leaving the Debugger

When you have finished your debugging session, enter the File Exit command or simply press *Alt-X*. The Debugger automatically closes all files and returns you to the DOS prompt.

The Debugger Display

Before you can make full use of *The CA-Clipper Debugger*, you must be familiar with the various windows, menus, and keys on which the debugging environment is based.

The Debugger Windows

The Debugger display is based on a series of windows, each with a unique purpose. The following table gives the name and a brief description of each of these windows:

The Debugger Windows

Window Name	Purpose
Callstack Window	Display the Callstack
Code Window	Display program code
Command Window	Display commands and their results
Help Window	Display help information
Monitor Window	Display monitored variables
Set Colors Window	Display and modify color settings
View Sets Window	Display and modify SET values
View Workareas Window	Display work area and (.dbf) information
Watch Window	Display Watchpoints and Tracepoints

Windows are used to display program code, enter commands, and offer help. The following sections explain the various windows and their functions.

Window Operations

This section describes the general behavior of windows that appear on the main Debugger screen (i.e., Code, Command, Monitor, Watch, and Callstack), including the processes of sizing and moving between windows.

The remaining windows, Help, View Sets, View Workareas, and Set Colors, are not part of the main Debugger display. Rather, they represent modes in which the Debugger displays a special type of window. When you open one of these windows using the appropriate Debugger command or function key, you enter a mode in which the window takes control of the screen and normal Debugger operation is temporarily suspended. When you are finished with the window, you

The Debugger Display

must close it using the *Esc* key before you can continue debugging. The operations in this section do not apply to these modal windows.

Navigating Between Windows

Of all the windows on the main screen, one is said to be the active window and the rest are inactive. When you select a window, you make it the active window. The active window is indicated by a highlighted border; inactive windows have a single-line border.

To select the next window on the screen

- Press *Tab*.

To select the previous window

- Press *Shift-Tab*.

These keystrokes are equivalent to the Window Next and Window Prev commands. The order of the windows is as follows: Code, Monitor, Watch, Callstack, Command.

When a window is active, any valid keystrokes affect only that window. For example, when the Monitor Window is active pressing *Dnarrow* moves the highlight bar to the next variable in the list of monitored variables but does not affect the cursor or highlight bar in any other window.

The exception to this rule is that a command can be typed and executed from any active window. For example if you type "List BP" while the Code Window is active, the command will appear in the Command Window next to the > prompt as you are typing. As soon as you press *Return*, the command will be executed and the result displayed in the Command Window. Furthermore, *Return* always executes the command pending in the Command Window (if any), taking precedence over the normal operation of the *Return* key for the active window.

Opening and Closing

Not all Debugger windows must be open on the Debugger screen at all times. The Debugger automatically controls the opening and closing of certain windows while leaving the control of others up to you. This section describes how to open and close each window.

The Code and Command windows are always open and cannot be closed by you or the Debugger.

The Debugger automatically opens the Watch Window when there are Tracepoints and/or Watchpoints defined. When all Tracepoint and Watchpoint definitions are deleted, the Debugger closes this window.

The Debugger automatically opens the Monitor Window when one or more variable classes is being monitored. When there are no monitored variables, the Debugger closes this window.

The Debugger Display

You control opening and closing of the Callstack Window using the View Callstack command.

Iconizing
Any window on the screen can be effectively put out of sight without actually closing the window. For example, if the Callstack Window is open and you are not interested in its contents for the moment, you can shrink it down so small that only the window name is visible.

To iconize a window

1. Press *Tab* or *Shift-Tab* until you activate the window.
2. Press *Alt-W I* to execute the Window Iconize command.
   The active window is replaced by an icon (its name) so that you can no longer see its contents.

When a window is iconized, certain window operations, such as sizing, are not available.

To return an iconized window to view

1. Press *Tab* or *Shift-Tab* until you activate the iconized window.
2. Press *Alt-W I* to execute the Window Iconize command.

Zooming
Windows can be zoomed to full-screen in order to view more information at one time.

To zoom a window to full-screen

1. Press *Tab* or *Shift-Tab* until you activate the window.
2. Press *F2*.

To return a zoomed window to its original window display

1. Press *Tab* or *Shift-Tab* until you activate the zoomed window.
2. Press *F2*.

F2 acts as a toggle between the full-screen and window display modes. It is equivalent to the Window Zoom command.

When a window is zoomed, other window operations such as moving and sizing are not allowed.

Sizing
The height and width of all Debugger windows is determined by the display mode and which windows are open at any given time. The size of a window, however, can be changed to suit your particular needs.

To size a window

1. Press *Tab* or *Shift-Tab* until you activate the window that you want to size.

The Debugger Display

2. Press *Alt-W S* to execute the Window Size command.
   The border of the window changes to a different pattern to indicate that size mode is active.

3. Use the cursor keys to change the size of the window.

4. Press *Return* to complete the sizing.

The following table lists some shortcut keys for changing the height of a window:

Sizing Windows

Key	Action
Alt-G	Grow active window by one line
Alt-S	Shrink active window by one line
Alt-D	Shrink Command Window by one line
Alt-U	Grow Command Window by one line

Note: You cannot size a window that is zoomed to full-screen.

Moving

The location of all Debugger windows is determined by the display mode and which windows are open at any given time and can be changed to suit your particular needs.

To move a window

1. Press *Tab* or *Shift-Tab* until you activate the window that you want to move.

2. Press *Alt-W M* to execute the Window Move command.
   The border of the window changes to a different pattern to indicate that move mode is active.

3. Use the cursor keys to move the window to the desired screen location.

4. Press *Return* to complete the moving process.

Note: You cannot move a window that is zoomed to full-screen.

Tiling

The Window Tile command is a quick way to clean up the screen. This command restores each window on the screen to its default location and size. Any windows that have been zoomed or iconized are also restored to the original window display mode.

To tile the Debugger windows

- Press *Alt-W T*.

The Code Window

The Code Window (shown in the figure below) is located underneath the Menu Bar and is used to display program code and header (.ch) files. Preprocessed output can be displayed beneath program code on a line-by-line basis using the Options Preprocessed command. The name of the file currently being displayed is shown at the top of the window.

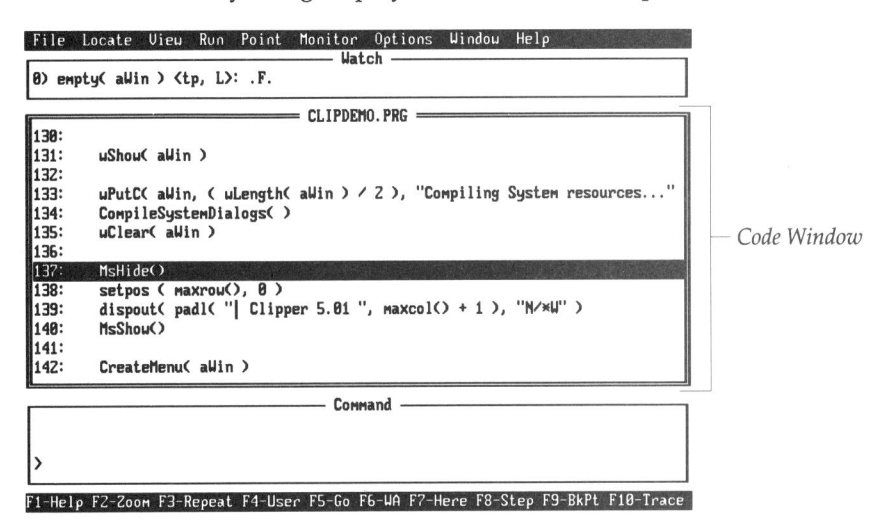

```
File  Locate  View  Run  Point  Monitor  Options  Window  Help
─────────────────────── Watch ───────────────────────
0) empty( aWin ) <tp, L>: .F.

═══════════════════════ CLIPDEMO.PRG ═══════════════════════
130:
131:    wShow( aWin )
132:
133:    wPutC( aWin, ( wLength( aWin ) / 2 ), "Compiling System resources..."
134:    CompileSystemDialogs( )
135:    wClear( aWin )
136:
137:    MsHide()
138:    setpos ( maxrow(), 0 )
139:    dispout( padl( "| Clipper 5.01 ", maxcol() + 1 ), "N/*W" )
140:    MsShow()
141:
142:    CreateMenu( aWin )

─────────────────────── Command ───────────────────────
>

F1-Help  F2-Zoom  F3-Repeat  F4-User  F5-Go  F6-WA  F7-Here  F8-Step  F9-BkPt  F10-Trace
```
— Code Window

Figure 10-1: The Code Window

The Code Window is initially set to display a certain number of lines but can be sized to display fewer or more lines (see the *Window Operations* section in this chapter for more information). The minimum number of display lines is zero, and the initial and maximum number of lines depend on the display mode.

Inside the Code Window is a highlight bar called the Execution Bar which is positioned on the line of code about to be executed. The Execution Bar moves as execution continues.

When the Code Window is active, the cursor appears in the window to indicate your current position in the file being viewed. Initially the cursor and Execution Bar appear on the same line, but the cursor can be moved up and down using the cursor keys. The cursor is used to show the result of Locate commands, to mark (or delete) a line of code as a Breakpoint, and to tell the Run To command where to stop.

To navigate within the Code Window, use the keys shown in the table below:

The Debugger Display

Code Window Active Keys

Key	Action
Uparrow/Ctrl-E	Move cursor up one line
Dnarrow/Ctrl-X	Move cursor down one line
Leftarrow/Ctrl-S	Pan left one character
Rightarrow/Ctrl-D	Pan right one character
Home/Ctrl-A	Pan left one screen
End/Ctrl-F	Pan right one screen
PgUp/Ctrl-R	Scroll window contents up
PgDn/Ctrl-C	Scroll window contents down
Ctrl-PgUp	Move cursor to first line
Ctrl-PgDn	Move cursor to last line
Return	Execute pending command
Tab	Activate next window
Shift-Tab	Activate previous window
F2	Toggle full-screen/window display
F3	Retype last command in command history

The Command Window

The Command Window, displayed at the bottom of the screen, shows the Debugger commands that you enter. The output of Debugger commands (if any) is also displayed in this window directly underneath the command.

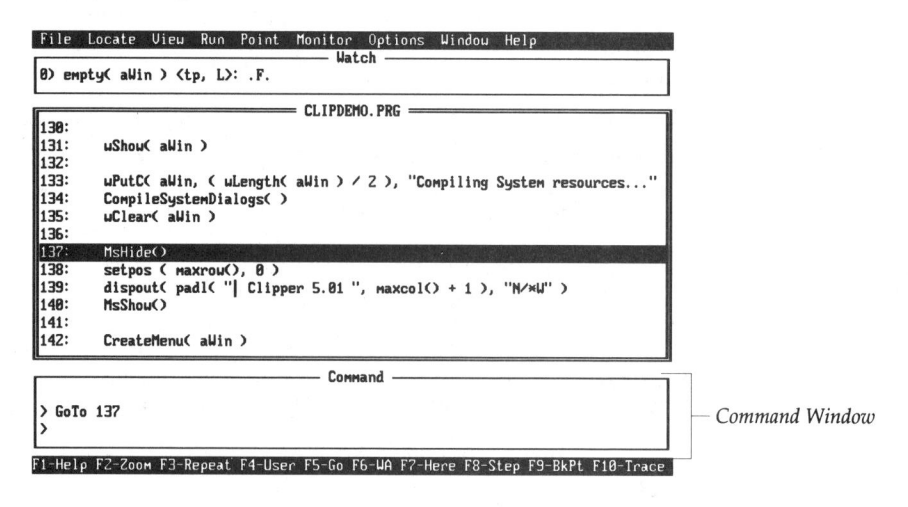

```
File  Locate  View  Run  Point  Monitor  Options  Window  Help
──────────────────────────── Watch ────────────────────────────
0) empty( aWin ) <tp, L>: .F.

═══════════════════════════ CLIPDEMO.PRG ═══════════════════════
130:
131:     wShow( aWin )
132:
133:     wPutC( aWin, ( wLength( aWin ) / 2 ), "Compiling System resources..." )
134:     CompileSystemDialogs( )
135:     wClear( aWin )
136:
137:     MsHide()
138:     setpos ( maxrow(), 0 )
139:     dispout( padl( "| Clipper 5.01 ", maxcol() + 1 ), "N/*W" )
140:     MsShow()
141:
142:     CreateMenu( aWin )

──────────────────────────── Command ────────────────────────────
> GoTo 137
>

F1-Help F2-Zoom F3-Repeat F4-User F5-Go F6-WA F7-Here F8-Step F9-BkPt F10-Trace
```
— *Command Window*

Figure 10-2: The Command Window

The Command Window is like most other Debugger windows in that it can be sized, moved, and zoomed to full-screen when it is the active window (see the Window Operations section in this chapter for more information). The window is made active by selecting it with *Tab* or *Shift-Tab*. When the Command Window is active, its border is highlighted.

Note, however, that the Command Window does not have to be active in order to enter commands. Commands are entered in the same manner no matter what window happens to be active at the time.

To execute a command

- Type the command and press *Return*.
      OR
- Make the equivalent menu selection.

The only difference between these two methods of command execution is that commands you type appear in the Command Window and are, therefore, available in the command history buffer discussed below—not so with menu selections.

See the Menu Command Reference section at the end of this chapter for an alphabetical list of Debugger commands. See The Debugger Menus section later on in this chapter for a brief description of each of the Debugger menus and information regarding their use.

The following table gives a summary of keys that are available when the Command Window is active:

Command Window Active Keys

Key	Action
Uparrow/Ctrl-E	Move cursor to previous line in command history
Dnarrow/Ctrl-X	Move cursor to next line in command history
Leftarrow/Ctrl-S	Move cursor one character to the left
Rightarrow/Ctrl-D	Move cursor one character to the right
Home/Ctrl-A	Move cursor to the beginning of line
End/Ctrl-F	Move cursor to the end of line
Ins/Ctrl-V	Toggle the insert mode on or off
Del/Ctrl-G	Delete character under cursor
Backspace/Ctrl-H	Delete character to the left of cursor

The Debugger Display

Command Window Active Keys (cont.)

Key	Action
Esc	Clear command line
Return	Execute pending command
Tab	Activate next window
Shift-Tab	Activate previous window
F2	Toggle full-screen/window display
F3	Retype last command in command history

History

Commands that you type are saved in a history buffer where they can be accessed when the Command Window is active.

To execute the last command in the history buffer

1. Press *Tab* or *Shift-Tab* until you activate the Command Window.

2. Press *F3* or *Uparrow* to redisplay the command next to the > prompt.

3. Press *Return* to execute the command.

To change a command in the history buffer before executing it

1. Press *Tab* or *Shift-Tab* until you activate the Command Window.

2. Press *Uparrow* or *Dnarrow* until the command that you are looking for appears next to the > prompt.

3. Edit the command to suit your needs.

4. Press *Return* to execute the edited command.

Overwrite and Insert Modes

When the Command Window is active, there are two data entry modes: overwrite and insert.

To toggle between overwrite and insert modes

1. Press *Tab* or *Shift-Tab* until you activate the Command Window.

2. Press *Ins*.
   The cursor changes to indicate the current mode.

Overwrite mode (the default) is indicated by the underscore cursor. In this mode, new characters that you type overwrite existing characters, causing one character to be deleted for each character typed.

Insert mode is indicated by the block cursor. In this mode, new characters that you type are inserted to the right of the cursor, pushing existing characters to the right.

The Watch Window

The Watch Window is displayed at the top of the screen (see Figure 10-3), underneath the Menu Bar and above the Code Window. It appears whenever a Watchpoint or Tracepoint is created.

In this window, the number and name of each Watchpoint and Tracepoint is displayed along with its data type, value, and storage class (e.g., LOCAL). Watchpoints and Tracepoints are distinguished using the abbreviations "wp" and "tp."

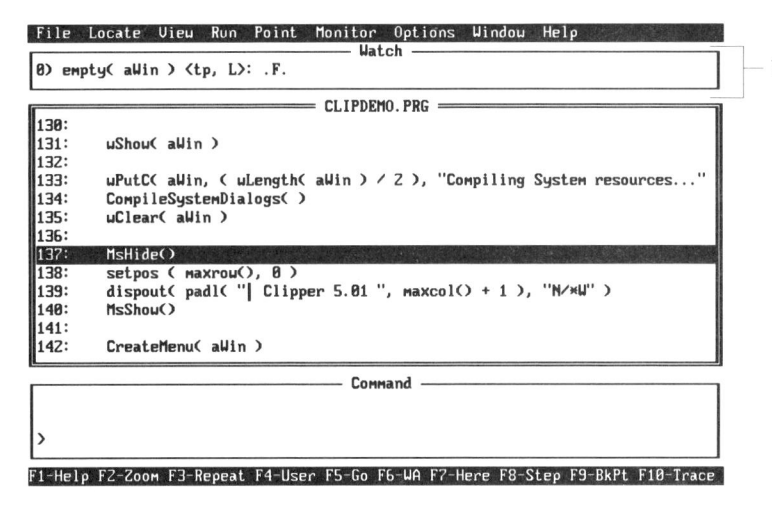

Figure 10-3: The Watch Window

Tracepoints and Watchpoints are created using the Point Tracepoint and Point Watchpoint commands, respectively. To remove a point definition, use the Point Delete command or press *Delete* when the point is highlighted in the Watch Window. To list Tracepoints and Watchpoints in the Command Window, use the List command.

When the Watch Window is active, you can inspect the value of almost any variable or expression that appears in the window. Code blocks cannot be inspected, and inspecting arrays and objects is slightly more complicated than other data types.

To inspect an array or object

1. Press *Tab* or *Shift-Tab* until you activate the Watch Window.

2. Press *Uparrow* or *Dnarrow* to highlight the array or object that you want to inspect.

The Debugger Display

3. Press *Return* to view the item in a dialog box.
   The dialog box shows an iconic representation of the array or object.

To inspect the array elements or object instance variables

1. Press *Return* again.
   Another dialog box opens in which the individual array or object components can be viewed and changed.
2. Press *Uparrow* or *Dnarrow* to highlight the element to change.
3. Press *Return* to enter edit mode or simply type a new value.
4. Press *Esc* twice to close both dialog boxes.
   All changes that you have made are saved and control returns to the Watch Window.

To inspect other data types

1. Press *Tab* or *Shift-Tab* until you activate the Watch Window.
2. Press *Uparrow* or *Dnarrow* to highlight the variable that you are interested in.
3. Press *Return* to view the value in a dialog box.
4. Press *Return* to enter edit mode or simply type a new value.
5. Press *Return* to accept the changes or *Esc* to abandon them.
   The dialog box closes and control returns to the Watch Window.

The following table summarizes the keys that are available when the Watch Window is active:

Watch Window Active Keys

Key	Action
Uparrow/Ctrl-E	Move highlight bar up one line
Dnarrow/Ctrl-X	Move highlight bar down one line
PgUp/Ctrl-R	Scroll window contents up
PgDn/Ctrl-C	Scroll window contents down
Ctrl-PgUp	Move highlight bar to first line
Ctrl-PgDn	Move highlight bar to last line
Return	Execute pending command or change selected item
Tab	Activate next window
Shift-Tab	Activate previous window
Del	Delete currently highlighted Watchpoint or Tracepoint
F2	Toggle full-screen/window display

The Monitor Window

The Monitor Window is similar to the Watch Window except that it is used to monitor variables of a particular storage class rather than variables that are set as Watchpoints and Tracepoints. The Monitor Window appears on the screen only when one or more of the storage classes in the Monitor menu is turned on (indicated by a check mark next to the storage class menu option).

Monitored variables relate to the program currently displayed in the Code Window. This is true regardless of whether the program is the current activation or a pending activation that is being viewed using the Callstack Window. The values displayed for monitored variables are the values they held when the routine in the Code Window was active. Furthermore, the only variables displayed in the Monitor Window are those that are visible to the routine in the Code Window.

Inspecting monitored variables with the ? | ?? command or specifying them as Watchpoints or Tracepoints yields the *current* value, which may be different from the value displayed in the Monitor Window.

To monitor a particular storage class (or stop monitoring one), use the associated Monitor command (e.g., Monitor Local for LOCAL variables). Each variable name that is being monitored is displayed along with its storage class, data type, and value.

By default, the variables are grouped by storage class. If Monitor Sort is on (indicated by a check mark beside the *Monitor:Sort* menu option), variables are listed in the Monitor Window in alphabetical order by variable name.

When the Monitor Window is active, you can change the value of almost any variable in the window by highlighting it and pressing *Return*. Editing variables in this window is identical to the method described above for the Watch Window and with the same exceptions—code blocks cannot be edited, and editing array values and objects is slightly more complicated than other data types. The keys for this window are summarized in the table below:

The Debugger Display

Monitor Window Active Keys

Key	Action
Uparrow/Ctrl-E	Move highlight bar up one line
Dnarrow/Ctrl-X	Move highlight bar down one line
PgUp/Ctrl-R	Scroll window contents up
PgDn/Ctrl-C	Scroll window contents down
Ctrl-PgUp	Move highlight bar to first line
Ctrl-PgDn	Move highlight bar to last line
Return	Execute pending command or change selected item
Tab	Activate next window
Shift-Tab	Activate previous window
F2	Toggle full-screen/window display

The Callstack Window

The Callstack Window (shown in Figure 10-4) appears on the righthand side of the screen and contains the names of all pending activations. This list is called the *Callstack*. The current activation is always at the top of the Callstack.

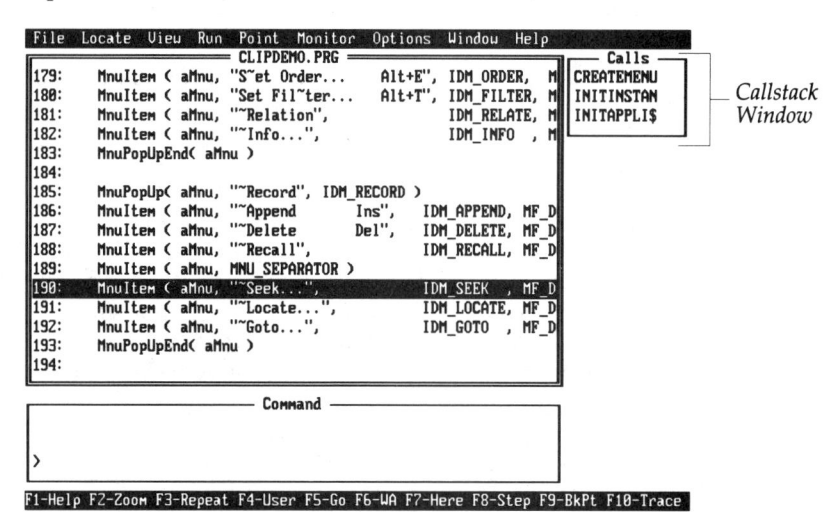

Figure 10-4: The Callstack Window

To open and select the Callstack Window

1. Press *Alt-V C* to execute the View Callstack command.

2. Press *Tab* or *Shift-Tab* until you highlight the window.

The following is a list of keys available when the Callstack Window is active:

Callstack Window Active Keys

Key	Action
Uparrow/Ctrl-E	Move highlight bar up one line
Dnarrow/Ctrl-X	Move highlight bar down one line
PgUp/Ctrl-R	Scroll window contents up
PgDn/Ctrl-C	Scroll window contents down
Ctrl-PgUp	Move highlight bar to first line
Ctrl-PgDn	Move highlight bar to last line
Return	Execute pending command
Tab	Activate next window
Shift-Tab	Activate previous window
F2	Toggle full-screen/window display

The Help Window

The Debugger offers online help in the form of a Help Window (shown in the figure below), which is divided into two panes: the left pane contains a list of topics for which help is available, and the right pane contains the help text for the currently highlighted topic. You can activate the Help Window using the Help menu, pressing *F1*, or entering the Help command.

Figure 10-5: The Help Window

The Debugger Display

There are several main topics of discussion in the Help Window, and when the window is first activated, one of these topics is highlighted on the left with its associated help text displayed on the right. The following table summarizes the keys used to navigate within the Help Window:

Help Window Active Keys

Key	Action
Uparrow/Ctrl-E	Move highlight bar up one line
Dnarrow/Ctrl-X	Move highlight bar down one line
PgUp/Ctrl-R	Scroll window contents up
PgDn/Ctrl-C	Scroll window contents down
Esc	Leave Help Window

The text that the Debugger displays in the Help Window is contained in a separate file, CLD.HLP, which must be present on your disk. When you ask for help, the Debugger will first search the current directory and then all directories specified in the DOS PATH. If CLD.HLP is not found in any of these locations, the Debugger will prompt you for a directory name. If CLD.HLP is still not found, you will not have access to help text within the Debugger.

The View Sets Window

When the View Sets Window is active, you can view and change the status of the *CA-Clipper* system settings.

To change a system setting using the View Sets Window

1. Press *Alt-V S* to execute the View Sets command.

2. Press *Uparrow* or *Dnarrow* to highlight the setting you want to change.

3. Press *Return* to enter edit mode or simply type a new value next to the setting.

4. Press *Return* to complete the editing process and move on to the next setting.

5. Press *Esc* to close the View Sets Window and continue debugging.

When you close the View Sets Window, the new settings are saved and take effect immediately in your program.

The View Workareas Window

The View Workareas Window allows you to view database (.dbf) and other work area information. The window is divided into three panes called Area, Status, and Structure.

To activate the View Workareas Window

- Press *F6* to execute the View Workareas command.

To move to the next window pane

- Press *Tab*.

To move to the previous window pane

- Press *Shift-Tab*.

To move the highlight up and down within the current window pane

- Press *Uparrow* and *Dnarrow*.

The Area pane displays the alias name for each open database file with the active file highlighted. Information regarding the currently highlighted file is shown in the other two window panes.

The Status pane shows the current database driver name and the status of most work area flag settings with information regarding the selected database file underneath. The database information is in the form of an outline that can be expanded and collapsed.

To expand or collapse an entry in the Status pane

1. Press *Uparrow* or *Dnarrow* until you highlight the entry.
2. Press *Return*.

Return acts as a toggle, expanding the entry if it is collapsed and collapsing it if it is expanded. For instance, highlight *Current Record* and press *Return*. The field values usually displayed underneath disappear. Press *Return* again, and the field values reappear. Expanding the *Workarea Information* heading displays additional work area settings that are not shown at the top of the window pane.

The Structure pane lists the structure of the selected database file.

To close the View Workareas Window and continue debugging

- Press *Esc*.

The Set Colors Window

When the Set Colors Window is active, you can view and change the status of the Debugger color settings.

The Debugger Display

To change a color setting using the Set Colors Window

1. Press *Alt-O C* to execute the Options Color command.

2. Press *Uparrow* or *Dnarrow* to highlight the color setting you want to change.

3. Press *Return* to enter edit mode or simply type a new value next to the setting.
   Each setting is a foreground/background color string enclosed in double quotes (see SETCOLOR() in the Reference guide for a list of colors).

4. Press *Return* to complete the editing process and move on to the next setting.

5. Press *Esc* to close the Set Colors Window and continue debugging.

When you close the Set Colors Window, the new colors take effect immediately but will be lost as soon as you exit the Debugger. To save the new colors in a script file for future use, execute the Options Save command.

Dialog Boxes

Several menu options and window selections require that you enter further information before continuing. Menu options that require more input are always indicated by an ellipsis (...) to the right of the option name. Some examples of window selections that require additional information are items in the Monitor and Watch Windows.

Dialog boxes are also used if you enter an incomplete command. For example, Locate Find with no search string opens a dialog box to ask for the search string.

In other words, anytime more information is required to proceed, a dialog box is displayed to prompt you.

The Debugger Display

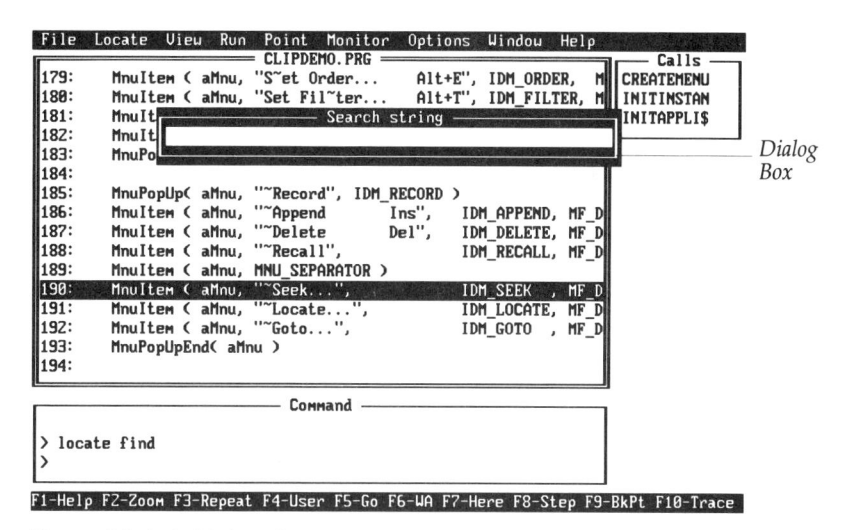

```
┌─File  Locate  View  Run  Point  Monitor  Options  Window  Help─────────┐
│╔════════════════ CLIPDEMO.PRG ════════════════╗┌── Calls ──┐
│║179:    MnuItem ( aMnu, "S~et Order...    Alt+E", IDM_ORDER, M║│CREATEMENU │
│║180:    MnuItem ( aMnu, "Set Fil~ter...   Alt+T", IDM_FILTER, M║│INITINSTAN │
│║181:    MnuIt┌──────────── Search string ────────────┐        ║│INITAPPLI$ │
│║182:    MnuIt│                                        │        ║│           │
│║183:    MnuPo│                                        │        ║│           │         Dialog
│║184:         └────────────────────────────────────────┘        ║│           │         Box
│║185:    MnuPopUp( aMnu, "~Record", IDM_RECORD )                ║│           │
│║186:    MnuItem ( aMnu, "~Append      Ins",  IDM_APPEND, MF_D  ║│           │
│║187:    MnuItem ( aMnu, "~Delete      Del",  IDM_DELETE, MF_D  ║│           │
│║188:    MnuItem ( aMnu, "~Recall",            IDM_RECALL, MF_D ║│           │
│║189:    MnuItem ( aMnu, MNU_SEPARATOR )                        ║│           │
│║190:    MnuItem ( aMnu, "~Seek...",           IDM_SEEK  , MF_D ║│           │
│║191:    MnuItem ( aMnu, "~Locate...",         IDM_LOCATE, MF_D ║│           │
│║192:    MnuItem ( aMnu, "~Goto...",           IDM_GOTO  , MF_D ║│           │
│║193:    MnuPopUpEnd( aMnu )                                    ║│           │
│║194:                                                           ║│           │
│╚═══════════════════════════════════════════════════════════════╝└───────────┘
│                        ┌──────── Command ────────┐
│                        │ > locate find            │
│                        │ >                        │
│                        └──────────────────────────┘
└F1-Help F2-Zoom F3-Repeat F4-User F5-Go F6-WA F7-Here F8-Step F9-BkPt F10-Trace─┘
```

Figure 10-6: A Dialog Box

The following table summarizes the keys that are available when a dialog box is open:

Dialog Box Active Keys

Key	Action
Leftarrow/Ctrl-S	Move cursor one character to the left
Rightarrow/Ctrl-D	Move cursor one character to the right
Home/Ctrl-A	Move cursor to the beginning of line
End/Ctrl-F	Move cursor to the end of line
Ins/Ctrl-V	Toggle the insert mode on or off
Del/Ctrl-G	Delete character under cursor
Backspace/Ctrl-H	Delete character to the left of cursor
Esc	Close dialog box without executing
Return	Execute and close dialog box

The Debugger Menus

The following section contains a description of the menus available in *The CA-Clipper Debugger* and how to access and execute options in those menus. Specific information on the operation of individual menu options, however, is not discussed here. Refer to the Menu Command Reference section at the end of this chapter for more detailed information on how menu options work.

The Debugger Display

Menu Operations

This section describes the process of accessing menus and selecting particular menu options. Also included are tables of menu system active keys.

Accessing a Menu

To access a menu, hold down the *Alt* key and press the first letter of the menu name.

For example, to access the View menu

- Press *Alt-V*.

Whenever a menu is displayed on the screen, any menu on the Menu Bar can be accessed.

To access the menu to the right of the current one

- Press *Rightarrow*.
  If you are positioned on the last menu, the cursor wraps around to the first menu on the Menu Bar.

To access the menu to the left of the current one

- Press *Leftarrow*.
  If you are positioned on the first menu, the cursor wraps around to the last menu on the Menu Bar.

To close the current menu

- Press *Esc*.

The following table summarizes the keys discussed in this section:

Menu Access Keys

Key	Action
Alt-<Menu first letter>	Activate designated menu
Leftarrow/Ctrl-S	Activate menu to the left; wrap if on first menu
Rightarrow/Ctrl-D	Activate menu to the right; wrap if on last menu
Esc	Close menu

Selecting a Menu Option

To select an option from an open menu

1. Press *Uparrow* or *Dnarrow* to highlight the option.
2. Press *Return* to select the option and close the menu.

When selecting a menu option, pressing *Uparrow* on the first option causes the highlight bar to wrap around to the last option; the reverse is true if you press *Dnarrow* on the last option. The following table lists the active keys within a menu:

10-28 CA-Clipper

The Debugger Display

Menu Option Keys

Key	Action
<Option letter>	Select designated option
Return	Select highlighted option
Uparrow/Ctrl-E	Move highlight bar up one option; wrap if on first option
Dnarrow/Ctrl-X	Move highlight bar down one option; wrap if on last option

Accelerator Keys

Once a menu is open, all options within the menu have an associated accelerator key that you can press to select the option. The accelerator key is a single letter which is usually the first letter of the option name.

Within a menu, the accelerator key for each option is highlighted within the option name. Typing the indicated key is equivalent to highlighting that option and pressing *Return*.

For example, to select *Point:Breakpoint* using its accelerator key

- Press *Alt-P B*.

In addition to accelerator keys, many options also have an associated hot key that allows you to select the option without first opening its associated menu. These keynames appear to the right of the option name in the menu.

For example, to select *File:Exit* using its hot key

- Press *Alt-X*.

All hot key and accelerator key equivalents are documented in the Menu Command Reference section at the end this chapter.

Menu Commands

Any menu option can be turned into a command that can be executed from the Command Window or from a script file. These commands, called menu commands, are formed using the menu name followed by the first word of the option name.

For example, to execute the *Monitor:Public* menu option type

```
Monitor Public ↵
```

Arguments can be specified following a menu command if the menu option requires further input. For example, selecting *Run:Speed* prompts you for the step delay using a dialog box.

To specify a delay speed of .5 seconds using a menu command

The Debugger Display

- `Run Speed 5 ↵`

Menu commands can usually be abbreviated down to one letter per word, but in some cases a second or third letter is required in the option keyword to distinguish it from another option that begins with the same letter(s).

For instance, instead of Run Speed 5 you can type

- `R Sp 5 ↵`

The abbreviation "Sp" distinguishes Run Speed from the Run Step command.

All of the Debugger commands are documented in detail in the Menu Command Reference section at the end of this chapter.

The Menu Bar

Menu selections appear on the Menu Bar at the top of the screen, as shown in Figure 10-7. Each menu contains a group of similar, commonly used options. For example, the Find, Next, Previous, Goto Line, and Case Sensitive options are found on the Locate menu.

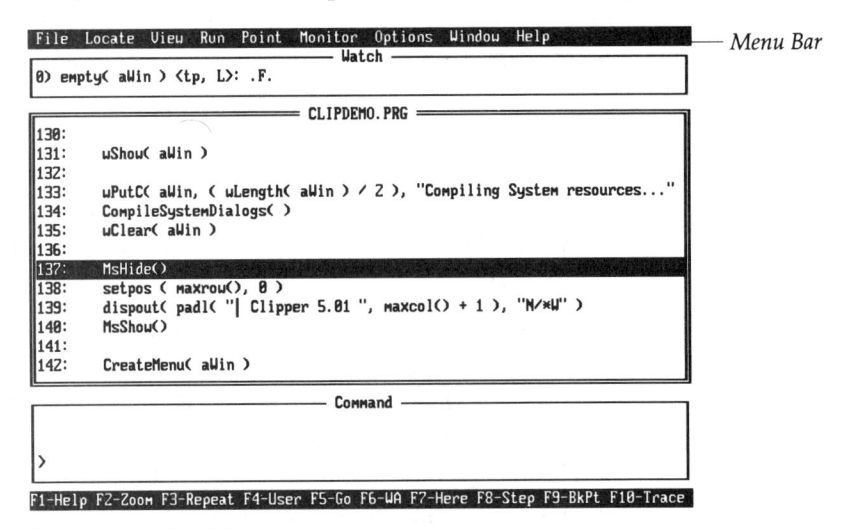

```
File Locate View Run Point Monitor Options Window Help        ── Menu Bar
─────────────────────── Watch ───────────────────────
0) empty( aWin ) <tp, L>: .F.

═══════════════════════ CLIPDEMO.PRG ═══════════════════════
130:
131:    wShow( aWin )
132:
133:    wPutC( aWin, ( wLength( aWin ) / 2 ), "Compiling System resources..."
134:    CompileSystemDialogs( )
135:    wClear( aWin )
136:
137:    MsHide()
138:    setpos ( maxrow(), 0 )
139:    dispout( padl( "| Clipper 5.01 ", maxcol() + 1 ), "N/*W" )
140:    MsShow()
141:
142:    CreateMenu( aWin )

─────────────────────── Command ───────────────────────

>

F1-Help F2-Zoom F3-Repeat F4-User F5-Go F6-WA F7-Here F8-Step F9-BkPt F10-Trace
```

Figure 10-7: The Menu Bar

Although each of the options has an associated command that performs the same function (in the above example: Locate Find, Locate Next, Locate Previous, Locate Goto, and Locate Case), the menu system provides a quick way for the beginner to learn how to use the Debugger.

Each of the Debugger menus is described briefly in the sections below. Refer to the Menu Command Reference section at the end of this chapter for detailed information on specific menu options.

The File Menu The File menu (*Alt-F*) contains options to view other files and allows you to access DOS without leaving the current program.

The Locate Menu The Locate menu (*Alt-L*) contains options that allow you to search for a character string in a program and move the cursor in the Code Window to a particular line number.

Note: If the Command Window is active when a Locate menu option is selected, you will not see the cursor move to its new location in the Code Window. You must select the Code Window in order to see the new cursor position in the file.

The View Menu The View menu (*Alt-V*) contains a set of options that allow you to view certain information that is not normally displayed as part of the Debugger screen.

The Run Menu The Run menu (*Alt-R*) allows you to run the current application using one of several Debugger execution modes. Additionally, it allows you to control how much of the application to run using the Code Window cursor and to restart the current application from the beginning.

The Point Menu The Point menu (*Alt-P*) contains options to set and delete Breakpoints, Watchpoints, and Tracepoints.

The Monitor Menu The Monitor menu (*Alt-M*) contains a set of options that control the display of PUBLIC, PRIVATE, STATIC, and LOCAL variables in the Monitor Window. Variables displayed using the options in this menu are known as *monitored* variables.

The Options Menu The Options menu (*Alt-O*) allows you to control the Debugger display options and to create and run Debugger script files.

The Window Menu The Window menu (*Alt-W*) allows you to perform certain Window operations including sizing and moving the active window. See the Window Operations section earlier in this chapter for more information on how negotiate the various Debugger windows.

The Help Menu The Help menu (*Alt-H*) allows you to activate the Help Window. The menu options select the help topic that is initially highlighted when the window is opened.

The Function Keys

Many of the function keys used in the Debugger are shortcuts to using a menu selection or command. For example, using *F9* to set a Breakpoint is slightly easier than selecting the Breakpoint option from the Point menu and much easier than entering the appropriate BP or Delete command. The following table shows a list of all the available functions keys and their actions:

Function Keys

Key	Action	Alternative
F1	Help	Help Menu, Help
F2	Full-screen/window toggle	Window Zoom
F3	Retype last command	None
F4	View application screen	View App
F5	Execute application	Run Go
Ctrl-F5	Execute until next activation	Run Next
F6	View Workareas Window	View Workareas
F7	Execute application to cursor	Run To
F8	Execute application in step mode	Run Step
F9	Set/Delete Breakpoint	Point Breakpoint
F10	Trace	Run Trace

Debugging a Program

So far, we have described some of the most commonly used features of *The CA-Clipper Debugger*. We have explained how to start the debugging process and have given a description of all the elements of the screen display. In order to make full use of this knowledge, this section provides a more detailed study of the most powerful features, together with suggestions on the most appropriate places to use them.

When you are told how to accomplish a particular task, this section often refers to the Debugger command without mentioning any of its shortcuts such as a function key equivalent. Keep in mind, however, that there are almost always menu and key equivalents that you can use to make Debugger command execution easier. The Debugger Display section in this chapter describes how commands are formed from menu options and discusses key equivalents. You can also use the Menu

Command Reference section at the end of this chapter to look up any command and quickly see its key equivalents.

Executing Program Code

The most basic function of a debugger is to execute an application and display the results. This section describes the various ways to run a program and explains how to control the speed and view the output.

Modes of Execution

The CA-Clipper Debugger provides several different ways to execute a program, called execution modes. The following table lists each of these modes together with the command and/or function key that is used to access the mode:

Program Execution Modes

Command/Function Key	Execution Mode
Run Animate	Animate
Run Go, *F5*	Run
Run Next, *Ctrl-F5*	Run Next
Run To, *F7*	Run to Cursor
Run Step, *F8*	Single Step
Run Trace, *F10*	Trace

Animate Mode To run a program in Animate Mode, use the Run Animate command (there is no function key equivalent). This execution mode allows you to execute large portions of an application, stopping when the value of a suspect variable or expression changes.

In Animate Mode, the Debugger executes the line of code currently highlighted by the Execution Bar, moves the Execution Bar to the next line of code, and continues executing sequentially in this manner until it reaches a Breakpoint or a Tracepoint.

The Options Exchange command controls whether or not the Debugger displays program output. By default Options Exchange is on, causing the Debugger to display the output of each line of code after it is executed.

The Run Speed command controls the speed at which Animate Mode executes.

Debugging a Program

Run Mode

Run Mode is the fastest mode of execution offered by the Debugger. In this mode, the Debugger runs your application in much the same manner as it would in Animate Mode. The difference between these two execution modes is that Run Mode does not return to the Debugger display after each line of code is executed.

If you have not set any Breakpoints or Tracepoints, when you execute an application in Run Mode the application runs just as if you had executed it without using the Debugger. The only difference is that when the application terminates, it returns control to the Debugger rather than the operating system.

If you have set Breakpoints or Tracepoints, the application runs until it reaches one and then returns control to the Debugger.

When a program is executing in Run Mode, you can voluntarily terminate the program at any time by pressing *Alt-D*, assuming this key has not been disabled with ALTD(0). Doing this returns control to the Debugger where the Execution Bar will be positioned on the next line of code to be executed.

There are three versions of Run Mode. The first is the one described above which you enter by pressing *F5* or using the Run Go command.

The second, called Run Next, is identical to the first except that the Debugger runs the application only until the start of the next activation. This mode is accessed by pressing *Ctrl-F5* or using the Run Next command and is equivalent to setting a Tracepoint of PROCLINE() == 0.

The third, called Run to Cursor, is identical to the first except that the Debugger runs the application only to the current cursor position in the Code Window rather than to the application's end. This mode is accessed by pressing *F7* or using the Run To command.

Single Step Mode

To run a program in Single Step Mode, press *F8* or use the Run Step command. This execution mode allows you to step slowly through each line of code, viewing the output and examining variables as you go. Ideally, you will use Single Step Mode in situations where the error you are looking for has been narrowed down to a few lines of code.

In Single Step Mode, the Debugger executes the line of program code highlighted by the Execution Bar, moves the Execution Bar to the next line of code to be executed, and stops. As functions are called by the current program, the Debugger displays the function code in the Code Window.

If a code block is evaluated in the course of executing a program in Single Step Mode, the Debugger moves the Execution Bar to the line of code where the block was created. This allows you to see the contents of the block (which are unavailable during normal inspection), and occurs regardless of whether the block was declared in the current routine. Press *F8* again to move past the code block definition to the next executable line.

You can control this behavior using the Options Codeblock command. By default, Options Codeblock is on and code block tracing is enabled as described above. Turn Options Codeblock off if you prefer not to view the code block definition.

Trace Mode To run a program in Trace Mode, press *F10* or use the Run Trace command. This execution mode is similar to Single Step Mode in that it allows you to execute one line of program code at a time. However, Trace Mode does not display the code for functions and procedures called by the current program nor does it trace code blocks back to their definition.

Finding Program Errors

As we have already mentioned, the Debugger provides several ways to execute a program. The method you use depends on how far you have progressed in your debugging effort. The following sections explain three common debugging stages and suggest an appropriate method for each stage.

Starting Out The easiest way to debug a program for the first time is to execute the application in Run Mode to show the program output in the context of the application. You will use the Run Go command (or simply press *F5*) to invoke the Run Mode.

Simple runtime errors can be identified using the Run Mode. When a runtime error is encountered, control is automatically returned to the Debugger where you can see the offending line of code. A runtime error normally causes the application to terminate prematurely because continued execution is prohibited by the error. Thus, you usually have to correct the error, recompile, and relink before you can continue debugging.

Cosmetic errors, such as mistakes in screen design, can also be identified in Run Mode, but it usually is not necessary to correct them before continuing since they do not affect the actual execution of the program. As soon as you notice a cosmetic error, press *Alt-D* to invoke the Debugger and make note of the code that needs to be changed.

Debugging a Program

After you have identified and noted all cosmetic errors, you will probably want to go ahead and edit your source code to make the necessary corrections using your text editor or word processor. Then, you must compile and link the application again so that the changes will take effect the next time you execute.

Note that all source code changes must take place outside of the context of the Debugger (i.e., you must leave the Debugger to make them) since *The CA-Clipper Debugger* has no facility for correcting errors on the fly.

For more complex problems such as mistakes in program logic or other runtime errors that do not have an obvious solution, continue to the next stage of debugging, taking a closer look.

Taking a Closer Look

When the cause of a problem is not immediately obvious, the next step is to examine the contents of suspected fields, memory variables, and expressions. Often, you can use the ? | ?? command to reveal an unexpected value for one of these items. If this is the case, specify the item as a Watchpoint or Tracepoint. This allows you to pinpoint exactly where in the application the erroneous value was stored.

If you suspect that several variables in a particular storage class may be at fault, use the options on the Monitor menu to display these variables in the Monitor Window. For example, to display PRIVATE variables, use the Monitor Private command. As you continue your debugging session, the PRIVATE variables visible to the routine displayed in the Code Window are updated in the Monitor Window.

After setting up the necessary Watchpoints and Tracepoints, use the Run Animate command to run the program slowly, displaying the code for each line executed (Animate Mode). Use the Run Speed command to control the speed of the display.

Continue this process until the problem has been confined to a few lines of code. Now it is time to move on to the final stage.

Final Stage

When the cause of an error has been narrowed down to a few lines of code, move the cursor in the Code Window to the first of these lines and use the *F9* key or the Point Breakpoint command to set a Breakpoint there. Then use the *F5* key to execute the application in Run Mode up to that Breakpoint.

When you have reached the suspect area of code, use the *F8* key to step through the code one line at a time (Single Step Mode). If you are certain that the error is in the current program and not in any subsidiary procedures or functions, use the *F10* key instead (Trace Mode). Remember that in Trace Mode the Debugger does not display the code for called routines.

Debugging a Program

Executing your application in Single Step or Trace Mode while watching the values of the items in the Watch Window is usually sufficient to uncover the cause of a problem. Having done that, simply alter your code, recompile, relink, and begin searching for the next error.

Viewing Program Output

There may be times when you want to look at the output of a program rather than the code. To do this, press *F4*, the function key equivalent for View App. When you do, the Debugger erases the screen and displays the current output exactly as the person running the application will see it. Press any key to return to the Debugger display.

Alternatively, you can view program output following execution of each individual line of code by running the application in Animate Mode (using the Run Animate command) with Options Exchange turned on.

Inspecting Data and Expressions

One of the most common uses for a debugger is examining (inspecting) and altering the values of variables and expressions. *The CA-Clipper Debugger* provides several ways to do this: the first of these is the ? | ?? command.

When you enter "?" followed by an expression, the expression is evaluated and the return value displayed in the Command Window. This is a useful feature that allows you to execute procedures and functions linked into your application and to make inline assignments to change existing variables or create new ones.

Another way to inspect data and expressions is to set up Watchpoints and Tracepoints. These are explained in the following sections.

Using Watchpoints and Tracepoints

Watchpoints and Tracepoints are two of the most valuable features of *The CA-Clipper Debugger*. They allow you to examine field variables, memory variables, and expressions and to pinpoint exactly where in your application their values change. This section gives a definition of Watchpoints and Tracepoints and explains how they are set, deleted, and inspected.

Definition of Watchpoints and Tracepoints

A Watchpoint is a field, memory variable, or expression whose current value is displayed in the Watch Window. The value or each Watchpoint is updated as your application executes.

Debugging a Program

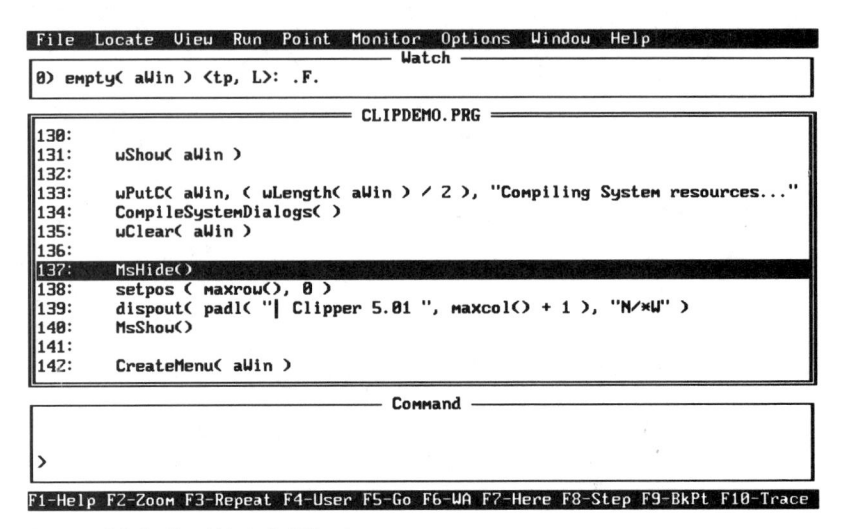

```
┌─File─Locate─View─Run─Point─Monitor─Options─Window─Help──────┐
│                              Watch                          │
│ 0) empty( aWin ) (tp, L): .F.                               │
│════════════════════════ CLIPDEMO.PRG ═══════════════════════│
│ ║130:                                                       ║│
│ ║131:    wShow( aWin )                                      ║│
│ ║132:                                                       ║│
│ ║133:    wPutC( aWin, ( wLength( aWin ) / 2 ), "Compiling System resources..." ║│
│ ║134:    CompileSystemDialogs( )                            ║│
│ ║135:    wClear( aWin )                                     ║│
│ ║136:                                                       ║│
│ ║137:    MsHide()                                           ║│
│ ║138:    setpos ( maxrow(), 0 )                             ║│
│ ║139:    dispout( padl( "| Clipper 5.01 ", maxcol() + 1 ), "N/xW" ) ║│
│ ║140:    MsShow()                                           ║│
│ ║141:                                                       ║│
│ ║142:    CreateMenu( aWin )                                 ║│
│────────────────────────── Command ──────────────────────────│
│                                                             │
│ >                                                           │
│                                                             │
│F1-Help F2-Zoom F3-Repeat F4-User F5-Go F6-WA F7-Here F8-Step F9-BkPt F10-Trace│
```

Figure 10-8: The Watch Window

A Tracepoint is similar to a Watchpoint; however, whenever its value changes, program execution pauses and control passes to the Debugger. In this respect, a Tracepoint is similar to a Breakpoint.

Setting Watchpoints and Tracepoints

To set Watchpoints and Tracepoints you can use the *Point:Watchpoint* and *Point:Tracepoint* menu options or their command equivalents. If you choose the menu options, the Debugger opens a dialog box where you are prompted to enter an expression or the name of a variable.

If you use the Point Watchpoint or Point Tracepoint commands, you can specify the name of the variable on the command line as in the following examples:

```
> Point Watchpoint nPageNum ↵

> Point Tracepoint nInvNum = 1000 ↵
```

In either case, the Debugger creates the Watchpoint or Tracepoint and displays it in the Watch Window, together with its type, value, and the abbreviation "wp" or "tp."

Debugging a Program

Deleting Watchpoints and Tracepoints

To delete an individual Watchpoint or Tracepoint, use the *Point:Delete* menu option. When the dialog box appears, enter the number that appears on the far lefthand side of the Watch Window.

Alternatively, you can use the Point Delete command and specify the Watchpoint or Tracepoint number on the command line or simply highlight the point in the Watch Window and press *Delete*. No matter which method you use, the specified entry is removed from the Watch Window and the numbers of all the remaining entries are updated.

To delete all Watchpoints and Tracepoints, use the Delete command as follows:

```
> Delete All WP ↵
```

Note that since Delete does not distinguish between Watchpoints and Tracepoints, you could have used Delete All TP just as effectively.

Warning! *Be careful not to use Delete All on its own unless you want to delete all Watchpoints, Tracepoints, and Breakpoints.*

Listing Watchpoints and Tracepoints

To display all Watchpoints and Tracepoints in the Command Window, use the List command as follows:

> List WP ↵

Note that since List does not distinguish between Watchpoints and Tracepoints, you could have used List TP just as effectively.

Inspecting Watchpoints and Tracepoints

To inspect a Watchpoint or Tracepoint and change its value, select the Watch Window. Then, move the highlight bar to the item you wish to inspect and press *Return*. The current value of the field, memory variable, or expression is displayed in a dialog box. Within the dialog box, you can either enter a new value and press *Return* or press *Esc* to leave the variable as it is.

Note that although you can inspect the value of almost any variable or expression that appears in the Watch Window, code blocks cannot be inspected, and inspecting arrays and objects is slightly more complicated than other data types.

When you are finished inspecting items in the Watch Window, you can select another window with *Tab* or *Shift-Tab*. For more information on the keys used to navigate the Watch Window and for specific instructions on inspecting arrays and objects, see the section entitled The Debugger Display in this chapter.

Debugging a Program

Creating New Variables

Sometimes, an error may be caused by something as simple as a misspelled variable name. In this case, you may want to create the variable from within the Debugger. Ultimately, you will have to address this type of error in your source code, but it is very convenient to make a temporary fix from within the Debugger so that you can continue the debugging process without having to recompile and relink right away.

Using the ? | ?? command, you can create a PRIVATE variable and assign a value to it. For example:

```
> ? myVar := 100 ↵
```

This creates a variable called *myVar* and assigns to it a value of 100. Remember that this variable is PRIVATE to the routine being executed when it was created.

Note: When attempting to create a variable in this manner, you must use the inline assignment operator (:=). If you use the = operator, the Debugger will perform a comparison and most likely return an error message.

Inspecting Program Code

As you debug an application, you may need to examine the code of programs other than the one currently being executed. This section describes how to view other programs and header files, how to use the Callstack to access programs that have already been executed, and how to use Breakpoints.

Using Breakpoints

Breakpoints are similar to (and just as powerful as) Watchpoints and Tracepoints. However, Breakpoints refer to actual lines of code rather than variables and expressions.

Breakpoints allow you to return control to the Debugger and execute problem areas of code in Single Step Mode (for an explanation of the modes of execution, see Executing Program Code in this chapter). This section explains what Breakpoints are and how to set, delete, and list them.

Debugging a Program

Definition of a Breakpoint A Breakpoint is a line of program code or a function or procedure call that, when encountered, causes the Debugger to halt execution of the program. In other words, a Breakpoint defines a physical breaking point in a program.

Setting Breakpoints To set a Breakpoint in the current program, move the cursor to the appropriate line in the Code Window and execute the Point Breakpoint command (or simply press *F9*). The selected line is highlighted, as shown in the figure below.

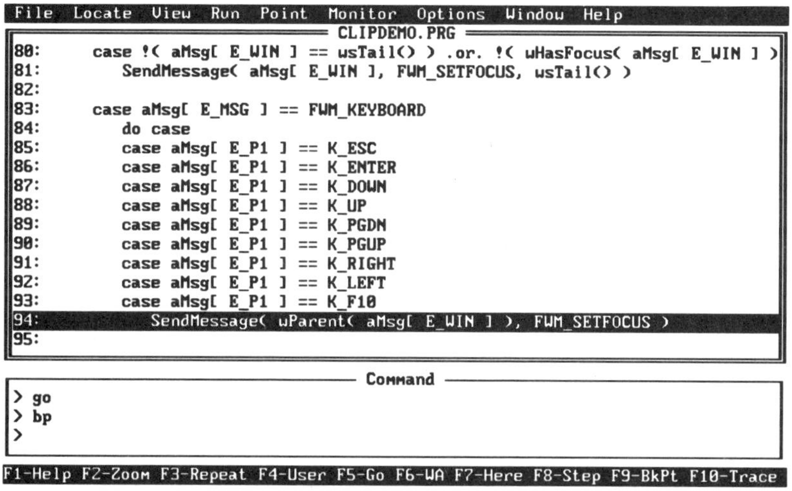

```
File  Locate  View  Run  Point  Monitor  Options  Window  Help
═══════════════════════════════ CLIPDEMO.PRG ═══════════════════════════════
80:       case !( aMsg[ E_WIN ] == usTail() ) .or. !( uHasFocus( aMsg[ E_WIN ] )
81:          SendMessage( aMsg[ E_WIN ], FWM_SETFOCUS, usTail() )
82:
83:       case aMsg[ E_MSG ] == FWM_KEYBOARD
84:          do case
85:             case aMsg[ E_P1 ] == K_ESC
86:             case aMsg[ E_P1 ] == K_ENTER
87:             case aMsg[ E_P1 ] == K_DOWN
88:             case aMsg[ E_P1 ] == K_UP
89:             case aMsg[ E_P1 ] == K_PGDN
90:             case aMsg[ E_P1 ] == K_PGUP
91:             case aMsg[ E_P1 ] == K_RIGHT
92:             case aMsg[ E_P1 ] == K_LEFT
93:             case aMsg[ E_P1 ] == K_F10
94:                SendMessage( uParent( aMsg[ E_WIN ] ), FWM_SETFOCUS )
95:
─────────────────────────────── Command ───────────────────────────────
> go
> bp
>
F1-Help F2-Zoom F3-Repeat F4-User F5-Go F6-WA F7-Here F8-Step F9-BkPt F10-Trace
```

Figure 10-9: Setting Breakpoints

To define a function or procedure as a Breakpoint, use the BP command instead of Point Breakpoint. These commands are similar in that they both define Breakpoints, but BP is more flexible and powerful. The following command defines the routine SayData as a Breakpoint. Whenever a call is made to this routine, the Debugger halts execution:

```
> BP SayData ↵
```

You can also set Breakpoints in programs other than the one currently displayed in the Code Window using the BP command. This example sets a Breakpoint at line 15 in the program OpenDbfs.prg:

```
> BP 15 In OpenDbfs.prg ↵
```

Deleting Breakpoints There are several ways to delete an individual Breakpoint. The first is to move the cursor to the appropriate line in the Code Window and execute the Point Breakpoint command (or simply press *F9* or issue BP command with no parameters). When the Breakpoint is deleted, the line returns to its normal color.

Debugging a Program

Another way to delete a Breakpoint is to use the Delete BP command followed by the Breakpoint number on the command line. Note that Breakpoint numbering begins with zero. If you do not know the number, use the List command (see Listing Breakpoints in this chapter).

To delete all Breakpoints, use the Delete command as follows:

```
> Delete All BP ↵
```

Warning! Delete All without any other arguments also deletes all Watchpoints and Tracepoints.

Listing Breakpoints

To display all Breakpoints, use the List BP command. This lists and numbers Breakpoints in the order they were entered, as shown in the figure below. Note that numbering begins with zero.

```
File  Locate  View  Run  Point  Monitor  Options  Window  Help
=============================== FILEMAN.PRG ==============================
107:      // Set the default values
108:      nMenuItem    := 1
109:      nTagged      := 0
110:      nFileItem    := 1
111:      nEl          := 1
112:      nRel         := 1
113:      lReloadDir   := .T.
114:      aFileMan     := {}
115:      aFileList    := {}
116:
117:      // Create the array
118:      aFileMan := ARRAY( FM_ELEMENTS )
119:
120:      // Resolve parameters
121:      IF nRowTop = NIL
=============================== Command ==================================
) List BP
    0) 108 FILEMAN.PRG
    1) 121 FILEMAN.PRG
)
F1-Help F2-Zoom F3-Repeat F4-User F5-Go F6-WA F7-Here F8-Step F9-BkPt F10-Trace
```

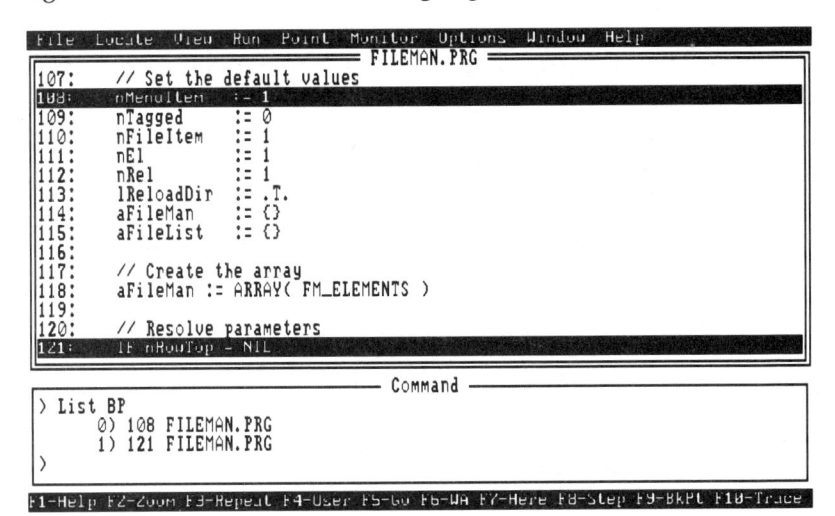

Figure 10-10: Listing Breakpoints

Navigating the Callstack

The Callstack is a list of pending activations, including procedures, functions, code blocks, and message sends. It is displayed in the Callstack Window on the righthand side of the screen, which is controlled using the View Callstack command.

When it is active, the Callstack Window allows you to highlight a program in the Callstack and view it in the Code Window. To select the window, press *Tab* until its border is highlighted.

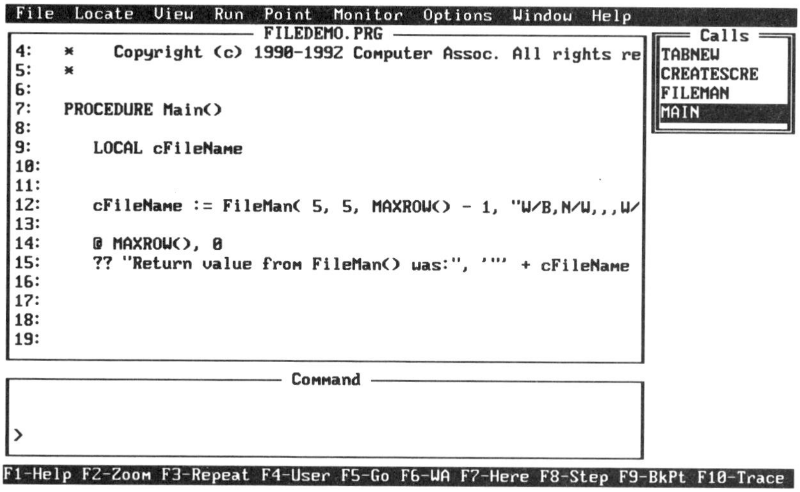

```
File  Locate  View  Run  Point  Monitor  Options  Window  Help
┌─────────────────────── FILEDEMO.PRG ───────────────────────┐┌─ Calls ──┐
│ 4:   *     Copyright (c) 1990-1992 Computer Assoc. All rights re││TABNEW    │
│ 5:   *                                                          ││CREATESCRE│
│ 6:                                                              ││FILEMAN   │
│ 7:   PROCEDURE Main()                                           ││MAIN      │
│ 8:                                                              │└──────────┘
│ 9:      LOCAL cFileName                                         │
│10:                                                              │
│11:                                                              │
│12:      cFileName := FileMan( 5, 5, MAXROW() - 1, "W/B,N/W,,,W/ │
│13:                                                              │
│14:      @ MAXROW(), 0                                           │
│15:      ?? "Return value from FileMan() was:", '"' + cFileName  │
│16:                                                              │
│17:                                                              │
│18:                                                              │
│19:                                                              │
└─────────────────────────────────────────────────────────────────┘
┌─────────────────────────── Command ────────────────────────────┐
│                                                                │
│ >                                                              │
└────────────────────────────────────────────────────────────────┘
F1-Help F2-Zoom F3-Repeat F4-User F5-Go F6-WA F7-Here F8-Step F9-BkPt F10-Trace
```

Figure 10-11: Inspecting the Callstack

When the Callstack Window is active, the highlight bar appears on the first item in the window. You can use *Uparrow* and *Dnarrow* to move to the activation whose code you wish to view. As the highlight is moved within the window, the Code Window is immediately updated to display the selected code.

If the activation you have chosen is a code block (denoted by a lowercase "b"), the name in the Callstack is the name of the routine in which the block was created. It is this code that appears in the Code Window.

To return to the program originally being debugged, use the File Resume command. For a list of the Callstack navigation keys, see the section entitled The Debugger Display in this chapter.

The Callstack is concerned only with executable code. If you want to view the contents of a header file you must use the procedure outlined in the next section.

Viewing Files

Although inspecting the Callstack provides an easy way to select and view program code, it does not display header files. If you already know the name of the program you want to view or you want to examine a header file, use the File Open command.

Debugging a Program

The following example displays the code for the header file Inkey.ch in the Code Window, as shown in the figure below.

```
> File Open C:\CLIPPER5\INCLUDE\Inkey.ch ↵
```

Notice that this example gives a file extension; if none is supplied, a (.prg) extension is assumed.

```
File  Locate  View  Run  Point  Monitor  Options  Window  Help
                      ═══ C:\CLIPPER5\INCLUDE\Inkey.ch ═══
1:  /***
2:  *        Inkey.ch
3:  *        Standard INKEY() key-code definitions
4:  *        Copyright (c) 1990-1992 Computer Assoc.  All rights reserved.
5:  */
6:
7:  // Cursor movement keys
8:  #define K_UP           5      // Uparrow, Ctrl-E
9:  #define K_DOWN         24     // Dnarrow, Ctrl-X
10: #define K_LEFT         19     // Leftarrow, Ctrl-S
11: #define K_RIGHT        4      // Rightarrow, Ctrl-D
12: #define K_HOME         1      // Home, Ctrl-A
13: #define K_END          6      // End, Ctrl-F
14: #define K_PGUP         18     // PgUp, Ctrl-R
15: #define K_PGDN         3      // PgDn, Ctrl-C
16: #define K_CTRL_LEFT    26     // Ctrl-Leftarrow, Ctrl-Z
17: #define K_CTRL_RIGHT   2      // Ctrl-Rightarrow, Ctrl-B

──────────────── Command ────────────────
> File Open C:\CLIPPER5\INCLUDE\Inkey.ch
>

F1-Help F2-Zoom F3-Repeat F4-User F5-Go F6-WA F7-Here F8-Step F9-BkPt F10-Trace
```

Figure 10-12: Viewing a Header File

You can also use the File Open command to view other program files in your application. This feature is convenient if you want to set a Breakpoint in another program but do not know the line number and cannot, therefore, set it using the BP command.

To return to the original program, use the File Resume command. The Debugger saves any Breakpoints that you have set and respects them the next time you run the application.

For a list of navigation keys used within the Code Window, see the section entitled The Debugger Display in this chapter.

Accessing DOS

There may be times when you wish to execute a DOS command without leaving the Debugger. For example, you may want to see a directory listing of all program (.prg) files. To do this, use the File DOS command. The DOS prompt will be displayed, and you can execute any DOS command you like.

Accessing DOS invokes a temporary copy of COMMAND.COM. Make sure that you have sufficient memory available to load this file and any others you may wish to execute. Remember that you are using a temporary copy of DOS, so any environment variables you set will be lost upon returning to the Debugger.

When you have finished and wish to continue executing your application, type "Exit" at the DOS prompt.

Menu Command Reference

This section is a complete reference to the syntax and operation of all *The CA-Clipper Debugger* commands. You can execute Debugger commands by selecting options from the Menu Bar or by entering commands directly into the Command Window. You have seen the former method described in The Debugger Display section and the latter method used in the examples in Debugging a Program.

The important thing to remember about *The CA-Clipper Debugger* is that any menu option can be expressed in the form of a command using the menu name followed by the first word of the option name. If required, arguments are placed at the end of the menu command line. Selecting a menu option and entering the equivalent menu command have the same effect.

All of the menu command equivalents and several other commands that have no menu equivalents are explained in this section. To execute a command, either type the command and press *Return* to execute, or make the equivalent menu selection if one is available. The only difference between these two methods of command execution is that commands you type appear in the Command Window and are, therefore, available in the command history buffer which is not so with menu selections.

? | ??

Display the value of an expression

Syntax

? | ?? <exp>

Arguments

<exp> is the expression to be displayed.

Description

? causes the value of the specified expression to be displayed in the Command Window. ?? displays the value in a popup window which is closed by pressing *Esc*. The expression used as the ? | ?? command line argument can be any valid *CA-Clipper* expression, including user-defined functions, arrays, and inline assignments.

When the display expression is a code block, the value cannot actually be displayed. For this data type, an icon is displayed instead.

If the expression is an array or an object, an icon is displayed if the value is displayed with either ? or ??. ??, however, will also display the array or object contents if you press *Return* in the first popup window. To close both popups, you must press *Esc* twice.

Both ? and ?? can be used to query individual array elements by subscripting the array name.

The ability to display the return value of a user-defined function allows you to extend the abilities of *The CA-Clipper Debugger*. For example, you might want to modify the structure of a database file while debugging an application. To do this, simply write a function to perform the modification, compile and link it with your application, and execute it using the ? | ?? command. Any feature that doesn't already exist can be written as a user-defined function and executed in this way.

Using the inline assignment operator (:=) with the ? | ?? command allows you to create new variables.

Examples

- This example displays the value of a numeric field in the Command Window:

```
> ? InvNum ↵
    465
```

- This shows the result of an expression in the Command Window:

  ```
  > ? X = Y ↵
    .T.
  ```

- Here, the return value of a *CA-Clipper* function is shown in the Command Window:

  ```
  > ? EOF() ↵
  *.F.
  ```

- In this example, the EVAL() function is used to execute a code block:

  ```
  > ? EVAL(bSortBlock) ↵
  ```

- This example creates a new PRIVATE variable called *nPageNo*, assigns it the value 2, and displays the variable in a popup window. To close the window, press *Esc*:

  ```
  > ?? nPageNo := 2 ↵
  ```

Animate

See: Run Animate

BP

Set or remove a Breakpoint

Syntax

```
BP [[At] [<lineNum>] [[In] <idProgramFile>]]]
BP <idFunction> | <idProcedure>
```

Arguments

<lineNum> is the line number where you want program execution to stop. *<lineNum>* acts as a toggle.

<idProgramFile> is the filename of the program in which to set the Breakpoint. If no extension is specified, (.prg) is assumed.

If neither *<lineNum>* not *<idProgramFile>* is specified, BP refers to the line indicated by the cursor in the Code Window.

<idFunction> | *<idProcedure>* is the name of the routine where you want program execution to stop. Function names are specified without parentheses. If no options are specified, BP toggles a breakpoint at the current line in the current source file.

Description

BP designates a line of program code or a call to a routine as a Breakpoint. When you execute this command, the line of code is indicated in a new color to distinguish it as a Breakpoint and execution will pause as soon as the line is encountered or the routine is called.

BP acts as a toggle so that if the specified line of code is already a Breakpoint, the Breakpoint is deleted and the color of the line returns to normal.

To delete a Breakpoint, use the Delete command. To see all Breakpoint settings in the Command Window, use the List BP command. For more information on the use of Breakpoints, see the section entitled Debugging a Program in this chapter.

Examples

- Entering:

    ```
    > BP 12 ↵
    ```

    highlights line 12 of the program displayed in the Code Window. Highlighted lines are those lines that have been designated as Breakpoints.

- This example highlights the current line and creates a breakpoint on that line. If the current line is already a BP, it will delete it.

  ```
  > BP
  ```

- This example inserts a Breakpoint at line 15 of File2.prg:

  ```
  > BP At 15 In File2 ↵
  ```

- Here, a user-defined function is used as a Breakpoint. Execution pauses as soon as the function is called:

  ```
  > BP TestUdf ↵
  ```

See also: Delete, List, Point Breakpoint, Run Go, Run Step

Callstack

Control the Callstack Window

Syntax

```
Callstack [on | Off]
```

Arguments

On opens the Callstack Window on the righthand side of the screen.

Off closes the Callstack Window.

If no options are specified, Callstack acts as a toggle by changing the current status of the Callstack Window (e.g., if it is On, Callstack turns it Off). Executed without an option in this manner, Callstack is functionally equivalent to the View Callstack command.

Description

Callstack toggles the display of the Callstack Window. This window contains the Callstack: a list of pending activations with the current activation at the top. For more information on the Callstack, see the section entitled Debugging a Program in this chapter.

Examples

- To open the Callstack Window, enter:

    ```
    > Callstack On ↵
    ```

- To close the window, type:

    ```
    > Callstack Off ↵
    ```

See also: View Callstack

Delete

Delete Debugger settings

Syntax

```
Delete All [BP | TP | WP]
Delete BP | TP | WP <number>
```

Arguments

All deletes all settings of a specified type or all settings if no type is specified.

BP specifies Breakpoints.

TP specifies Tracepoints and Watchpoints.

WP specifies Watchpoints and Tracepoints.

<*number*> is the number of the individual Watchpoint, Tracepoint, or Breakpoint to delete. If not specified, Delete opens a dialog box to prompt you for the item number.

Description

Delete removes Watchpoints, Tracepoints, and Breakpoints, either individually or as a whole.

To delete all items of a particular type, use the Delete All form of the command.

To perform a single deletion, specify the number and type of the item to be deleted. For Watchpoints and Tracepoints, this is the number that appears on the lefthand side of the Watch Window. For Breakpoints the List command must be used to determine the number.

Note: Tracepoints and Watchpoints are not distinguished by the Delete command. Thus, Delete TP and Delete WP are synonymous as are Delete All TP and Delete All WP.

For a detailed explanation of Breakpoints, Tracepoints, and Watchpoints, see the section entitled Debugging a Program in this chapter.

Delete

Examples

- To delete a Breakpoint, you must first find out where it falls in the list of Breakpoints:

    ```
    > List BP ↵
    ```

    If the Breakpoint you want to delete is number three on the list, type:

    ```
    > Delete BP 3 ↵
    ```

    and the Breakpoint is deleted.

- This example deletes all Watchpoints (and Tracepoints):

    ```
    > Delete All WP ↵
    ```

- Here, all Breakpoints, Tracepoints, and Watchpoints are deleted:

    ```
    > Delete All ↵
    ```

See also: BP, List, Point Breakpoint, Point Delete, Point Tracepoint, Point Watchpoint

DOS

See: File DOS

File DOS

Access DOS without leaving the current application

Syntax

```
File DOS
```

File DOS Command Equivalents

Type	Equivalent
Accelerator key	Alt-F D
Abbreviation	F D
Synonym	DOS

Description

File DOS loads a temporary copy of COMMAND.COM, allowing you to enter DOS commands without leaving the current application. To return to the Debugger, type "Exit" at the DOS prompt.

When using this command, you must make sure sufficient memory is available to load COMMAND.COM and any additional programs you want to execute. If the amount of memory is insufficient, an error message will be displayed.

File Exit

Exit the Debugger

Syntax

```
File Exit
```

File Exit Command Equivalents

Type	Equivalent
Hot key	*Alt-X*
Accelerator key	*Alt-F X*
Abbreviation	F E
Synonym	Quit

Description

File Exit terminates the Debugger, closes all files and returns to DOS.

File Open

Examine a file during the current debugging session

Syntax

```
File Open <idFileName>
```

File Open Command Equivalents

Type	Equivalent
Accelerator key	Alt-F O
Abbreviation	F O

Arguments

<idFileName> is the name of the file you want to view. If no extension is specified, (.prg) is assumed. If *<idFileName>* is not specified on the command line, File Open opens a dialog box to prompt you for a filename.

Description

File Open allows you to look at other files without leaving the current debugging session. It can also be used to view header files specified with the #include preprocessor directive. Breakpoints can be set and are saved when you return to the original program.

Unless you have defined a source file search path with the Options Path command, the filename that you specify is searched for in the current directory only. If, however, a search path has been defined, the directories in the path are searched in order until the file is found.

When you view a file with File Open, the current program is cleared from the Code Window and the new file is displayed in its place. To continue with the original program, use the File Resume command.

Note: File Open is almost identical to the View command. The only difference between these two commands is that File Open assumes a (.prg) extension whereas View assumes no file extension.

Examples

- To view ListDbfs.prg from within the current program (EnterData.prg), type:

  ```
  > File Open ListDbfs ↵
  ```

- It is possible to continue viewing other programs without returning to the original program. For example:

  ```
  > File Open PrintDbfs ↵
  ```

  loads PrintDbfs.prg in place of ListDbfs.prg. Typing File Resume returns the Debugger to EnterData.prg—the original program.

See also: File Resume, Options Path, View

File Resume

Return from viewing a file

Syntax

```
File Resume
```

File Resume Command Equivalents

Type	Equivalent
Accelerator key	Alt-F R
Abbreviation	F R
Synonym	Resume

Description

File Resume clears the file being viewed with the File Open command (or using the Callstack Window) and redisplays the file originally shown in the Code Window. Any Breakpoints which have been set are saved.

Examples

- Suppose the program currently being debugged is called PrintData.prg. The following line temporarily removes PrintData.prg from the screen and displays ReportHead.prg in its place:

  ```
  > File Open ReportHead ↵
  ```

- To return to PrintData.prg, type the following:

  ```
  > File Resume ↵
  ```

See also: File Open

Find

See: Locate Find

Go

See: Run Go

Goto

See: Locate Goto

Help

Activate the Help Window

Syntax

```
Help [Keys | Windows | Menus | Commands]
```

Help Command Equivalents

Type	Equivalent
Hot key	F1
Accelerator keys	Alt-H K, Alt-H W, Alt-H M, Alt-H C
Abbreviations	H K, H W, H M, H C

Arguments

Keys highlights the Keys topic when the Help Window is activated.

Windows highlights the Window topic when the Help Window is activated.

Menus highlights the Menus topic when the Help Window is activated.

Commands highlights the Commands topic when the Help Window is activated.

If issued without an argument, Help highlights the About Help topic when the Help Window is activated.

Description

The Debugger Help Window is divided into two panes: the left pane contains a list of topics for which help is available, and the right pane contains the help text for the currently highlighted topic. When the Help Window is activated with the Help command, one of the topics in the left pane is highlighted, and a general discussion of the topic is displayed in the right pane of Help Window.

Moving the highlight bar using *Uparrow* or *Dnarrow* selects a new help topic. *PgUp* and *PgDn* page through the associated help text on the right. *Esc* closes the Help Window and returns you to the main Debugger screen.

The text that the Debugger displays in the Help Window is contained in a separate file, CLD.HLP, which must be present on your disk. When you ask for help, the Debugger will first search the current directory and then all directories specified in the DOS PATH. If CLD.HLP is not found in any of these locations, the Debugger will prompt you for a directory name. If CLD.HLP is still not found, you will not have access to help text within the Debugger.

For more information on the Help Window, see the section entitled The Debugger Display in this chapter.

Input

See: Options Restore

List

List Watchpoints, Tracepoints, and Breakpoints

Syntax

```
List BP | TP | WP
```

Arguments

BP specifies Breakpoints.

TP specifies Tracepoints and Watchpoints.

WP specifies Watchpoints and Tracepoints.

Description

List displays all Watchpoints, Tracepoints, or Breakpoints in the Command Window. This command is useful when you want to Delete settings by number.

Note: Tracepoints and Watchpoints are not distinguished by the List command. Thus, List TP and List WP are synonymous. The Watch Window can also be used to determine the number of these items.

For more information on the setting and deleting of these items, see the section entitled Debugging a Program in this chapter.

Examples

- If there are Breakpoints at lines 10 and 15 of AddData.prg and lines 2 and 6 of ViewData.prg, List BP displays the following:

    ```
    0)  10 ADDDATA.PRG
    1)  15 ADDDATA.PRG
    2)  2 VIEWDATA.PRG
    3)  6 VIEWDATA.PRG
    ```

- To list all Watchpoints (and Tracepoints), type:

    ```
    > List WP ↵
    ```

See also: BP, Delete, Point Breakpoint, Point Delete, Point Tracepoint, Point Watchpoint

Locate Case

Toggle search case-sensitivity setting on and off

Syntax

```
Locate Case
```

Locate Case Command Equivalents

Type	Equivalent
Accelerator key	*Alt-L C*
Abbreviation	L C

Description

Locate Case changes the current case-sensitivity setting. This setting is used by all the Locate search commands (i.e., Locate Find, Locate Next, and Locate Previous) to determine whether to ignore or respect the case of the letters in the search string when testing for a match.

If case-sensitivity is on, a string must match the contents and case of the specified search string. If off (the default), only the contents need be the same for a successful match.

The *Locate:Case sensitive* menu option can be used to check the status of the case-sensitivity setting. An on setting is indicated by a check mark to the left of the menu option. An off setting is indicated by the absence of the check mark in the menu.

Examples

- Suppose you want to locate all occurrences of the USE command in your program, but are not always consistent in using uppercase letters for commands. If case-sensitivity is currently on, Locate Find USE will not find all occurrences of the command but only those in which all uppercase letters are used. To turn the case-sensitivity setting off and locate the USE command in all its possible forms:

  ```
  > Locate Case ↵
  > Locate Find USE ↵
  ```

See also:

Locate Find, Locate Next, Locate Previous

Locate Find

Locate a character string

Syntax

```
Locate Find <searchString>
```

Locate Find Command Equivalents

Type	Equivalent
Accelerator key	Alt-L F
Abbreviation	L F
Synonym	Find

Arguments

<searchString> is the character string you want to locate. If <searchString> is not specified on the command line, a dialog box opens on the screen to prompt you for a search string.

Description

Locate Find searches the file displayed in the Code Window for a specified search string.

The search always begins at the first line of code and moves down through the file, regardless of the current cursor position in the Code Window. If the search string is found, the Debugger moves the cursor to the line containing the first occurrence of the string in the Code Window; otherwise, the cursor remains at its current location.

Note: If the Command Window is active when Locate Find is executed, you will not see the cursor move to its new location in the Code Window. You must select the Code Window in order to see the new cursor position in the file.

This command obeys the *Locate:Case sensitive* menu setting. If this menu setting is on (indicated by a check mark), Locate Find looks for a character string that matches the contents and case of the specified search string. If off, only the contents need be the same for a successful match. The Locate Case command is used to toggle this menu setting on and off.

Locate Find

Examples

- If you suspect that the error you are trying to eliminate relates to a database (.dbf) file that is not open, search for the keyword "USE":

  ```
  > Locate Find USE ↵
  ```

See also: Locate Case, Locate Next, Locate Previous

Locate Goto

Move the cursor to a specified line in the Code Window

Syntax

```
Locate Goto <lineNum>
```

Locate Goto Command Equivalents

Type	Equivalent
Accelerator key	Alt-L G
Abbreviation	L G
Synonym	Goto

Arguments

<lineNum> is the number of the line where you want to move the cursor. If *<lineNum>* is not specified on the command line, Locate Goto opens a dialog box to prompt you for a line number.

Description

Locate Goto moves the cursor from its current position in the Code Window to the specified line number.

This command works regardless of whether or not line numbers are currently displayed in the Code Window. However, for the sake of readability we recommend that line numbers be displayed. For more information refer to the Options Line command in this section.

Note: If the Command Window is active when Locate Goto is executed, you will not see the cursor move to its new location in the Code Window. You must select the Code Window in order to see the new cursor position in the file.

Examples

- This example moves the cursor to line 30 of the program currently being debugged:

    ```
    > Locate Goto 30 ↵
    ```

See also: Num, Options Line

Locate Next

Locate the next occurrence of a character string

Syntax

```
Locate Next
```

Locate Next Command Equivalents

Type	Equivalent
Accelerator key	Alt-L N
Abbreviation	L N
Synonym	Next

Description

Locate Next locates the next occurrence of the most recently defined search string. A search string is defined using the Locate Find command or the *Locate:Find* menu option. If no search string has been defined, Locate Next opens a dialog box to prompt you for one.

The search begins at the current cursor position in the Code Window and moves down through the file. If a match is found, the Debugger moves the cursor to the line containing the next occurrence of the string; otherwise, the cursor remains at its current location.

Note: If the Command Window is active when Locate Next is executed, you will not see the cursor move to its new location in the Code Window. You must select the Code Window in order to see the new cursor position in the file.

Like Locate Find, Locate Next obeys the current case-sensitivity setting. This setting can be determined and changed using the *Locate:Case sensitive* menu option.

See also: Locate Case, Locate Find, Locate Previous

Locate Previous

Locate the previous occurrence of a character string

Syntax

```
Locate Previous
```

Locate Previous Command Equivalents

Type	Equivalent
Accelerator key	Alt-L P
Abbreviation	L P
Synonym	Prev

Description

Locate Previous searches for the previous occurrence of the most recently defined search string. A search string is defined using the Locate Find command or the *Locate:Find* menu option. If no search string has been defined, Locate Previous opens a dialog box to prompt you for one.

The search begins at the current cursor position in the Code Window and moves up through the file. If a match is found, the Debugger moves the cursor to the line containing the previous occurrence of the string; otherwise, the cursor remains at its current location.

Note: If the Command Window is active when Locate Previous is executed, you will not see the cursor move to its new location in the Code Window. You must select the Code Window in order to see the new cursor position in the file.

Like Locate Find, Locate Previous obeys the current case-sensitivity setting. This setting can be determined and changed using the *Locate:Case sensitive* menu option.

See also:
Locate Case, Locate Find, Locate Next

Monitor All

Toggle the display of variables in all storage classes in the Monitor Window

Syntax

```
Monitor All
```

Monitor Local Command Equivalents

Type	Equivalent
Accelerator key	*Alt-M A*
Abbreviation	M A

Description

Monitor All acts as a toggle by changing the current display status of all storage classes (i.e., local, private, public, and static) in the Monitor Window. The default setting for Monitor All is off.

Monitored variables relate to the routine currently displayed in the Code Window. When a pending activation is viewed, the values displayed for the monitored variables are the values that they held when the pending routine was active.

Inspecting the variables with the ? | ?? command or specifying them as Watchpoints or Tracepoints yields the *current* value, which may be different from the value displayed in the Monitor Window. Any existing variables that are not visible to the activation in the Code Window do not appear in the Monitor Window.

Examples

- To view all variable in all storage classes in the Monitor Window, enter:

```
> Monitor All ↵
```

See also:

Monitor Local, Monitor Private, Monitor Public, Monitor Sort, Monitor Static

Monitor Local

Toggle the display of Local variables in the Monitor Window

Syntax

```
Monitor Local
```

Monitor Local Command Equivalents

Type	Equivalent
Accelerator key	*Alt-M L*
Abbreviation	M L

Description

Monitor Local acts as a toggle by changing the current display status of Local variables in the Monitor Window (e.g., if Local variables are not being monitored, Monitor Local begins monitoring them). The default setting for Monitor Local is off, meaning that Local variables are not monitored.

The Monitor Window indicates the display status of Local variables in its window title. The word "Local" in the window title indicates that Local variables are being monitored.

Monitored variables relate to the routine currently displayed in the Code Window. When a pending activation is viewed, the values displayed for the monitored variables are the values that they held when the pending routine was active.

Inspecting the variables with the ? | ?? command or specifying them as Watchpoints or Tracepoints yields the *current* value, which may be different from the value displayed in the Monitor Window. Any existing variables that are not visible to the activation in the Code Window do not appear in the Monitor Window.

Examples

- To change the display status of Local variables in the Monitor Window, enter:

  ```
  > Monitor Local ↵
  ```

See also:

Monitor All, Monitor Private, Monitor Public, Monitor Sort, Monitor Static

Monitor Private

Toggle the display of Private variables in the Monitor Window

Syntax

```
Monitor Private
```

Monitor Private Command Equivalents

Type	Equivalent
Accelerator key	*Alt-M V*
Abbreviation	M Pr

Description

Monitor Private acts as a toggle by changing the current display status of Private variables in the Monitor Window (e.g., if Private variables are not being monitored, Monitor Private begins monitoring them). The default setting for Monitor Private is off, meaning that Private variables are not monitored.

The Monitor Window indicates the display status of Private variables in its window title. The word "Private" in the window title indicates that Private variables are being monitored.

Monitored variables relate to the routine currently displayed in the Code Window. When a pending activation is viewed, the values displayed for the monitored variables are the values that they held when the pending routine was active.

Inspecting the variables with the ? | ?? command or specifying them as Watchpoints or Tracepoints yields the *current* value, which may be different from the value displayed in the Monitor Window. Any existing variables that are not visible to the activation in the Code Window do not appear in the Monitor Window.

Examples

- To change the display status of Private variables in the Monitor Window, enter:

    ```
    > Monitor Private ↵
    ```

See also:

Monitor All, Monitor Local, Monitor Public, Monitor Sort, Monitor Static

Monitor Public

Toggle the display of Public variables in the Monitor Window

Syntax

```
Monitor Public
```

Monitor Public Command Equivalents

Type	Equivalent
Accelerator key	Alt-M P
Abbreviation	M P

Description

Monitor Public acts as a toggle by changing the current display status of Public variables in the Monitor Window (e.g., if Public variables are not being monitored, Monitor Public begins monitoring them). The default setting for Monitor Public is off, meaning that Public variables are not monitored.

The Monitor Window indicates the display status of Public variables in its window title. The word "Public" in the window title indicates that Public variables are being monitored.

Monitored variables relate to the routine currently displayed in the Code Window. When a pending activation is viewed, the values displayed for the monitored variables are the values that they held when the pending routine was active.

Inspecting the variables with the ? | ?? command or specifying them as Watchpoints or Tracepoints yields the *current* value, which may be different from the value displayed in the Monitor Window. Any existing variables that are not visible to the activation in the Code Window do not appear in the Monitor Window.

Examples

- To change the display status of Public variables in the Monitor Window, enter:

  ```
  > Monitor Public ↵
  ```

See also: Monitor All, Monitor Local, Monitor Private, Monitor Sort, Monitor Static

Monitor Sort

Control the order in which variables are displayed in the Monitor Window

Syntax

```
Monitor Sort
```

Monitor Sort Command Equivalents

Type	Equivalent
Accelerator key	Alt-M O
Abbreviation	M So

Description

Monitor Sort controls the order in which items are displayed in the Monitor Window.

When Monitor Sort is on (indicated by a check mark to the left of the *Monitor:Sort* menu option), items in the Monitor Window are displayed in alphabetical order according to variable name.

When Monitor Sort is off (the default), the monitored variables are grouped by storage class.

Examples

- To sort monitored variables by name if they are currently grouped by storage class, enter:

  ```
  > Monitor Sort ↵
  ```

See also: Monitor All, Monitor Local, Monitor Private, Monitor Public, Monitor Static

Monitor Static

Toggle the display of Static variables in the Monitor Window

Syntax

```
Monitor Static
```

Monitor Static Command Equivalents

Type	Equivalent
Accelerator key	Alt-M S
Abbreviation	M S

Description

Monitor Static acts as a toggle by changing the current display status of Static variables in the Monitor Window (e.g., if Static variables are not being monitored, Monitor Static begins monitoring them). The default setting for Monitor Static is off, meaning that Static variables are not monitored.

The Monitor Window indicates the display status of Static variables in its window title. The word "Static" in the window title indicates that Static variables are being monitored.

Monitored variables relate to the routine currently displayed in the Code Window. When a pending activation is viewed, the values displayed for the monitored variables are the values that they held when the pending routine was active.

Inspecting the variables with the ? | ?? command or specifying them as Watchpoints or Tracepoints yields the *current* value, which may be different from the value displayed in the Monitor Window. Any existing variables that are not visible to the activation in the Code Window do not appear in the Monitor Window.

Examples

- To change the display status of Static variables in the Monitor Window, enter:

    ```
    > Monitor Static ↵
    ```

See also:

Monitor All, Monitor Local, Monitor Private, Monitor Public, Monitor Sort

Next

See: Locate Next

Num

Toggle the display of line numbers

Syntax

    Num [On | off]

Arguments

On displays line numbers at the beginning of each line of code.

Off removes the line numbers.

If no options are specified, Num acts as a toggle by changing the line number status. Executed without an option in this manner, Num is functionally equivalent to the Options Line command.

Description

Num toggles the display of line numbers at the beginning of each line of code in the Code Window. When you first start the Debugger, Num is On and line numbers are displayed. This is particularly useful when using the Locate Goto command to move the cursor to a certain line in the Code Window.

See also: Locate Goto, Options Line

Options Codeblock

Control the tracing of code blocks during Single Step Mode

Syntax

```
Options Codeblock
```

Options Codeblock Command Equivalents

Type	Equivalent
Accelerator key	Alt-O B
Abbreviation	O C

Description

Options Codeblock acts as a toggle to control whether or not the Debugger traces code blocks in Single Step Mode.

By default, Options Codeblock is on (indicated by a check mark next to the *Options:Codeblock Trace* menu option), causing Single Step Mode to trace a code block back to its definition each time the code block is evaluated. It does this by moving the Execution Bar to the line where the code block was defined, allowing you to see the code block definition.

Tracing code blocks involves an extra step each time a code block is evaluated because you also have to step over the line of code defining the code block. If you do not want to trace code block definitions, turn Options Codeblock off.

For more information on modes of execution, see the section entitled Debugging a Program in this chapter.

Examples

- To change the current code block trace status, enter:

```
> Options Codeblock ↵
```

See also: Run Step

Options Color

Open the Set Colors Window

Syntax

```
Options Color
```

Options Color Command Equivalents

Type	Equivalent
Accelerator key	*Alt-O C*
Abbreviation	O Col

Description

Options Color activates the Set Colors Window where you can inspect the Debugger display colors. Press *Esc* to close the window and continue debugging. See The Debugger Display section for more information on the Set Colors Window.

Note that although Options Color expects additional input, the color settings cannot be specified on the command line as with other menu commands.

Note: When the display mode is monochrome (see Options Mono), the Options Color command is not operational.

Examples

- To inspect the Debugger color settings, enter:

    ```
    > Options Color ↵
    ```

See also: Options Mono, View Sets, View Workareas

Options Exchange

Control the display of program output while in Animate Mode

Syntax

```
Options Exchange
```

Options Exchange Command Equivalents

Type	Equivalent
Accelerator key	Alt-O E
Abbreviation	O E

Description

Options Exchange acts as a toggle to control the display of program output while in Animate Mode.

By default, this option is on (indicated by a check mark next to the *Options:Exchange Screens* menu option), causing Animate Mode to display the application output for each line of code executed.

If Options Exchange is off, the Debugger displays the application screen only when input is required. For more information on modes of execution, see the section entitled Debugging a Program in this chapter.

Examples

- To change the current program output display status, enter:

  ```
  > Options Exchange ↵
  ```

See also: Options Swap, Run Animate, View App

Options Line

Toggle the display of line numbers in the Code Window

Syntax

```
Options Line
```

Options Line Command Equivalents

Type	Equivalent
Accelerator key	Alt-O L
Abbreviation	O L

Description

Options Line toggles the display of line numbers at the beginning of each line of code in the Code Window (e.g., if line numbers are being displayed, Options Line turns their display off). The default setting for Options Line is on, meaning that line numbers are displayed.

Having line numbers in the code listing is particularly useful when using the Locate Goto command to move the cursor to a certain line in the Code Window.

Options Line is functionally equivalent to the Num command specified without an option.

Examples

- To change the display status of line numbers in the Code Window, enter:

    ```
    > Options Line ↵
    ```

See also: Locate Goto, Num

Options Menu

Toggle the Debugger Menu Bar display

Syntax

```
Options Menu
```

Options Menu Command Equivalents

Type	Equivalent
Accelerator key	Alt-O M
Abbreviation	O M

Description

Options Menu toggles the display of the Debugger Menu Bar (e.g., if the Menu Bar is currently visible, Options Menu hides it from view). The default setting for Options Menu is on, meaning that the Menu Bar is visible.

Note: Menu options can be executed using the appropriate accelerator key even when the Menu Bar is hidden from view. The Menu Bar will be displayed while the menu selection is being made but will disappear after the option is executed.

Examples

- To change the display status of the Menu Bar, enter:

  ```
  > Options Menu ↵
  ```

Options Mono

Toggle the Debugger display mode between color and monochrome

Syntax

```
Options Mono
```

Options Mono Command Equivalents

Type	Equivalent
Accelerator key	Alt-O D
Abbreviation	O Mo

Description

Options Mono toggles the Debugger display mode between color and monochrome (e.g., if the current display mode is color, Options Mono changes it to monochrome). This command is effective for color monitors only.

The default setting for Options Mono depends on the kind of monitor you are using (i.e., for color monitors the default is off, whereas for monochrome monitors it is on).

Note: When the display mode is set to monochrome, the Options Color command is not available.

Examples

- To change the Debugger display mode, enter:

  ```
  > Options Mono ↵
  ```

See also: Options Color

Options Path

Define the search path for source files

Syntax

```
Options Path <idPathList>
```

Options Path Command Equivalents

Type	Equivalent
Accelerator key	Alt-O F
Abbreviation	O Pa

Arguments

<idPathList> is a list of one or more directory names separated by semicolons. This option specifies the path used by the Debugger to search for source files if they cannot be found in the default directory. If *<idPathList>* is not specified on the command line, Options Path opens a dialog box to prompt you for a search path.

Description

Options Path allows you to specify one or more alternative directories to be searched if a particular source file cannot be found in the default directory. The search path that you specify pertains to source files only (i.e., the View and File Open commands). Other file searches are not affected.

Examples

- To specify the DBU and RL source code directories, enter:

```
> Options Path C:\CLIPPER5\SOURCE\DBU;C:\CLIPPER5\SOURCE\RL ↵
```

See also: File Open, View

Options Preprocessed

Toggle the display of preprocessed code in the Code Window

Syntax

```
Options Preprocessed
```

Options Preprocessed Command Equivalents

Type	Equivalent
Accelerator key	Alt-O P
Abbreviation	O P

Description

Options Preprocessed acts as a toggle by changing the current display status of preprocessed code in the Code Window (e.g., if the preprocessed code is not being displayed, Options Preprocessed displays it). By default, Options Preprocessed is off which means preprocessed code is not displayed.

When Options Preprocessed is on, preprocessed code for the current program appears underneath each line of source code. Since the preprocessed code is taken from the corresponding (.ppo) file, the program in the Code Window must have been compiled with the /P option.

The Code Window indicates the display status of preprocessed code in its window title as well as in the window itself. The presence of the corresponding (.ppo) filename in the window title indicates that preprocessed code is being displayed.

Examples

- To change the display status of preprocessed code in the Code Window, enter:

    ```
    > Options Preprocessed ↵
    ```

Options Restore

Read commands from a script file

Syntax

```
Options Restore <idScriptFile>
```

Options Restore Command Equivalents

Type	Equivalent
Accelerator key	Alt-O R
Abbreviation	O R
Synonym	Input

Arguments

<idScriptFile> is the name of the script file from which to read commands. If no extension is specified, (.cld) is assumed. If *<idScriptFile>* is not specified on the command line, Options Restore opens a dialog box to prompt you for a filename.

Description

Options Restore causes the Debugger to read and execute all commands in the specified script file and then resume accepting input from the keyboard.

Script files can be created automatically using the Options Save command and can also be written using a text editor or word processor. For a full explanation of script files and their uses, see the section entitled Starting the Debugger in this chapter.

Examples

- To execute the commands in the script file ViewBug.cld, enter the following:

    ```
    > Options Restore ViewBug ↵
    ```

See also: Options Save

Options Save

Save the current Debugger settings to a script file

Syntax

```
Options Save <idScriptFile>
```

Options Save Command Equivalents

Type	Equivalent
Accelerator key	*Alt-O S*
Abbreviation	O Sa

Arguments

<idScriptFile> is the name of the script file to which you want to save the current Debugger settings. If no extension is specified, (.cld) is assumed. If *<idScriptFile>* is not specified on the command line, Options Save opens a dialog box to prompt you for a filename.

Description

Options Save writes the current Debugger settings to a script file. The settings are written to the script file using standard menu commands.

The script file can be executed using the Options Restore command or from the CLD command line the next time you execute the Debugger. For more information on script files, see the section entitled Starting the Debugger in this chapter.

Examples

- To save the current Debugger settings to a file named ViewBug.cld, enter:

```
> Options Save ViewBug ↵
```

See also:

Options Restore

Options Swap

Control the display of the application screen when input is required

Syntax

```
Options Swap
```

Options Swap Command Equivalents	
Type	**Equivalent**
Accelerator key	*Alt-O I*
Abbreviation	O S

Description

If Options Exchange is on, Options Swap has no effect. If, however, Options Exchange is off, Options Swap acts as a toggle to control whether or not the application screen is displayed when input is required.

By default, Options Swap is on (indicated by a check mark next to the *Options:Swap on Input* menu option), which causes Run Animate to swap to the application screen when input is required. If Options Swap is off, the application screen is not displayed during Animate Mode.

For more information on modes of execution, see the section entitled Debugging a Program in this chapter.

Examples

- To change the current application screen display status, enter:

  ```
  > Options Swap ↵
  ```

See also: Options Exchange, Run Animate, View App

Options Tab

Options Tab

Set the tab size for the Code Window

Syntax

```
Options Tab <tabSize>
```

Options Tab Command Equivalents

Type	Equivalent
Accelerator key	Alt-O T
Abbreviation	O T

Arguments

<tabSize> is a numeric value indicating the size to which the tabs are expanded when displayed in the Code Window. The default tab size is 4. If *<tabSize>* is not specified on the command line, Options Tab opens a dialog box to prompt you for the tab size.

Description

Options Tab allows you to set the tab size for the Code Window. This command is effective only if the file you are viewing contains tabs. Lines that are indented with spaces are not affected.

Examples

- To change the tab size to ten, enter:

  ```
  > Options Tab 10 ↵
  ```

See also: Options Color, Options Mono

Programming and Utilities Guide   10–93

Output

See: View App

Point Breakpoint

Set or remove a Breakpoint at the current cursor position

Syntax

```
Point Breakpoint
```

Point Breakpoint Command Equivalents

Type	Equivalent
Hot key	F9
Accelerator key	Alt-P B
Abbreviation	P B

Description

Point Breakpoint designates the line of code indicated by the cursor in the Code Window as a Breakpoint. When you execute this command, the line of code is indicated in a new color to distinguish it as a Breakpoint.

Point Breakpoint acts as a toggle so that if the current line of code is already a Breakpoint, the Breakpoint is deleted and the color of the line returns to normal.

To see all Breakpoint settings in the Command Window, use the List BP command. For more information on the use of Breakpoints, see the section entitled Debugging a Program in this chapter.

Examples

- To set the current line of code as a Breakpoint, enter:

    ```
    > Point Breakpoint ↵
    ```

See also:

BP, Delete, List, Point Tracepoint, Point Watchpoint

Point Delete

Delete a Tracepoint or Watchpoint setting

Syntax

```
Point Delete <number>
```

Point Delete Command Equivalents

Type	Equivalent
Accelerator key	Alt-P D
Abbreviation	P D
Synonyms	Delete TP, Delete WP

Arguments

<*number*> is the number of the individual Watchpoint or Tracepoint to delete. If <*number*> is not specified on the command line, Point Delete opens a dialog box to prompt you for a number.

Description

Point Delete allows you to delete a single Watchpoint or Tracepoint by specifying its associated number. The number that you specify appears to the left of the Watchpoint or Tracepoint definition in the Watch Window. Alternatively, you could use List TP or List WP to show the numbers and point definitions in the Command Window.

For a detailed explanation of Tracepoints and Watchpoints, see the section entitled Debugging a Program in this chapter.

Examples

- This example deletes the first Watchpoint or Tracepoint defined in the Watch Window:

  ```
  > Point Delete 0 ↵
  ```

See also:

Delete, List, Point Tracepoint, Point Watchpoint

Point Tracepoint

Specify a variable or expression as a Tracepoint

Syntax

```
Point Tracepoint <exp>
```

Point Tracepoint Command Equivalents

Type	Equivalent
Accelerator key	Alt-P T
Abbreviation	P T
Synonym	TP

Arguments

<exp> is the field variable, memory variable or expression to be traced. It can be of any data type. If *<exp>* is not specified on the command line, Point Tracepoint opens a dialog box to prompt you for an expression.

Description

Point Tracepoint designates the specified variable or expression as a Tracepoint. The Tracepoint is then added to the Watch Window along with its type and value, and noted as "tp."

A Tracepoint can be thought of as a conditional Breakpoint. The two are very similar, except that a Tracepoint halts program execution as soon as its value changes. For more information on Tracepoints, see the section entitled Debugging a Program in this chapter.

Examples

- This example halts execution as soon as the value of the variable *nInvNum* changes:

    ```
    > Point Tracepoint nInvNum ↵
    ```

- This pauses when end of file is reached:

    ```
    > Point Tracepoint EOF() ↵
    ```

- This example halts as soon as the value of *i* exceeds 10, and is very useful in looping operations:

    ```
    > Point Tracepoint i > 10 ↵
    ```

See also:

Delete, Point Breakpoint, Point Delete, Point Watchpoint

Point Watchpoint

Specify a variable or expression as a Watchpoint

Syntax

```
Point Watchpoint <exp>
```

Point Watchpoint Command Equivalents

Type	Equivalent
Accelerator key	Alt-P W
Abbreviation	P W
Synonym	WP

Arguments

<exp> is the field variable, memory variable or expression to be watched and can be of any data type. If <exp> is not specified on the command line, Point Watchpoint opens a dialog box to prompt you for an expression.

Description

Point Watchpoint designates the specified variable or expression as a Watchpoint. The Watchpoint is then added to the list in the Watch Window together with its type and value and the abbreviation "wp." As each line of the current program is executed, the value of the Watchpoint is updated in the Watch Window.

A Watchpoint is identical to a Tracepoint, except that it does not cause a break in program execution every time its value changes. For more information on Watchpoints, see the section entitled Debugging a Program in this chapter.

Examples

- This example specifies the field *CustNum* as a Watchpoint:

  ```
  > Point Watchpoint CustNum ↵
  ```

- In this example, the current record number is displayed as each line of code is executed:

  ```
  > Point Watchpoint RECNO() ↵
  ```

See also: Delete, Point Breakpoint, Point Delete, Point Tracepoint

Prev

See: Locate Previous

Quit

See: File Exit

Restart

See: Run Restart

Resume

See: File Resume

Run Animate

Run application in Animate Mode

Syntax

```
Run Animate
```

Run Animate Command Equivalents

Type	Equivalent
Accelerator key	Alt-R A
Abbreviation	R A
Synonym	Animate

Description

The Run Animate command runs the application in Animate Mode. This means that the Debugger executes a single line, moves the Execution Bar to the next line, executes it, and so on. If the *Options:Exchange Screens* menu option is on (indicated by a check mark), the output of each line is displayed after the line has been executed.

Execution continues in this manner until a Breakpoint or Tracepoint is reached. To control the speed of this process, use the Run Speed command. Press any key to stop execution.

For more information on execution modes, see the section entitled Debugging a Program in this chapter.

See also: Run Speed

Run Go

Execute the application in Run Mode

Syntax

```
Run Go
```

Run Go Command Equivalents

Type	Equivalent
Hot key	F5
Accelerator key	Alt-R G
Abbreviation	R G
Synonym	Go

Description

Run Go displays your application screen and executes the application until a Breakpoint or Tracepoint is reached, or until the Debugger is deliberately invoked (i.e., by pressing *Alt-D* or using ALTD()). At that point, the Debugger screen is redisplayed and program execution halts.

This is known as Run Mode. Using Run Go again causes the application to be executed from the current position to the next Breakpoint or Tracepoint. To reload the application so that it can be executed from the beginning, use Run Restart.

For more information on modes of execution and the use of Breakpoints and Tracepoints, see section entitled Debugging a Program in this chapter.

Examples

- Suppose Breakpoints have been set at lines 15 and 30 of the current program. Typing

    ```
    > Run Go ↵
    ```

    executes the application as far as line 15. Entering Run Go again executes up to line 30, and so on.

See also: Run Next, Run Restart, Run To

Run Next

Execute the application in Run Mode up to the start of the next activation

Syntax

```
Run Next
```

Run Go Command Equivalents

Type	Equivalent
Hot key	*Ctrl-F5*
Accelerator key	*Alt-R N*
Abbreviation	R N

Description

Run Next displays your application screen and executes the application until it reaches line zero of the next activation (i.e., function, procedure, code block, or message send). The application is executed in Run Mode until that point is reached. This is equivalent to setting a Tracepoint of "PROCLINE() == 0."

Examples

- Suppose the current program contains its first function call on line 15. Typing

  ```
  > Run Next ↵
  ```

  executes the application as far as line 15, the function call. Entering Run Next again executes up to the next activation.

See also:

Run Go, Run Restart, Run To

Run Restart

Reload the current application

Syntax

```
Run Restart
```

Run Restart Command Equivalents

Type	Equivalent
Accelerator key	*Alt-R R*
Abbreviation	R R
Synonym	Restart

Description

Run Restart reloads the current application in preparation to be reexecuted, keeping intact all Debugger settings. This command is the only way to execute an application which has already been run.

Examples

- In this example, the current application is executed:

    ```
    > Run Go ↵
    ```

- If errors are discovered, you may want to make changes and run the application again:

    ```
    > Run Restart ↵
    ```

See also: Run Next, Run Go

Run Speed

Set step delay for Animate Mode

Syntax

```
Run Speed <delay>
```

Run Speed Command Equivalents

Type	Equivalent
Accelerator key	Alt-R P
Abbreviation	R Sp
Synonym	Speed

Arguments

<delay> is the increment of delay for animation in tenths of seconds. If *<delay>* is not specified, Run Speed opens a dialog box which displays the current setting and allows you to enter a new one.

Description

Run Speed controls the speed of display while in Animate Mode (for an explanation of this mode, see section entitled Debugging a Program in this chapter). Remember that this setting is expressed in tenths of seconds so that smaller settings are faster than larger ones.

See also: Run Animate

Run Step

Execute the current program in Single Step Mode

Syntax

```
Run Step
```

Run Step Command Equivalents

Type	Equivalent
Hot key	F8
Accelerator key	Alt-R S
Abbreviation	R S
Synonym	Step

Description

Run Step executes the current program in Single Step Mode. This means that it executes the line of code at the Execution Bar, moves the Execution Bar to the next line, and stops. As functions are called by the current program, their code is displayed in the Code Window.

Examples

- This example executes the next line of the application being debugged:

    ```
    > Run Step ↵
    ```

    The Debugger stops with the Execution Bar on the next line to be executed.

See also: Options Codeblock, Run Trace

Run To

Execute the current program up to the current cursor position

Syntax

```
Run To
```

Run To Command Equivalents

Type	Equivalent
Hot key	F7
Accelerator key	Alt-R C
Abbreviation	R To

Description

Run To executes only those lines of code up to the line indicated by the current cursor position in the Code Window. The application is executed in Run Mode until that line is reached.

If the line indicated by the current cursor position is never executed, the Debugger continues to the end of the application.

See also: Run Go, Run Next

Run Trace

Execute the current program in Trace Mode

Syntax

```
Run Trace
```

Run Trace Command Equivalents

Type	Equivalent
Hot key	*F10*
Accelerator key	*Alt-R T*
Abbreviation	R T

Description

The Run Trace command is similar to Run Step in that it executes one line of program code at a time. However, Run Trace does not display the code for functions called by the current program.

See also: Run Step

Speed

See: Run Speed

Step

See: Run Step

TP

See: Point Tracepoint

View

Examine a file during the current debugging session

Syntax

```
View <idFileName>
```

Arguments

<idFileName> is the name of the file you want to examine. Unless one is explicitly specified as part of *<idFileName>*, no extension is assumed. If *<idFileName>* is not specified on the command line, View opens a dialog box to prompt you for a filename.

Description

View allows you to look at other files without leaving the current debugging session. It can also be used to view header files specified with the #include preprocessor directive. Breakpoints can be set and are saved when you return to the original program.

Unless you have defined a source file search path with the Options Path command, View searches for *<idFileName>* in the current directory only. If, however, a search path has been defined the directories in the path are searched in order until the file is found.

Note: View is almost identical to the File Open command. The only difference between these two commands is that File Open assumes a (.prg) extension whereas View assumes no file extension.

Examples

- To view ListDbfs.prg from within the current program (EnterData.prg), type:

    ```
    > View ListDbfs.prg ↵
    ```

- It is possible to continue viewing other programs without returning to the original program. For example:

    ```
    > View PrintDbfs.prg ↵
    ```

    loads PrintDbfs.prg in place of ListDbfs.prg. Typing File Resume returns the Debugger to EnterData.prg—the original program.

See also:

File Open, File Resume, Options Path

View App

Display program output

Syntax

```
View App
```

View App Command Equivalents

Type	Equivalent
Hot key	F4
Accelerator key	Alt-V A
Abbreviation	V A
Synonym	Output

Description

View App temporarily clears the Debugger screen and displays your application screen in its place. This allows you to see the output of the current program in the context of the application itself. To return to the original screen, press any key.

View Callstack

Control the Callstack Window

Syntax

```
View Callstack
```

View Callstack Command Equivalents
Type	Equivalent
Accelerator key	Alt-V C
Abbreviation	V C

Description

View Callstack acts as a toggle by changing the current status of the Callstack Window (e.g., if the window is open, View Callstack closes it). By default, View Callstack is off which means that the Callstack Window is closed.

The Callstack Window contains a list of pending activations with the current activation at the top. For more information on the Callstack, see the section entitled Debugging a Program in this chapter.

View Callstack is functionally equivalent to the Callstack command specified without an option.

Examples

- To change the current status of the Callstack Window, enter:

    ```
    > View Callstack ↵
    ```

See also: Callstack

View Sets

Display the View Sets Window

Syntax

```
View Sets
```

View Sets Command Equivalents

Type	Equivalent
Accelerator key	*Alt-V S*
Abbreviation	V S

Description

View Sets activates the View Sets Window. When this window is active, you can view and change the status of the *CA-Clipper* system settings.

Use *Uparrow* and *Dnarrow* to move the highlight up and down in the list of settings. To change a setting, highlight it and press *Return*. After changing the value, press *Return* again and move on to the next setting.

To close the View Sets Window and continue debugging, press *Esc*. The new settings are saved and take effect immediately in your program.

See also: Options Color, View Workareas

View Workareas

Display the View Workareas Window

Syntax

```
View Workareas
```

View Workareas Command Equivalents

Type	Equivalent
Hot key	F6
Accelerator key	*Alt-V W*
Abbreviation	V W

Description

View Workareas activates the View Workareas Window. This window allows you to view information regarding all database files that are currently in use. To close the window and continue debugging, press *Esc*.

See also: Options Color, View Sets

Window Iconize

Toggle active Debugger window between icon and window display

Syntax

```
Window Iconize
```

Window Iconize Command Equivalents

Type	Equivalent
Accelerator key	*Alt-W I*
Abbreviation	W I

Description

Window Iconize reduces the active window to an icon (its name). This command acts as a toggle between the icon and window display modes so that when the active window is iconized, executing Window Iconize resumes the original window display.

A window that is iconized remains open, but you cannot see the window contents. Certain window operations such as moving, however, are possible.

Examples

- To iconize the active window, enter:

  ```
  > Window Iconize ↵
  ```

See also: Window Move, Window Size, Window Tile, Window Zoom

Window Move

Move the active Debugger window

Syntax

```
Window Move
```

Window Move Command Equivalents	
Type	Equivalent
Accelerator key	*Alt-W M*
Abbreviation	W M

Description

Window Move allows you to move the active window around on the screen. When you execute this command, the border of the active window changes to a different pattern and the cursor keys are used to move the window. *Return* completes the moving process.

Note: You cannot move a window that is zoomed to full-screen.

Examples

- To move the active window, enter:

  ```
  > Window Move ↵
  ```

See also: Window Iconize, Window Size, Window Tile, Window Zoom

Window Next

Select the next Debugger window

Syntax

```
Window Next
```

Window Next Command Equivalents

Type	Equivalent
Hot key	*Tab*
Accelerator key	*Alt-W N*
Abbreviation	W N

Description

Window Next selects the next window on the main Debugger screen. The order of the windows is as follows: Code, Monitor, Watch, Callstack, Command. Thus, if the Callstack Window is active, Window Next selects the Command Window.

Examples

- To select the next window, enter:

  ```
  > Window Next ↵
  ```

See also: Window Prev

Window Prev

Select the previous Debugger window

Syntax

```
Window Prev
```

Window Prev Command Equivalents

Type	Equivalent
Hot key	*Shift-Tab*
Accelerator key	*Alt-W P*
Abbreviation	W P

Description

Window Previous selects the previous window on the main Debugger screen. The order of the windows is as follows: Code, Monitor, Watch, Callstack, Command. Thus, if the Callstack Window is active, Window Prev selects the Watch Window.

Examples

- To select the previous window, enter:

  ```
  > Window Prev ↵
  ```

See also: Window Next

Window Size

Change the size of the active Debugger window

Syntax

```
Window Size
```

Window Size Command Equivalents

Type	Equivalent
Accelerator key	*Alt-W S*
Abbreviation	W S

Description

Window Size allows you to change both the height and the width of the active window. When you execute this command, the border of the active window changes to a different pattern and the cursor keys are used to change the size of the window. *Return* completes the sizing process.

Note: You cannot size a window that is zoomed to full-screen.

Examples

- To size the active window, enter:

  ```
  > Window Size ↵
  ```

See also: Window Iconize, Window Move, Window Tile, Window Zoom

Window Tile

Restore Debugger windows to default size and location

Syntax

```
Window Tile
```

Window Tile Command Equivalents

Type	Equivalent
Accelerator key	*Alt-W T*
Abbreviation	W T

Description

Window Tile provides a quick way to clean up the screen by restoring each window on the screen to its default location and size. Any windows that have been zoomed or iconized are also restored to the original window display mode.

Examples

- To restore Debugger windows, enter:

    ```
    > Window Tile ↵
    ```

See also: Window Iconize, Window Move, Window Size, Window Zoom

Window Zoom

Toggle active Debugger window between window and full-screen display

Syntax

```
Window Zoom
```

Window Zoom Command Equivalents

Type	Equivalent
Hot key	F2
Accelerator key	*Alt*-W Z
Abbreviation	W Z

Description

Window Zoom allows you to zoom the active window to full-screen. This command acts as a toggle between the full-screen and window display modes so that when the active window is zoomed to full-screen, executing Window Zoom resumes the original window display.

When a window is zoomed, some window operations such as moving and sizing are not allowed.

Examples

- To zoom the active window to full-screen, enter:

  ```
  > Window Zoom ↵
  ```

See also: Window Iconize, Window Move, Window Size, Window Tile

WP

See: Point Watchpoint

Chapter 11
Program Maintenance
RMAKE.EXE

CA-Clipper provides a program maintenance facility with RMAKE.EXE. RMAKE is a powerful tool for keeping programs involving several source, header, object, and library files up-to-date. It does this by comparing the date and time stamps of files related to one another and performing a series of actions if the date and time stamps do not match.

A make facility is generally used to speed the process of compiling a program system composed of several source files by compiling only those files that have changed since the last program build. *.RTLink*, the *CA-Clipper* linker, provides a similar facility for the link phase with incremental linking.

In This Chapter

This chapter describes the operation of RMAKE. The following general subjects are covered:

- Invoking RMAKE
- RMAKE options
- The RMAKE environment variable
- The RMAKE return code
- How RMAKE works
- How RMAKE searches for files
- The make file
- Examples
- Notes

Invoking RMAKE

If you installed the default configuration of the *CA-Clipper* development system, RMAKE.EXE is located in the \CLIPPER5\BIN directory. RMAKE.EXE has the following general syntax:

```
RMAKE [<makeFile list>][<macroDef list>][<option list>]
```

<makeFile list> is a list of one or more make files to process, separated by a space. If no extension is specified as part of the filename, (.RMK) is assumed. Any filename may optionally include a drive designator and a path reference. A maximum of 16 files can be specified in the *<makeFile list>*.

<macroDef list> is a list of one or more macro definitions in the form *<macroName>=<value>*. Each macro definition specified is separated by a space. Macros specified like this take precedence over macros defined in the make file.

<option list> is a list of one or more RMAKE options. Options may be specified either as upper or lowercase characters and must be prefaced by either a slash (/) or a dash (-) character. No other separator is required between options.

If executed with no arguments, RMAKE displays a list of all available options with a short description.

RMAKE Options

The options described in this section control the behavior of RMAKE. When specifying options, they can be either upper or lowercase and must be prefaced by either a slash (/) or a dash (-) character. Options may be specified in any order. In addition, options may be specified either on the RMAKE command line or in the RMAKE environmental variable. Command line options take precedence if there is a conflict.

Some options have arguments. If an option has arguments, they are specified after the option, and no space is allowed between the option and any of its arguments.

/B   **Debugging Information**

Displays debugging information.

/D   Define a Macro

   `/D<macroName>[:<value>]`

   Defines a macro and an optional value. If the value is not supplied, the macro is defined with a null value. This option provides an alternative to defining a macro on the RMAKE command line with *<macroName>=<value>*. It is available for use as part of the RMAKE environment variable since the equal sign has special meaning when used with the DOS SET command.

/I   Ignore Execution Errors

   Causes RMAKE to continue if a command executed from the make file causes an error (i.e., produces a DOS return code greater than zero). By default, errors of this nature halt the make process. Note that fatal errors generated by RMAKE always terminate the make process, regardless of /I.

/N   Null Make

   Displays the commands that would be executed without actually executing them.

/S   Search Subdirectories

   Normally, RMAKE searches the current directory (or explicitly specified directories) for the make files you specify. This option causes it to search all subdirectories of the current directory (or of explicitly specified directories) for make files not found. For example, RMAKE MYMAKE /S will search for MYMAKE.RMK in the current directory and all subdirectories of the current directory, whereas RMAKE \PROJECT\MYMAKE /S will search the \PROJECT directory and all of its subdirectories.

/U   Enable (#) Comment Indicator

   Enables the # character as a comment indicator. If this option is specified, make file directives must be preceded by the exclamation (!) symbol instead of the hash (#) symbol.

/W   Show Warnings

   Displays warning messages while processing.

/XS   Set Symbol Table Size

/XS<numSymbols>

Sets the size of the internal symbol table. The default size is 500 symbols.

/XW   Set Internal Workspace Size

/XW<numBytes>

Sets the size of the internal workspace. The default size is 2048 bytes.

The RMAKE Environment Variable

RMAKE uses the environment variable, RMAKE, to allow the specification of command line arguments and options without supplying them on the RMAKE command line. To define the RMAKE environment variable, use a DOS SET command like the following:

```
SET RMAKE=[<option list>]
```

When defining the RMAKE environment variable, options are specified just as they would be on the RMAKE command line. Note that options specified in the environment are scanned first, before any command line options. This means command line options override those specified in the environment.

In order to save yourself from having to enter this SET command repeatedly, it can be placed in your AUTOEXEC.BAT file where it will be processed automatically each time you reset your computer.

Example

The following example causes RMAKE to display warning messages and debugging information:

```
SET RMAKE=/W /B
```

Note: DOS does not allow the equal sign to be specified as part of an environment variable string, making it impossible to define macros in RMAKE as you would normally on the command line. To overcome this problem, use the /D option to define macros in the RMAKE environment string. Like other command line macro definitions, macros defined in the RMAKE environment string cannot be redefined.

The RMAKE Return Code

RMAKE sets the DOS return code to 0 if the make was successful (i.e., there were no errors) and 1 if any errors were encountered during processing. The RMAKE return code can then be tested within a batch file using the DOS ERRORLEVEL condition. For example:

```
RMAKE RL
IF ERRORLEVEL 1 GOTO ERROR
.
.<commands>
.
GOTO END
:ERROR
.
.<commands>
.
:END
```

How RMAKE Works

If you are familiar with UNIX Make, you will probably find RMAKE easy to understand and use. Generally, it is upwardly compatible with UNIX Make and will, in most cases, successfully process UNIX make files; however, RMAKE provides several additional features.

In general, you describe your program in terms of *dependency rules* using a text file, called a *make file*. Then, instead of determining the compiling and linking steps to build the program each time you make a change to a source file, you use RMAKE with one or more make files designed for that particular program. RMAKE performs only the compiling and linking steps necessary to bring the executable version of the program up-to-date.

The basic element of a make system is the *dependency rule* consisting of a *dependency statement* that establishes a relationship between a *target file* and a series of *dependent files*. The dependency statement is then followed by a series of *actions* to update the target file if any of its dependent files are more recent.

RMAKE is a two pass system. The first pass parses the make files and is called the *parsing phase*. The second pass examines the dependency rules and performs the necessary actions and is called the *make phase*.

When invoked, RMAKE first performs the parsing phase by processing each make file sequentially in the order encountered. Files specified on the RMAKE command line are processed from left to right. All file searching is performed during the parsing phase, and all inference rules are applied. If macros or rules are redefined, then the new definitions apply to the dependencies specified below them.

After the parsing phase, RMAKE performs the make phase during which all of the rules defined in all of the files are applied in order to produce the target files. For each rule in the make file, the date of the target file is compared to the date of each dependent file. If any dependent file has a more recent date and time stamp than its associated target file, the accompanying actions are executed. If none of the dependent files have been changed since the target file was last updated, the rule is bypassed and the process is repeated for the next rule.

RMAKE will correctly handle the case where two input files define dependencies (and possibly actions) for the same target. It always builds all target files to ensure that multiple-specified targets (e.g., .LIB files) are not time stamped before all make files get a chance to specify dependencies. Although no particular order is guaranteed during the make phase, RMAKE does guarantee that no target will be built until all of its dependencies have been built.

When RMAKE terminates, it sets the DOS return code to indicate the state of the make process at termination.

How RMAKE Searches for Files

There are only a few types of files that serve as input to RMAKE. This section describes each input file type and any special rules governing how RMAKE searches for files of that type.

Make Files

When RMAKE is first executed, it attempts to locate the make files specified on the command line and in the RMAKE environment variable. By default, these files are assumed to have an extension of .RMK which you can override by specifying another extension as part of the make filename. To indicate a file with no extension, include a period at the end of the filename.

A full or partial path may be specified as part of any make filename in order to force RMAKE to look for the file in a particular drive and/or directory. If no path is specified, MAKE searches only the current directory. The /S option causes RMAKE to search subdirectories as well.

Target and Dependency Files

In addition to finding make files, RMAKE must also find the target and dependency files referenced in dependency statements in the make file. By default, RMAKE searches the current directory only for these files. However, there are special macros that you can define to specify where target and dependency files are to be found and/or created. The form of this special macro definition is:

```
makepath[.<extension>] = <pathSpec>
```

<extension> specifies the file type and <pathSpec> specifies one or more paths to be searched when trying to find or create files of the given type. To avoid problems with the backslash (\) character in a make file, enclose the path list in quotes as in the examples below.

If a *makepath* macro is defined for a particular file type, the current directory is not searched unless it is explicitly included in the <pathSpec>. Only the paths specified in the path list are searched, and they are searched in the order specified during the parsing phase.

If a file is not found along any of those paths, RMAKE will attach the path at the front of the list to the filename. For a target file, this means that the first path in the list for its file type determines where the file is created when the make actions are performed. For dependent files, it means that RMAKE will not be able to find the file and will result in an error.

makepath macros may be specified on the RMAKE command line, in the RMAKE environment variable (using the /D option), and in make files. Some examples follow:

```
makepath[.h]   = "C:\CLIPPER5\SOURCE;C:\CLIPPER5\INCLUDE"
makepath[.obj] = "C:\CLIPPER5\OBJ"
makepath[.lib] = "C:\CLIPPER5\LIB;C:\MSC\LIB"
makepath[.exe] = "C:\CLIPPER5\BIN"
```

To augment a path list, use the immediate macro assignment operator (:=) rather than the assignment operator (=):

```
makepath[.h] := "C:\MSC\INCLUDE;$(makepath[.h])"
```

Note: Actions are executed during the make phase and thus have no effect on file searches. This means, for example, that you cannot have an action line that places a file in a subdirectory specified with a path name with the intent of triggering a subsequent dependency rule.

For target and dependency files specified with a path as part of the filename, RMAKE searches only the explicitly named directory regardless of the existence of an appropriate *makepath* macro. For more information on macros and make files, see *The Make File* section in this chapter.

The Make File

The fundamental element of a make system is the make file, an ASCII text file created with a standard text editor. This file defines which files update other files and the update actions to perform when files are out-of-date. With a typical *CA-Clipper* program, a make file lists the program and header files required to create each object file and all the object files necessary to create the executable file. The make file also lists the commands required to build each file.

A make file consists of the following basic components:

- *Dependency rules* that define how each file is built
- *Inference rules* that define what to do if a dependency rule has no accompanying actions
- *Macros* that allow variable information to be used when constructing rules
- *Directives* that allow you to further control the manner in which the make file is processed

Each of these components is described below.

Using Quotation Marks

There are instances when quote marks are necessary to force RMAKE *not* to interpret certain symbols as commands. Either single or double quotes may be used.

For instance, when you specify a filename with a drive and/or path specification in a dependency or inference rule, both the colon and the backslash have special meanings to RMAKE and, therefore, the path specification must be delimited by quote marks. Another instance is a macro name definition containing these same or other special symbols. As a general principle, use quotes whenever you are in doubt since they are not interpreted literally.

Note: Using quotation marks does not affect the manner in which RMAKE interprets the macro character ($). Macros are expanded both inside and outside of quotes.

Line Continuation

Any comment, rule, or action in a make file can be continued to the next line with a backslash (\) character at the end of a line. If there is a comment on a continued line, it must occur after the continuation (\) symbol. For example:

```
TEST.OBJ: Test1.prg, Test2.prg, Test3.prg\   // line continued
Test4.prg
    CLIPPER @Test.clp
```

Comments

Comments may be specified in a make file using C-style inline comments (/* ... */) or C-style line comments (//) or, if the /U option is used, UNIX make-style comments (#). In a make file, a comment can be either on a line by itself or embedded at the end of a command line in a dependency rule. For example:

```
/* This is a valid comment line */
TEST.OBJ: Test.prg
    CLIPPER Test.prg    // This is a valid embedded comment.
```

If the /U option is specified, the hash (#) symbol can be used as a comment indicator. Comments cannot be nested.

Dependency Rules

The key component of the make file is the *dependency rule* which consists of a dependency statement followed by one or more actions. The dependency statement establishes a relationship between a target file and one or more dependent files. If any of the dependent files have a date and time stamp newer than the target file, the actions that follow the dependency statement are performed to bring the target file up-to-date.

A dependency rule has the following basic form:

```
<targetFile>: <dependentFile list>
   [<action>]
   [<action>]...
```

The Make File

You can specify up to 128 dependent files in the <dependentFile list>. You can specify up to 32 <action> lines per dependency rule, provided that the maximum number of bytes in all <action> lines does not exceed 1024.

Any filename specified as part of a dependency rule may be specified with a path and drive reference. If a path or drive is not specified, RMAKE assumes the current directory or the path list defined by the appropriate *makepath* macro (see the *How RMAKE Searches for Files* section above). Quotes may be used to delimit a filename that contains special characters—since the colon has special meaning in a dependency rule, a filename with a drive letter should always be enclosed in quotes. The filenames must include file extensions.

The <targetFile> is the name of the file to build followed by a colon and a space. For example, the target file specified here is built in the current directory:

```
MYAPP.EXE: MYOBJ.OBJ "\CLIPPER5\OBJ\IBM BIOS.OBJ"
```

and this target file is built in the specified directory:

```
"C:\CLIPPER5\APPS\MYAPP.EXE": MYOBJ.OBJ "\CLIPPER5\OBJ\IBM BIOS.OBJ"
```

The <dependentFile list> is a list of one or more filenames separated by spaces. The following dependency statement defines two dependent files:

```
MYAPP.EXE: MYOBJ.OBJ "\CLIPPER5\OBJ\IBM BIOS.OBJ"
```

RMAKE searches for MYOBJ.OBJ in the current directory and for IBM BIOS.OBJ in the \CLIPPER5\OBJ directory of the current drive.

The <action> statements are one or more valid DOS command lines with arguments. Each <action> must be placed on a line by itself and must be indented using spaces or tab characters. The action lines are assumed to end at the first line that is not indented but may also be terminated using a blank line.

It is not necessary for each <action>'s primary focus to be the creation of the target file. In fact, after the actions are executed, RMAKE does not verify that the target file was successfully created. However, RMAKE does check the DOS ERRORLEVEL variable upon completion of each item in the <action> list. If it contains a nonzero value, RMAKE will halt processing and produce an error message. The compiler and .RTLink both properly set ERRORLEVEL so that RMAKE will halt in the event that any errors are encountered.

This default error processing can be disabled using the /I option. When specified from the command line, RMAKE will no longer stop when

DOS's ERRORLEVEL is set. Instead, it will continue processing as if nothing had happened.

The <actions> listed are executed if any dependent files in the <dependentFile list> have a more recent date and time stamp than the <targetFile>. If no actions are specified, there should be a corresponding inference rule located somewhere in the make file that defines the <actions> to take.

As an example, if a target object file, TEST.OBJ, depends on Test.prg, then RMAKE should recompile Test.prg if someone has changed it since it was last compiled. The following dependency rule accomplishes this:

```
TEST.OBJ: Test.prg
   CLIPPER Test
```

Likewise, the target .EXE file can be created with a similar rule. To demonstrate the use of a dependent file list, this example assumes that WINDOW.OBJ must also be linked to create the executable file. The rule to create TEST.EXE when either TEST.OBJ or WINDOW.OBJ is more recent is as follows:

```
TEST.EXE: TEST.OBJ WINDOW.OBJ
   RTLINK FI TEST, WINDOW
```

Inference Rules

An *inference rule* specifies a series of actions for a dependency statement not followed by an action list. It can be used as a shortcut if you want to perform the same set of actions for several different dependency statements. The basic form of an inference rule is as follows:

```
.<dependentExtension>.<targetExtension>:
   [<action>]
   [<action>]...
```

In comparison with a dependency statement, an inference rule is stated with the order of the target and dependent file reversed. You can specify up to 32 <action> lines per inference rule, provided that the maximum number of bytes in all <action> lines does not exceed 1024.

When RMAKE encounters a dependency statement without an action list, it searches for an inference rule that matches the extensions of the target and dependent files in the dependency statement. If an inference rule is found, RMAKE performs the accompanying actions, including expanding any macros. As in a dependency rule, each action must be placed on a line by itself and must be indented using spaces or tab characters. The action lines are assumed to end at the first line that is not indented but may also be terminated using a blank line.

The Make File

When searching for an inference rule, the actual filenames in the dependency statement are insignificant—only the extensions are compared. If a dependency statement refers to more than one dependent file, the extension of the first file in the list is used. If the extensions match, the rule applies.

For example, the following inference rule compiles a single (.prg) file defined in dependency statements to an .OBJ file:

```
.prg.obj:
    CLIPPER $** /M
```

The $** symbol on the compiler command line is a predefined RMAKE macro described in the next section. Basically, it causes the dependent filename to be substituted on the command line. For the following dependency statement:

```
TEST.OBJ: Test.prg
```

the resulting action would be CLIPPER Test.prg /M.

If an action line in either a dependency or an inference rule contains a DOS SET command, it will be interpreted directly by RMAKE during the make phase rather than being passed to the DOS command processor like other action lines. This allows the setting of environment variables as part of an action. For example:

```
.prg.obj:
    SET CLIPPERCMD=$(flagList)
    CLIPPER $**
```

Understanding DOS SETs from Within RMAKE

Each program that runs under DOS receives a local copy of the DOS environment variables. RMAKE allows you to make changes to its copy so that programs in action lines can inherit a modified environment. Any changes made are discarded once RMAKE terminates. (Like most programs, RMAKE is unable to change its parent's environment.)

If RMAKE sees a DOS command in an action list, it simply shells to execute the command. The DOS shell will not inherit the modified environment, so the lone SET command below:

```
.prg.OBJ:
    SET INCLUDE=$(makepath[.ch])
    SET                              // <<--- THIS COMMAND
    CLIPPER $<
```

will display the INCLUDE setting prior to running the RMAKE. The SET command is internal to DOS; there is no Set.COM or Set.EXE file. Therefore, SET can only access the initial (DOS) copy of the environment. However, *CA-Clipper* will see the new setting of

INCLUDE because it is an application in its own right and will inherit a copy of RMAKE's modified environment.

If you are distributing a make file, you don't have to require that *.RTLink* be set to positional mode, or that certain *CA-Clipper* compiler options be set in the CLIPPERCMD variable. Instead, set up everything you need and assume absolutely nothing about the DOS environment.

DOS's environment variables can be used *at any time* in a make file by surrounding them with parentheses and preceding them with a dollar sign. The following code shows how to temporarily append the makepath[.ch] list to the current DOS INCLUDE variable:

```
.prg.OBJ:
   SET INCLUDE=$(include);$(makepath[.ch])
   CLIPPER $<
```

Directives

Several directives are provided for use in make files. Each directive must be specified on a separate line in the make file—directives cannot be used in dependency or inference rule action lines. Macros encountered in directive arguments are always expanded immediately. Directives are handled during the parsing phase of the make.

The syntax for all directives in this section includes an initial hash (#) symbol that must be specified. Note, however, that the exclamation (!) symbol can be used interchangeably with the hash (#) symbol unless the /U option is specified. In this instance, the ! symbol must be used or the directives will be treated as comments.

Conditional Directives

Conditional directive structures allow you to include or exclude sections of a make file evaluation based on the existence or value of a macro. The macro can be defined either earlier in the make file, on the RMAKE command line, or in the RMAKE environment variable (using the /D option). A conditional directive structure consists of an #if directive followed by a series of statements. Each construct is terminated with an #endif directive. Alternative statements may be executed within the construct if the current #if condition fails by specifying an #else directive.

Process If Macro Exists

```
#ifdef <macroName>
   <statements>...
[#else]
   <statements>...
#end[if]
```

#ifdef processes subsequent lines if the named macro exists. Do not use a dollar sign in front of the macro name unless you are trying to test for the existence of the macro whose name will result from the expansion of the macro you are specifying. For example:

```
#ifdef prepath
   makepath[.prg] := "$(prepath);$(makepath[.prg])"
#endif
```

Process If Macro Does Not Exist

```
#ifndef <macroName>
   <statements>...
[#else]
   <statements>...
#end[if]
```

#ifndef is the same as #ifdef except that #ifndef processes subsequent lines if the named macro *does not* exist. For example:

```
#ifndef "makepath[.prg]"
   makepath[.prg] := "$(src)"
#endif
```

Process If Two Words Are Identical

```
#ifeq <word1> <word2>
   <statements>...
[#else]
   <statements>...
#end[if]
```

#ifeq processes subsequent lines if the two words are identical. The words must be enclosed in quotes if they contain any spaces since a space is used to separate them. Note that the comparison is not case-sensitive. For example:

```
#ifeq $(os) windows
.obj.exe:
   RTLINK $**, $@;
#endif
```

Process If Specified File Exists

```
#iffile <fileSpec>
   <statements>...
[#else]
   <statements>...
#end[if]
```

#iffile processes subsequent lines if the specified file exists. Wildcards are allowed in the *fileSpec* so that you can test to see, for example, whether there are any files in a particular directory. Specifying the directory name by itself, however, will fail. For example:

```
#iffile "C:\MYDIR\*.*"
   prepath="C:\MYDIR"
#endif
```

Remember that directives are handled during the parsing phase—you cannot use #iffile to check for the presence of a file as part of an action. Also, *makepath* macros do not apply when searching for the #iffile.

Undefine Macro

```
#undef <macroName>
```

Removes any previous definition of the macro, including one supplied on the command line or in the RMAKE environment string. There is no way to undefine an environment variable. You can, however, hide it by defining a macro with the same name and then make it visible again using #undef.

Include File

```
#include "<fileName>"
```

Inserts and processes the contents of the specified file in place of the #include directive as part of the make file. The filename may include a path and/or extension, but no defaults are assumed. For example:

```
#include "RMAKE.INI"
```

The #include directive does not use the INCLUDE environment variable nor will it use a *makepath* macro when searching for the specified file. Unless a path is explicitly stated as part of the filename, only the current directory is searched.

The Make File

Display Error Message

```
#stderr "<text>"
```

Writes the *<text>* to the standard error file or device. For example:

```
#stderr "Error in compile step."
```

Display Message

```
#stdout "<text>"
```

Writes the *<text>* to the standard output file or device. For example:

```
#stdout "Making new version..."
```

Execute Action

```
#!<action>
```

Directly executes the *<action>* during the parsing phase. Like other actions specified in make files, this one can be any valid DOS command with arguments. For example:

```
#!DIR *.OBJ
```

Macros

In order to allow you to use variable information when constructing dependency and inference rules, RMAKE provides a macro facility. This allows you to associate a string of characters with a macro name. Whenever the name is subsequently encountered in the make file, it is substituted with the associated string. As you might already realize, it works similarly to the #define preprocessor directive when compiling a program (.prg) file.

Macros are used in a number of situations including the following:

- To control the location and command line options of the compiler and linker for a particular project
- To define dependent file lists used more than once in a make file

With RMAKE, there are two types of macros: user-defined and predefined as described below.

User-Defined Macros

In a make file, a macro definition occurs on a line by itself and takes the following form:

`<macroName>=<value>`

The macro *<value>* can be any string of characters including embedded spaces or a null string. Macro definitions can be nested so that the macro *<value>* may reference another macro name. This allows you to assign a macro to another macro.

Warning! *Almost any character may be used in a macro name via the use of quotes, but the following characters should not be used: dollar sign, parentheses, colon, period, and the percent sign.*

To subsequently use the macro, refer to it like this:

`$(<macroName>)`

When RMAKE encounters a $(*<macroName>*), it replaces the name with the associated string. If the macro is not defined, it is replaced with a null string. A macro remains in effect until it is redefined or undefined, or until RMAKE terminates.

As an example, the following statement creates a macro definition for a dependent file list:

`files=TEST.OBJ TEST1.OBJ TEST2.OBJ`

Later in the make file you can use the macro as a substitute for the file list in a rule, like this:

`TEST.EXE: $(files)`

There are two cases in which macros are not expanded when they are encountered. First, macros in inference rule actions are not expanded until those actions are attached to a dependency rule. Second, macros within macro definitions are not expanded until the macro which contains them is expanded. For example:

```
mac1 = $(cfile)            // cfile is not defined yet
cfile = Xfile.prg
XFILE.OBJ:  $(mac1)        // XFILE.OBJ depends on Xfile.prg
```

Delayed expansion makes it possible for a command line macro to refer to a make file macro, as long as the make file macro has been defined by the time the command line macro is used. However, since this behavior causes difficulty in some situations, macros can be assigned with immediate expansion using the immediate macro assignment (:=) operator. For example:

```
mac1 := $(cfile)           // Won't work unless cfile exists
```

Note: Macros may be defined on the RMAKE command line using the same syntax as in a make file and in the RMAKE environment variable using the /D option. Since they cannot be redefined, command line and environment macro definitions have higher precedence than those defined in the make file.

Predefined Macros

In addition to user-defined macros, there are several predefined macros allowing you to access the target and dependent file information from the last dependency statement encountered. Predefined macros are used when specifying commands associated with a dependency statement or inference rule. The predefined macros are listed in the following table:

Predefined RMAKE Macros

Macro	Meaning
$*	Expands to the target filename without a path or extension
$@	Expands to the target filename including path and extension
$**	Expands to the complete list of full dependency filenames
$<	Expands to the full name of the first file in the dependent file list
$?	Expands to a list of dependencies that have a more recent date and time stamp than the target file

As an example, the following inference rule uses the $** macro to compile a single program file (.prg) dependent on an object file (.OBJ) in a corresponding dependency statement:

```
.prg.obj:
    CLIPPER $** /M
```

This command line will work only for dependency statements that define a single dependent file because *CA-Clipper* compiles only one file at a time and $** returns the entire list of dependent files. See the *Notes* section at the end of this chapter for information on compiling more than one file.

There are also special *makepath* macros that specify paths to use when RMAKE is searching for and creating files. These are discussed in the How RMAKE Searches for Files section in this chapter.

Examples

The following example is a make file designed to maintain RL.EXE, the *CA-Clipper* report and label utility. This utility program consists of the following source code modules:

- Rlfront.prg
- Rlback.prg
- Rldialog.prg

To build RL.EXE, each of these source files is compiled into an object file, and the resulting object files are linked. The following make file defines these dependencies and issues the appropriate compile and link commands:

```
// RL.RMK
// Define macros
objs=RLFRONT, RLBACK, RLDIALOG
//
// Inference rule for compiling (.prg) to .OBJ files
.prg.obj:
   CLIPPER $** /M
//
// Dependency statements for .OBJ files
RLFRONT.OBJ: Rlfront.prg
RLBACK.OBJ: Rlback.prg
RLDIALOG.OBJ: Rldialog.prg
//
// Dependency rule for linking .OBJ files to a .EXE file
RL.EXE: RLFRONT.OBJ RLBACK.OBJ RLDIALOG.OBJ
   RTLINK OUTPUT $@ FI $(objs)
```

Assuming this make file is called RL.RMK, rebuild RL.EXE when source code changes are made by issuing the following command at the DOS prompt:

```
RMAKE RL ↵
```

Notes

- *DOS Command line Length:* Since RMAKE generates executable commands based on statements that include macros, there is a danger that an executable command might exceed the 128 character DOS command line limit. This is particularly a problem if an executable file has a large number of dependent object files making the action line to link the object files too long. In this case, create an *.RTLink* script file (.LNK) containing the list of dependent object files as the files to link, and use the script file in place of the object file list on the linker action line.

For more information on using a script file with the linker, refer to the *CA-Clipper Linker—RTLINK.EXE* chapter in this book.

- ***Using an Inference Rule with a Compiler Command line:*** When using an inference rule to execute the *CA-Clipper* compiler, it is important not to specify more than one dependent file in the accompanying dependency statement. For example, specifying the $** predefined macro expands to the entire dependent file list and the compiler command line may fail because *CA-Clipper* accepts only one source file argument.

  To compile more than one source file into the same object file with the *CA-Clipper* compiler, use a script file (.clp) to list all source files to compile into the target object file. To use this technique, give the script file the same name as the object file but with a (.clp) extension. Then when you create your make file, define the inference and dependency rules for the compile operation like this:

  ```
  // Standard inference rule for compiling with a script file
  .prg.obj:
      CLIPPER @$*
  //
  // Typical dependency rule to maintain an object file
  TEST.OBJ: Test1.prg Test2.prg Test3.prg
  ```

  The file, Test.clp, lists the three dependent (.prg) files to compile into TEST.OBJ. The compiler command line is then specified using the predefined macro, $*, to substitute the root portion of the target filename. This results in the command CLIPPER @TEST when the inference rule is evaluated. The at (@) symbol tells *CA-Clipper* to read the source filenames from Test.clp.

  For more information on compiler script (.clp) files, see the *CA-Clipper Compiler* chapter in this book. For more information on predefined macros, refer to the sections above.

- ***Technical Specifications:*** The following table summarizes the known RMAKE limitations:

  RMAKE Technical Specifications

Component	Limitation
Maximum number of make files	16
Maximum number of dependent files	128
Maximum number of action lines in an inference or a dependency rule	32
Total maximum length of action lines in an inference or a dependency rule	1024 bytes

Summary

In this chapter, you learned how to use RMAKE to maintain your *CA-Clipper* programs. It is a very convenient tool, especially for maintaining complex applications with many source files; however, it has other uses that may not be as apparent. For example, you can use RMAKE to maintain a library file or to make printouts and backups of source files as they are updated. As you use RMAKE, you will discover more and more of its power and flexibility. To look at other make file examples, check the \CLIPPER5\SOURCE subdirectory for files ending with an .RMK extension.

Chapter 12
CA-Clipper Program Editor PE.EXE

CA-Clipper gives you the ability to create and modify *CA-Clipper* source code and header files by supplying a simple program editor, PE.EXE. PE is a stand-alone application written in *CA-Clipper* that provides basic editing features for text and source files.

In This Chapter

This chapter provides a brief overview of the *CA-Clipper* program editor and general usage information. The following topics are covered:

- Invoking the program editor
- Navigation and editing
- Leaving the program editor
- The PE system architecture

Invoking the Program Editor

If you installed the default configuration of the *CA-Clipper* development system, PE.EXE is located in \CLIPPER5\BIN. The general syntax for invoking the program editor is as follows:

```
PE [ <filename>]
```

<filename> is the name of the text file to edit and can include a drive, path, and/or extension. If no extension is specified, a (.prg) extension is assumed. If the specified filename exists, it is loaded for editing; otherwise, a new file is created and you are presented with an empty edit window. If a filename is not specified, you are queried for the name of a file to edit.

Navigation and Editing

Like all other *CA-Clipper* utilities, PE can be executed from any drive or directory since the appropriate directory was placed in the PATH list by the *CA-Clipper* installation program.

Navigation and Editing

When you load PE, the file specified on the command line is either loaded or created and you are presented with the edit window. Once in the edit window, the contents of the file are displayed. You can now add new or modify existing text by moving the cursor to the location of the change and typing just as you would with another text editor or word processor. PE, however, is fairly simple and does not implement any block operations such as select, cut, copy, paste, or replace. While you are editing, you can save changes with *Alt-W*.

The following table lists the editing keys unique to PE. Navigation keys are the same as those used by MEMOEDIT(). See the MEMOEDIT() entry in the *Language Reference* chapter of the *Reference* guide for a complete list of these keys.

PE Editing Keys

Key	Action
Uparrow, Ctrl-E	Line up
Dnarrow, Ctrl-X	Line down
Leftarrow, Ctrl-S	Character left
Rightarrow, Ctrl-D	Character right
Ctrl-Leftarrow, Ctrl-A	Word-left
Ctrl-Rightarrow, Ctrl-F	Word-right
Home	Beginning of line
End	End of line
Ctrl-Home	Top of window
Ctrl-End	End of window
PgUp	Previous window
PgDn	Next window
Ctrl-PgUp	Top of file
Ctrl-PgDn	End of file
Return	Begin next line
Delete	Delete character
Backspace	Delete character left

PE Editing Keys (cont.)

Key	Action
Tab	Insert tab/spaces
Ctrl-Y	Delete line
Ctrl-T	Delete word-right
Alt-H, F1	Display help screen
Ctrl-W	Save and exit
Alt-W	Save and continue
Alt-O	New output filename
Alt-X, Esc	Exit
Alt-F	Display filename
Alt-S	Search
Alt-A	Search again
Alt-I, Ins	Toggle insert mode

Leaving the Program Editor

To exit the edit window and save your most recent changes, press *Ctrl-W*. To exit the edit window without saving changes, press *Esc*. If you have made changes to the file without saving them, you are prompted to save changes before the edit window is closed.

In either case, the edit window closes and you are returned to the initialization screen. On the initialization screen, you can either specify another file to edit or return to DOS. To edit another file, enter its name at the *File To Edit* prompt. Otherwise, press *Esc* to return to DOS.

The PE System Architecture

Like the DBU and RL utilities, PE is provided in source code form. Having access to the source code allows you to modify the program to suit your own needs and serves as a model for programming the MEMOEDIT() function. If you installed *CA-Clipper* in the default configuration, the source files are located in the \CLIPPER5\SOURCE\PE directory.

To facilitate maintenance of the PE program files, the make file, PE.RMK, is also installed with the PE source files. Provided you do not add new source files or make changes in the existing file dependencies, issuing the following command at the DOS prompt:

```
RMAKE PE ↵
```

will compile and link the PE source files to create a new PE.EXE executable file. See the *Program Maintenance—RMAKE.EXE* chapter in this guide for more information about make files.

Summary

In this chapter, you have learned about PE and its basic functionality. This program is useful for editing program files (.prg) if you do not already own a professional program editor. It also serves as an example illustrating how the MEMOEDIT() function may be programmed to create text editing systems for your application programs. PE, however, is limited and is not intended to replace professional program editing systems.

Chapter 13
Database Utility
DBU.EXE

In order to provide you with an interactive database design environment, *CA-Clipper* includes the DBU system. The system is a stand-alone application written in *CA-Clipper* that allows you to build database files, add data to the files, browse existing data, create and attach index files, and construct views using a completely menu-driven system. This chapter explains how to use all of the features provided in DBU.

In This Chapter

In this chapter, the following topics are discussed:

- Invoking DBU
- The main DBU screen
- Leaving DBU
- The DBU menus

Invoking the Database Utility

DBU exists on the disk as the executable file, DBU.EXE. The general syntax for executing the program is as follows:

```
DBU [/<colorString>][<filename>] /e
```

The DBU command line arguments may be specified in any order.

<*colorString*> determines whether the DBU screen display is color or monochrome. Color is specified using /C and monochrome using /M. If

a color option is not specified, the default color mode is the result of ISCOLOR().

<filename> is the name of a view file (.vew) previously created in DBU or a database file (.dbf). If a view and database file have the same name, the view file is assumed unless the database file is explicitly specified with an extension.

/e opens the file EXCLUSIVE. The parameter is case-insensitive.

Specifying *<filename>* on the DBU command line causes the named file to be opened and browsed after DBU is loaded. See the explanation for *F5 Browse* in The DBU Menus section of this chapter for information on how to use the Browse window. Otherwise, the main screen is active where you can open files and access the DBU menu bar.

If you installed the default configuration of the *CA-Clipper* development environment, DBU.EXE and DBU.HLP were installed in \CLIPPER5\BIN, and this directory was added to the PATH statement in your AUTOEXEC.BAT. You should, therefore, be able to access DBU from any drive and/or directory.

The Main DBU Screen

If you invoke DBU without a *<filename>* argument, the first thing you see is the main screen which gives you a visual image of the current view.

A *view* consists of one or more database files open in separate work areas, their associated index files, the relationships between the files, the active fields, and a logical record filter for each work area. In fact, it only takes a single open database file to comprise a view—all other parts of the view are optional. As you open and close files, the main screen is updated to show the current view definition.

While on the main screen items may be added, changed, or deleted in one of three ways:

- Typing the name of a file or a field
- Pressing a command key (*Return*, *Ins*, or *Del*)
- Pressing a menu key to activate one of the dropdown menus represented on the menu bar

This section describes the various components of the DBU screen and how each one operates.

The Menu Bar

The *menu bar* at the top of the screen consists of several menu names with menu keys that you press to activate associated dropdown menus. The menu bar is always available except when there is a pending prompt. Take advantage of this by accessing the menu bar and exploring it for available options whenever you are located in a window or dialog box. You may be quite surprised at the number of operations available.

The *menu names* on the menu bar are designed to give you a brief description of what operations you can perform with the associated dropdown menu.

The *menu key* is a function key whose name is displayed directly above the menu name. Pressing it activates the dropdown menu.

The *dropdown menu* is a vertical menu that is displayed directly below the menu name. When one is active, you can see all of the *menu items* that it has to offer. Available items are displayed in high intensity, while items that are disabled appear dimmer.

The *highlight* marks the current item and does not appear in the menu if all of the items are disabled. The figure below shows the DBU menu bar and all of its components.

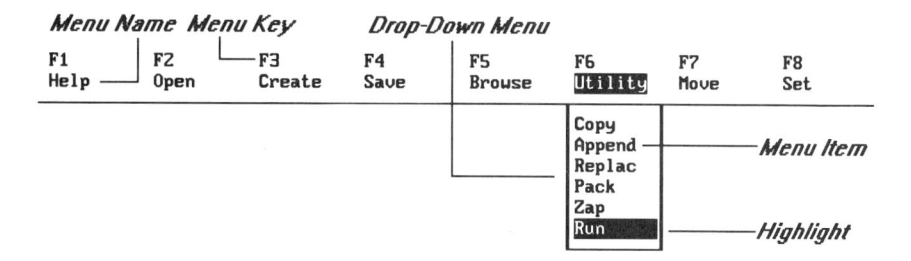

```
Menu Name  Menu Key        Drop-Down Menu
F1      F2      F3      F4      F5      F6       F7     F8
Help    Open    Create  Save    Browse  Utility  Move   Set
                                        Copy
                                        Append ———————— Menu Item
                                        Replac
                                        Pack
                                        Zap
                                        Run    ———————— Highlight
```

Figure 13-1: The DBU Menu Bar

When a menu is active, *Uparrow* and *Dnarrow* move the highlight within the menu and *Return* selects, or executes, the currently highlighted item. When using the arrow keys to navigate within a menu, navigation is not circular (e.g., pressing *Dnarrow* on the last menu item does not wrap-around to the first item).

A shortcut for navigating to a particular menu item is to type the first letter of its name. When using the first letter shortcut, menu navigation is circular which means that the highlight will move past the last item by wrapping around to the top of the menu if necessary.

Leftarrow and *Rightarrow* activate the previous and next menu on the menu bar. Pressing *Esc* terminates the menu and returns control to the previous operation.

Note: Pressing the same menu key twice activates the specified menu and selects the currently highlighted menu item allowing you to perform the same operation repetitively.

The Message and Prompt Area

The message and prompt area is located in the upper lefthand area of the main screen just below the menu bar. This area is used to display status messages while DBU performs certain operations and error messages when an error condition is encountered. Error messages remain on the screen until a key is pressed.

In addition to messages, this area of the screen is used for *prompts* which are messages requiring a user response. Generally, these are warnings requiring you to verify that you want to perform a potentially destructive action. Pressing *Esc* or *N* terminates a prompt without performing the action. Pressing *Y* executes the selected action.

Dialog Boxes

Most selections that you make from the menu bar cause a box to open on the main screen which is called a *dialog box*. Its purpose is to prompt you for additional information that is needed to complete your selection.

Dialog boxes are similar to dropdown menus in that you use the arrow keys to move the highlight from one item to another and use *Return* to make a selection. They are different in that selecting an item does not necessarily close the dialog box since it is likely that you will want to make several selections. The items in a dialog box are called *controls*. The figure below illustrates the components of a typical DBU dialog box.

The Main DBU Screen

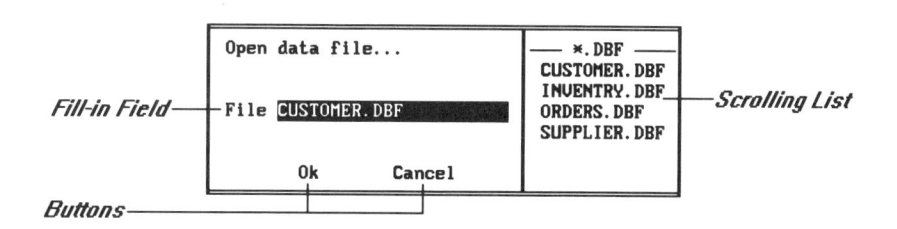

```
                    Open data file...         ┌── *.DBF ──┐
                                              │CUSTOMER.DBF│
                                              │INVENTRY.DBF│
    Fill-in Field ──── File CUSTOMER.DBF      │ORDERS.DBF  │──── Scrolling List
                                              │SUPPLIER.DBF│
                              Ok    Cancel
         Buttons ─────────────┘       └─┘
```

Figure 13-2: A Typical Dialog Box

Note: When a dialog box is open on the screen, you may only perform the actions available in the box. An attempt to activate a menu (with the exception of F1 Help which is always available) will reveal that all options are disabled.

All dialog boxes in DBU are similar in their behavior but may be slightly different in appearance depending on the information needed. This section describes each type of control that you will see. Not all dialog boxes will have each type of control described but, when a particular control is present, it will always behave in the manner described.

Buttons

Buttons are controls in a dialog box that are usually grouped together and displayed side-by-side. In DBU, there are *push buttons* to perform some sort of action and *radio buttons* to make one of several mutually exclusive selections.

Ok and *Cancel* are examples of push buttons. Selecting *Ok* confirms the choices you have made within the dialog box and performs the associated menu action. Selecting *Cancel* cancels the dialog box as well as the associated menu action.

SDF and *DELIMITED* are examples of radio buttons which do not perform an action, but rather set some parameter. Selecting a radio button causes it to be surrounded by a box as an indicator and deselects any other one that might have been selected. A button that is selected may also be deselected in the same manner as it was selected, thereby leaving all buttons in the group unselected.

Fill-in Fields

A *fill-in field* is a control in which you type text from the keyboard into a data entry area. Data is entered in the same manner as all data entry is performed in *CA-Clipper*. For example, the arrow keys are used to move the cursor around within the entry area, *Ins* is used to toggle

The Main DBU Screen

insert/overwrite mode, *Del* is to delete the current character, and *Esc* undoes any changes that were made. Fill-in fields are used, for example, to enter filenames and expressions.

Unlike other controls, you do not have to press *Return* to select a fill-in field—you may just start typing once it is highlighted. If the fill-in field already has contents, it is overwritten. To edit a fill-in field, however, you must press *Return* before you begin to type, move the cursor to the correct location, and make the change.

In many cases, the information you enter into a fill-in field will be a logical condition or some other type of expression. In these cases, the expression must be valid according to the rules established by the *CA-Clipper* language. For more information on expressions, see the *Basic Concepts* chapter in this guide.

Scrolling Lists

A *scrolling list* is a control that displays a list of items. Selecting a scrolling list item places the item in the currently highlighted fill-in field. The list is called a scrolling list because if there are more items than will fit in the dialog box (there are small arrows to indicate this situation), the list will scroll to bring those items into view.

There are several examples of scrolling lists in the DBU interface, including file and field lists. Navigation and selection in a scrolling list are the same as in a dropdown menu.

Closing a Dialog Box

To close a dialog box, you can select either the *Ok* or the *Cancel* button at the bottom of the box. Selecting *Ok* closes the dialog box and performs the action which activated it using all of the controls that you have set. Selecting *Cancel* (or pressing *Esc*) closes the box without performing the action. Both selections return you to the main screen.

Windows

Other menu selections open a window on the main screen. A *window* is similar in appearance to a dialog box but has a special purpose. Several windows are available in the DBU interface (e.g., *F3 Create:Database* and *F5 Browse:Database*), and each behaves more or less independently in order to accomplish its specific task. Unlike a dialog box, certain menu items other than *F1 Help* are available when a window is active. The figure below shows a typical DBU window.

The Main DBU Screen

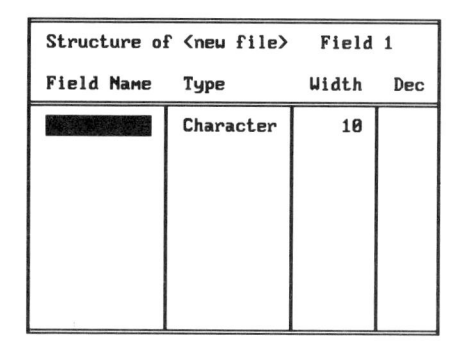

```
Structure of <new file>    Field 1

Field Name    Type       Width    Dec

▒▒▒▒▒▒▒▒     Character    10
```

Figure 13-3: A Typical DBU Window

How you navigate and what you can do inside a window depends on the window. Each window is described individually in The DBU Menus section of this chapter under the menu selection that you use to open it.

Work Areas

The screen is divided into columns with each column representing a *CA-Clipper work area*. Up to six work areas are available in DBU. Each work area is divided into three separate sections on the screen which identify its attributes and the database file, an index file list, and a field list. The files from all work areas, taken together with relations and filters comprise the current view. The figure below shows the DBU work areas with two open database files.

The Main DBU Screen

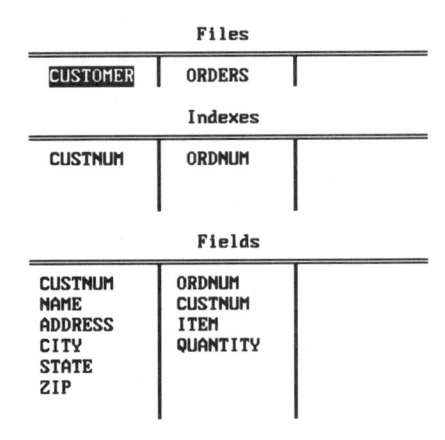

	Files	
CUSTOMER	ORDERS	

	Indexes	
CUSTNUM	ORDNUM	

	Fields	
CUSTNUM NAME ADDRESS CITY STATE ZIP	ORDNUM CUSTNUM ITEM QUANTITY	

Figure 13-4: DBU Work Areas

Files

The Files area is for opening and closing database files (.dbf). Pressing *Return* when the Files area is highlighted allows you to open an existing database file (.dbf) either by selecting a filename from a scrolling list or by typing the filename in a dialog box. Up to six database files can be open at one time in DBU.

When you open a database file, an unused Files column is also opened to represent the next available work area (unless all six work areas are already occupied). When the highlight is located in the unused work area, you cannot move it to the Indexes or Fields areas, nor can you select any menu item that requires an open database file. Other operations such as opening or creating a database file, however, are allowed.

Indexes

Highlighting the Indexes area allows you to open up to seven index files per work area. Opening an index file is accomplished in the same manner as opening a database file. To open more than one index file, move the highlight below the current index filename and press *Return*.

Opening database and index files in this manner (i.e., with *Return*) closes any other open files that happened to occupy that work area. *Ins*, on the other hand, allows you to open a file by pushing the other files across one column—if necessary, closing the file occupying the last work area to make room for the new one. Opening database and index files using

Ins is equivalent to using the *F2 Open* menu. *Del* closes the file in the currently highlighted work area.

Fields

Highlighting the Fields area allows you to delete, insert, and overwrite fields in the field list for the active database file. Operations on the Fields area do not affect the database file structure—only the field list that is used when you edit with the *F5 Browse* menu.

To delete a field from the list, highlight the field name and press *Del*.

To insert a field, highlight the proper location for the field and press *Ins*. The field dialog box opens allowing you to select the field to insert from a scrolling list. This action is equivalent to selecting *F8 Set:Fields* from the menu bar.

You can combine the actions of inserting a new field and deleting the current one by highlighting the field you want to delete and pressing *Return*. Similar to pressing *Ins*, this action opens the field dialog box; however, pressing *Return* causes the field that you select to overwrite the current field—thereby deleting the current field and inserting a new one. Using a combination of these actions, you can construct a field list for the current database file that is reflected when you select *F5 Browse:Database*.

Note: Besides using *Ins*, *Del*, and *Return* to manipulate files and fields, you can type file and field names directly on the main screen in the appropriate work area sections. Typing names in this manner is equivalent to selecting them from a dialog box.

Navigation on the Main Screen

The following table summarizes the keys that are available for use when operating on the main screen in DBU. These keys apply only when the main screen is active and you are dealing directly with the file and field work area attributes. In short, they do not apply when a menu, dialog box, or window is active.

Leaving DBU

DBU Main Screen Navigation Keys

Key	Action	Work Area Attribute
Leftarrow	Previous column	All
Rightarrow	Next column	All
Uparrow	Previous row	All
Dnarrow	Next row	All
PgUp	First row or previous section	All
PgDn	Last row or next section	All
Ctrl-PgUp	First row in current column/section	All
Ctrl-PgDn	Last row in current column/section	All
Home	First column in current section	All
End	Last column in current section	All
Return	Open file and close current Select field and delete current	Files, Indexes Fields
Ins	Open file without closing current Select field without deleting current	Files, Indexes Fields
Del	Close file Delete field	Files, Indexes Fields
Esc	Quit DBU	All
Menu Key	Activate dropdown menu	All

Leaving DBU

When you are finished with DBU, you can exit the utility by pressing *Esc* and responding with *Y* at the prompt. All files that you created and any data that you added or changed are automatically saved. A view, on the other hand, must be saved to disk if you plan to use it again without having to set up all of the components in the next DBU session.

Once a view is set up on the main screen with all the necessary files and fields represented, you can save the image to a view file (.vew). To do this, select *F4 Save:View* and enter the requested information into the dialog box. Later on, you can select *F2 Open:View* and enter the view filename as it was saved. All files in the view will be opened and all field lists established. For more information on view files, see *F4 Save* in The DBU Menus section in this chapter.

The DBU Menus

The DBU menu bar has a menu name and function key for each menu in the system. The menus are always available and are activated by pressing the corresponding menu key to activate a dropdown menu.

A highlight indicates the current menu item. Items that appear in bold are available for selection, while dimmer items are disabled. If all items are disabled, no highlight appears and no selection may be made from that menu. The following table summarizes the keys used to navigate and make selections when a menu is active:

Menu Navigation and Selection Keys

Key	Action
Return	Select current item
Esc	Abandon menu without selecting
Uparrow	Highlight previous item
Dnarrow	Highlight next item
Leftarrow	Activate previous menu
Rightarrow	Activate next menu
Home	Highlight first item
End	Highlight last item

In addition to these keys, you can navigate to a particular menu item by typing the first letter of its name. If there are several items in the menu beginning with the same first letter, this method navigates to the next one, wrapping back around to the top of the menu if necessary.

DBU uses the highlight on the main screen to make decisions about how certain menu selections operate. For example if you select *F6 Utility:Copy*, DBU assumes you want to copy from the currently highlighted database file.

Note: A menu key may be pressed when another menu is active to quickly open another menu. Pressing the same menu key twice in succession activates the menu and selects the currently highlighted item, allowing you to perform the same operation repetitively.

F1 Help

The *F1 Help* menu contains only a single menu item, also named *Help*. Selecting it gives you help information that is context-sensitive, depending on what you are trying to do when you make the menu

The DBU Menus

selection. For instance, if you are trying to open a database file and the open dialog box is active on the screen, requesting help gives you information on opening a database file, as shown in the figure below.

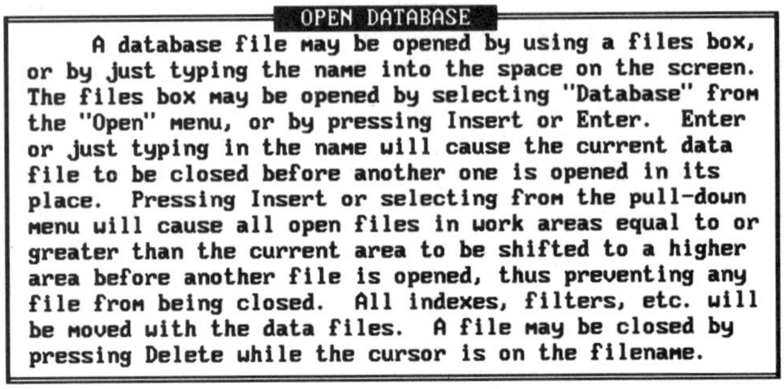

```
┌──────────────── OPEN DATABASE ────────────────┐
│     A database file may be opened by using a files box, │
│ or by just typing the name into the space on the screen. │
│ The files box may be opened by selecting "Database" from │
│ the "Open" menu, or by pressing Insert or Enter.  Enter │
│ or just typing in the name will cause the current data │
│ file to be closed before another one is opened in its │
│ place.  Pressing Insert or selecting from the pull-down │
│ menu will cause all open files in work areas equal to or │
│ greater than the current area to be shifted to a higher │
│ area before another file is opened, thus preventing any │
│ file from being closed.  All indexes, filters, etc. will │
│ be moved with the data files.  A file may be closed by │
│ pressing Delete while the cursor is on the filename. │
└─────────────────────────────────────────────┘
```

Figure 13-5: The Help Window

The help text is displayed in a window. If there is more information than will fit in the window, use *Uparrow, Dnarrow, PgUp,* and *PgDn* to scroll through the text. Press *Esc* when you have finished and want to close the window.

F2 Open

F2 Open allows you to open files. The menu contains the following items:

- Database
- Index
- View

All menu items activate an open dialog box similar to the one in the following figure. The scrolling list contains a different file type depending on the item you select.

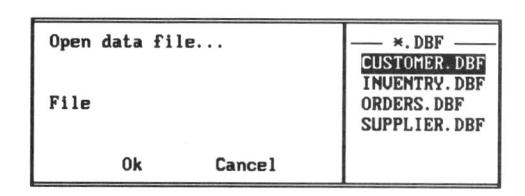

```
┌──────────────────────┬─── *.DBF ────┐
│ Open data file...    │ CUSTOMER.DBF │
│                      │ INVENTRY.DBF │
│ File                 │ ORDERS.DBF   │
│                      │ SUPPLIER.DBF │
│     Ok      Cancel   │              │
└──────────────────────┴──────────────┘
```

Figure 13-6: The Open Database Dialog Box

If the highlight on the main screen is located in the Indexes area, *Index* will be the current item when you activate the *Open* menu. Otherwise, the current item will be *Database*.

For all *Open* menu items, selecting an already open file or entering the name of a nonexistent file causes an error.

Database

Database opens a database file (.dbf) in the currently highlighted work area on the main screen. Selecting this menu item activates an open dialog box where you may either select a database filename from a scrolling list or type the filename directly into a fill-in field.

When you select *Ok*, the file is opened in the currently highlighted work area, pushing all other open files to the right by one work area. Any file occupying work area six is closed.

The open database dialog box may be activated by pressing *Return* or *Ins* instead of using the menu. If you know the name of the file that you want to open, you may also type it directly on the main screen.

Index

Index opens an index file (.ntx) in the currently highlighted work area. This menu item is available only if there is a corresponding open database file. Selecting *Index* activates an open dialog box where you may either select an index filename from a scrolling list or type the filename directly into a fill-in field.

When you select *Ok*, the file is opened in the currently highlighted slot, pushing all other open files across one slot. Any file occupying slot seven is closed.

The open index dialog box may be activated by pressing *Return* or *Ins* instead of using the menu. If you know the name of the file that you want to open, you may also type it directly on the main screen. The index file at the top of the column is the controlling index.

View

View opens a view file (.vew) by opening all its associated database and index files, establishing the field lists for each database file, establishing the relations between the files, and activating filters for each work area. Selecting this menu item activates an open dialog box where you may either select a view filename from a scrolling list or type the filename directly into a fill-in field.

Note: A view file can be designed, saved, and opened only in DBU—it cannot be used subsequently in a *CA-Clipper* program the way database and index files can.

F3 Create

F3 Create allows you to create and modify database and index files. The menu contains the following items:

- Database
- Index

If the highlight on the main screen is located in the Indexes area, *Index* will be the current item when you activate the *Create* menu. Otherwise, the current item will be *Database*.

Database

Selecting *Database* opens the Structure window in which you enter field definitions to define a database file structure. If a database file is open in the current work area on the main screen, its structure will be displayed in the window for you to modify, as shown in the following figure. Otherwise, the window will be empty under the assumption that you want to create a new file.

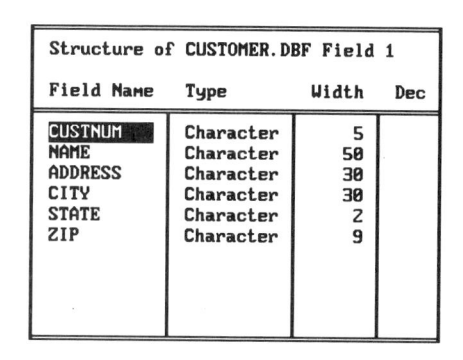

```
Structure of CUSTOMER.DBF Field 1

Field Name    Type         Width   Dec

CUSTNUM       Character        5
NAME          Character       50
ADDRESS       Character       30
CITY          Character       30
STATE         Character        2
ZIP           Character        9
```

Figure 13-7: The Structure Window

The following table summarizes the keys that are available when the Structure window is active:

Structure Window Keys

Key	Action
Return	Enter input mode Go to next column
Esc	Abandon input mode without saving Exit without saving structure
F4	Exit and save structure
SpaceBar	Select next value in Type column
Uparrow	Go to previous field definition
Dnarrow	Go to next field definition
Leftarrow	Go to previous column
Rightarrow	Go to next column
Ins	Insert new field definition
Del	Delete current field definition
Home	Go to first column
End	Go to last column
Ctrl-Home	Go to first column
Ctrl-End	Go to last column
PgUp	Scroll field definitions back
PgDn	Scroll field definitions forward
Ctrl-PgUp	Go to first field definition
Ctrl-PgDn	Go to last field definition

For each field in the database file, you must define a field name and data type. Then, depending on the data type, you may also have to define the field width and number of decimal places.

A field definition is checked for validity when you attempt to move the highlight to another field (e.g., by pressing *Dnarrow* to move to the next field). If you have made any mistakes an error message is displayed in the message and prompt area, and the highlight is moved to the erroneous column.

Field Name

To create a new field in the Structure window, move the highlight to an empty row in the Field Name column and type the field name. Field names may be up to ten characters in length, must begin with a letter, and may contain letters, numbers, and the underscore character only.

To edit an existing field name, highlight it and press *Return* before typing. This places you in input mode where data is entered in the same manner as all data entry is performed in *CA-Clipper*. For example, the arrow keys are used to move the cursor around within the entry area, *Ins* is used to toggle insert/overwrite mode, and *Esc* undoes any changes that were made. *Return* is used to terminate input mode. Field Name is essentially the same as a fill-in area in a dialog box.

Type

The Type column defines the data type of the field. In *CA-Clipper*, the valid field types are as follows:

- Character
- Date
- Logical
- Memo
- Numeric

To select a data type for a field, highlight the Type column of the field and type the first letter of the type you want. As an alternative, you can press the *SpaceBar* to move through the available data types until the one that you want is displayed.

Width

The Width column defines the length of the field. For date, logical, and memo fields, DBU assigns the column width automatically as 8, 1, and 10, respectively.

To enter a field width for a character or numeric field, move the highlight to the Width column and type a number. For character fields the number must be between one and 1024, and the default is ten. For numeric fields the width must be between one and 19, and the default is also ten.

To edit an existing field width, highlight it and press *Return* before typing. This places you in input mode where data is entered in the same manner as all data entry is performed in *CA-Clipper*. Width is essentially the same as a fill-in area in a dialog box.

Decimals

The Decimals column is for numeric fields only. It defines the number of decimal places allowed in the field.

The number that you enter in this column must be at least two less than the field width. Thus, it can range between one and 17, but the actual range depends on the current field width. Like Field Name and Width, Decimals behaves like a fill-in area in a dialog box—just start typing for a new value or press *Return* first to edit the current value.

Inserting and Deleting Fields

To insert a new field when you have one or more field definitions already in the Structure window, highlight the field before which you want to insert the new field and press *Ins*. An empty row will open in the window and all existing fields will be moved down one space.

To delete a field definition, highlight it and press *Del*. The field is deleted and all fields following it are moved up to close the space.

Saving the File Structure

When you have made all of the changes that you want to the database file structure, you can choose to either save the changes or abandon them altogether.

To abandon all changes and return to the main screen, press *Esc*. A prompt will be displayed in the message and prompt area for you to confirm your choice.

The DBU Menus

To save the database file structure, select *F4 Save:Struct*. A save dialog box will open on the screen where you can enter the name of a new database file or accept the current filename. If you select *Ok*, the file is created and opened in the currently highlighted work area. This dialog box is described under *F4 Save* in this section.

Warning! *When you are modifying a database file structure, certain changes will cause data to be lost when you save the file. In all cases, DBU attempts to save as much data as possible and may prompt you if it has a decision to make regarding whether or not to preserve data. To make sure that you do not lose anything important, make backup copies of the database file before modifying its structure significantly.*

Index

Index allows you to create a new index file or modify the key expression of an existing one for the database file in the current work area. To create a new file, move the highlight to an unused slot in the Indexes column; to modify an existing index key expression, make sure the index file is open in one of the Indexes columns and highlight its filename. Selecting this menu item activates a dialog box where an index filename and key expression may be entered as shown in the following figure.

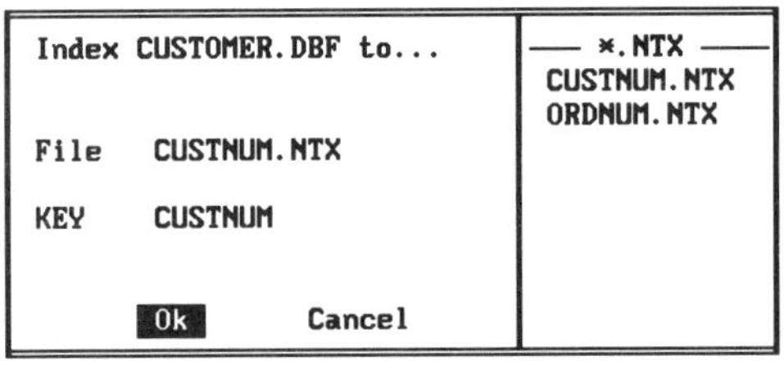

```
┌─────────────────────────────┬──────────────┐
│ Index CUSTOMER.DBF to...    │ ──── *.NTX ──│
│                             │ CUSTNUM.NTX  │
│                             │ ORDNUM.NTX   │
│   File    CUSTNUM.NTX       │              │
│                             │              │
│   KEY     CUSTNUM           │              │
│                             │              │
│            ▓Ok▓    Cancel   │              │
└─────────────────────────────┴──────────────┘
```

Figure 13-8: Create Index Dialog Box

The filename may be selected from a scrolling list or typed directly into a fill-in field. If an open index file is currently highlighted on the main screen, its name (and key expression) automatically appear in the dialog box. If you change the filename, a new index file will be created with that name; otherwise, the current index file will be recreated. If there is no current index file when you make the menu selection, no filename appears and you must provide one.

The index key is typed into a fill-in field. If there is already an index key expression present, it may be edited by pressing *Return* before you begin to type.

13-18   CA-Clipper

When you select *Ok* in the Create Index dialog box, the file is created and opened in the currently highlighted slot.

Note: Create:Index can be used to reindex a database file. To do this, go through the steps as if you were going to modify the key expression without changing anything. Simply select Ok, and the index will be recreated.

F4 Save

F4 Save allows you save a newly created or modified database file structure to a (.dbf) file or to save the view defined on the main screen to a (.vew) file. The menu contains the following items:

- View
- Struct

View

View saves the database environment defined on the main screen, including open database and index files, relations, filters, and field lists. Selecting this menu item opens a dialog box as shown in the figure below. You can select a filename from a scrolling list or type it directly into a fill-in field. Using an existing filename overwrites the file, while typing a new filename creates a new file.

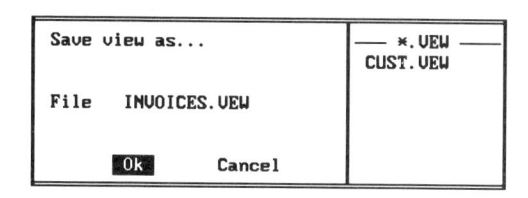

```
Save view as...                ── *.VEW ──
                               CUST.VEW
File    INVOICES.VEW

        [ Ok ]    Cancel
```

Figure 13-9: The Save View Dialog Box

When you select *Ok*, a view file (.vew) is created which can be opened and modified in DBU. You cannot, however, use a view file outside of DBU in a *CA-Clipper* program.

Struct

Struct saves the database file structure that you are actively creating or modifying in the Structure window. Selecting this menu item opens a

The DBU Menus

dialog box in which you can select a filename from a scrolling list or type it directly into a fill-in field.

Selecting *Ok* saves the named database file (.dbf) and appends the original data back into the file, if necessary.

Struct is not available unless you have an open Structure window on the screen (see *F3 Create* in this section).

F5 Browse

F5 Browse allows you to view and edit data in a database file or a view. The menu contains the following items:

- Database
- View

Each of these menu items opens a Browse window on the screen, and this section describes how to operate within that window. To close the window, press *Esc*. Any changes that you made will automatically be saved.

Database

Database opens a Browse window for the database file in the currently highlighted work area, as shown in the figure below. Only those fields shown in the Fields area on the main screen are available for editing. You can make changes to the existing record simply by typing over the old information, and you can add new records by moving the highlight below the last record in the file. Pressing *Del* marks the current record for deletion or reinstates the record if it is already marked.

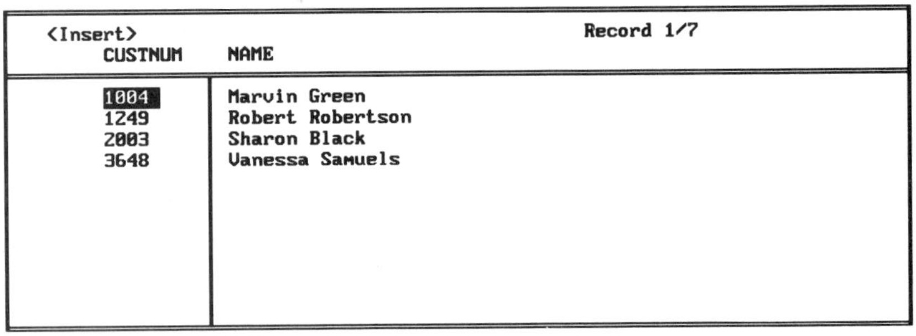

```
<Insert>                                    Record 1/7
CUSTNUM    NAME
1004       Marvin Green
1249       Robert Robertson
2003       Sharon Black
3648       Vanessa Samuels
```

Figure 13-10: The Browse Window

The following table summarizes the keys that you can use when a Browse window is active. These keys apply to both the *Database* and the *View* menu items.

Browse Window Keys

Key	Action
Return	Enter input mode Go to next field
Esc	Abandon input mode without saving Exit Browse
Uparrow	Go to previous record
Dnarrow	Go to next record
Leftarrow	Go to previous field
Rightarrow	Go to next field
Ctrl-Leftarrow	Pan screen to the right
Ctrl-Rightarrow	Pan screen to the left
Ins	Toggle insert mode
Del	Toggle record delete status
Home	Go to first field onscreen
End	Go to last field onscreen
Ctrl-Home	Go to first field
Ctrl-End	Go to last field
PgUp	Scroll records back
PgDn	Scroll records forward
Ctrl-PgUp	Go to first record
Ctrl-PgDn	Go to last record

Search functions are also available for quick navigation within the Browse window. See the explanation for the *F7 Move* menu in this section.

View

View opens a Browse window for the entire view defined on the main screen. All fields in the Fields area for all files in the Files area are available for editing. The window behaves as if you were editing a database file, except that you can neither edit data from multiple files nor append new records.

Editing Memo Fields

Memo fields are edited in a popup window. To open the window, highlight the memo field and press *Return* or just start typing. The following table summarizes all of the editing and navigation keys available when a memo editing window is active.

Memo Editing Keys

Key	Action
Return	Go to next line
Esc	Exit without saving
Uparrow	Go to previous line
Dnarrow	Go to next line
Leftarrow	Go to previous character
Rightarrow	Go to next character
Ctrl-Leftarrow	Go to first character of previous word
Ctrl-Rightarrow	Go to first character of next word
Ins	Toggle insert mode
Del	Delete current character
Backspace	Delete previous character
Home	Go to beginning of line
End	Go to end of line
Ctrl-Home	Go to beginning of file
Ctrl-End	Go to end of file
PgUp	Scroll screen back
PgDn	Scroll screen forward
Ctrl-PgUp	Go to beginning of screen
Ctrl-PgDn	Go to end of screen
Ctrl-B	Reformat current paragraph
Ctrl-N	Insert blank line
Ctrl-T	Delete next word
Ctrl-W	Exit and save
Ctrl-Y	Delete current line

F6 Utility

F6 Utility provides you with several batch database operations. Each of the menu items listed below directly corresponds to a *CA-Clipper* command of the same name:

- Copy

- Append
- Replace
- Pack
- Zap
- Run

Copy

Copy allows you to copy the current database file to another database file or a text file. Any filter condition that is in effect is respected, but the field list is ignored as are any file relations. Only the active database file is copied, and it is copied in order according to the first index file in the Indexes column, if any.

To copy a file, move the highlight to the database file work area that you want to copy; *Copy* cannot be selected if the *Utility* menu is activated when the unused work area is highlighted.

Selecting this menu item activates a dialog box (shown in the figure below) where a filename, FOR and WHILE conditions, and a record scope may be entered. SDF or DELIMITED may also be selected to copy to a text file instead of a database file.

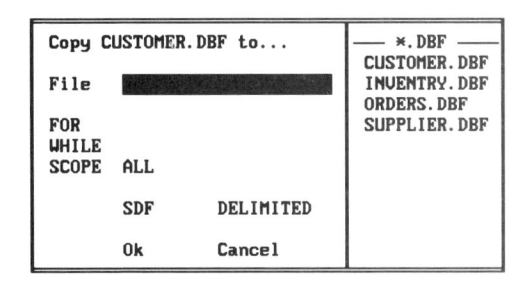

```
Copy CUSTOMER.DBF to...          ── *.DBF ──
                                 CUSTOMER.DBF
File   ▮▮▮▮▮▮▮▮▮▮▮▮▮▮            INVENTRY.DBF
                                 ORDERS.DBF
FOR                              SUPPLIER.DBF
WHILE
SCOPE  ALL

       SDF        DELIMITED

       Ok         Cancel
```

Figure 13-11: The Copy Dialog Box

The filename may be selected from a scrolling list or typed directly into a fill-in field. If SDF or DELIMITED is selected, the scrolling list will show text files (.txt); otherwise, it will show database files (.dbf).

SDF and DELIMITED are radio buttons, only one of which may be selected at a time. If one is selected, it is surrounded by a box as an indicator. To select (or deselect) one of these buttons, highlight it and press *Return*.

SDF indicates a text file that is undelimited and without separators between fields. DELIMITED indicates a text file in which fields are separated by commas and character fields are enclosed in quotation marks.

The remaining controls in the dialog box are used to designate a subset to operate on in lieu of all records in the current database file.

The SCOPE can be either ALL records or a specific number of records. This control acts like a kind of specialized fill-in field. If you select it, the ALL changes to NEXT with a 0 to the right of it, and you are expected to enter a positive number. Pressing *Return* with the number set at zero will return the scope to the default, ALL records. Any other number changes the scope to NEXT *n*, causing the operation to begin with the current record and continue until *n* records have been processed or until the end of file is encountered.

FOR and WHILE are both fill-in fields where you enter expressions to extract a logical subset of records from the file. The value that you enter must be a valid logical expression.

A FOR condition will be evaluated as each new record is considered, and those records that evaluate to false (.F.) will not be processed. This process continues until the end of file is reached.

A WHILE condition causes processing to begin with the current record and continue only while the condition evaluates to true (.T.). If both FOR and WHILE conditions are entered, WHILE takes precedence over FOR.

After all controls are set, selecting *Ok* creates the new file but does not open it on the main screen.

Append

Append allows you to add records to the active database file from either a database file or a text file.

To add records from one file to another, move the highlight to the work area containing the file that you want to append to; *Append* cannot be selected if the *Utility* menu is activated when the unused work area is highlighted.

Selecting this menu item activates a dialog box that is almost identical to the *Copy* dialog box. In it a filename, FOR and WHILE conditions, and a record scope may be entered. SDF or DELIMITED may also be selected to append from a text file instead of a database file.

The filename may be selected from a scrolling list or typed directly into a fill-in field. If SDF or DELIMITED is selected, the scrolling list will show text files (.txt); otherwise, it will show database files (.dbf).

The remaining controls were described previously in the discussion of the *Copy* dialog box and may be set to control which records are added. Selecting *Ok* appends the designated records to the end of the current file and updates all open index files in the same work area.

Replace

Replace does conditional or global field replacements in the active database file. In DBU, only one field at a time can be replaced.

To replace a field in a database file, move the highlight to the appropriate work area; *Replace* cannot be selected if the *Utility* menu is activated when the unused work area is highlighted.

Selecting this menu item activates the dialog box shown in the figure below. To perform the replace, you must select a field name from the scrolling list and enter a WITH expression in the fill-in field. FOR and WHILE conditions and a record scope may be designated also.

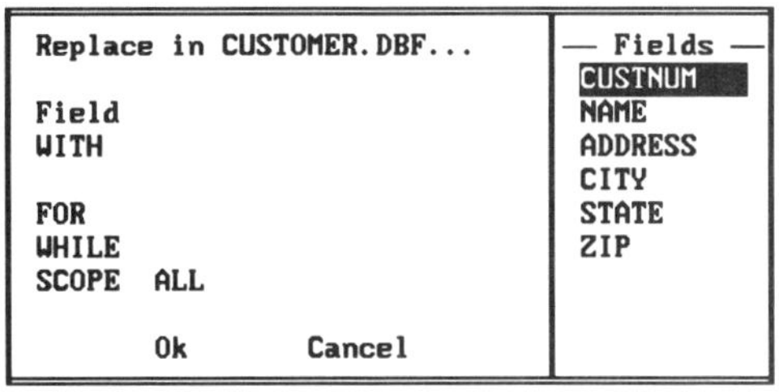

```
 Replace in CUSTOMER.DBF...      ── Fields ──
                                 CUSTNUM
 Field                           NAME
 WITH                            ADDRESS
                                 CITY
 FOR                             STATE
 WHILE                           ZIP
 SCOPE   ALL

           Ok        Cancel
```

Figure 13-12: The Replace Dialog Box

The Field area in this dialog box is not a fill-in field; you must select the field from the scrolling list. This is done as a precaution to make sure that only valid field names are used.

WITH is a fill-in field, and the value that you enter must be a valid expression with the same data type as the field you select.

The remaining controls were described previously in the discussion of the *Copy* dialog box and may be set to control which records are replaced. Selecting *Ok* replaces the indicated field in the specified records and updates all index files in the same work area.

The DBU Menus

Pack

Pack permanently removes records that are marked for deletion in the active database files. In DBU, records are marked for deletion in the Browse window using the *Del* key.

To pack a database file, move the highlight to the appropriate work area; *Pack* cannot be selected if the *Utility* menu is activated when the unused work area is highlighted.

Selecting this menu item causes a prompt to be displayed in the message and prompt area which you must answer before the operation will proceed. Typing *Y* packs the database file and updates all open index files. Typing *N* or pressing *Esc* abandons the operation without packing the file.

Zap

Zap permanently removes all records from the active database file, regardless of their delete status. This item is useful for getting rid of test data in a file.

To zap a database file, move the highlight to the appropriate work area; *Zap* cannot be selected if the *Utility* menu is activated when the unused work area is highlighted.

Selecting this menu item causes a prompt to be displayed in the message and prompt area which you must answer before the operation will proceed. Typing *Y* zaps the database file and updates all open index files. Typing *N* or pressing *Esc* abandons the operation.

Run

Run allows you to execute another program from within DBU. In order to use this menu item, you must have enough available memory to load the other program.

Selecting *Run* causes a prompt to be displayed at the bottom of the screen. To the right of the prompt is a large fill-in field where you can enter any DOS command just as if the DOS prompt were on the screen. Press *Return* to execute the command.

After the command that you enter has completed execution, you are returned to DBU with a new prompt at the bottom of the screen. At this point, you can execute another command or press *Esc* to go back to the main screen.

F7 Move

F7 Move gives you a quick way to move around in a Browse window. The menu items are as follows:

- Seek
- Goto
- Locate
- Skip

None of the items in this menu are available unless you have an open Browse window on the screen (see *F5 Browse* in this section). All record pointer movement is performed in the current work area or the first work area if you are browsing a view.

Seek

Seek allows you to locate a particular index key value in the Browse window. This menu item is available only if there is an open index file.

When you select *Seek*, a dialog box (shown in the figure below) opens and allows you to enter an expression into a fill-in field. The value that you enter must be a valid expression and must match the data type of the controlling index key expression. Do not use quotes when seeking a character expression.

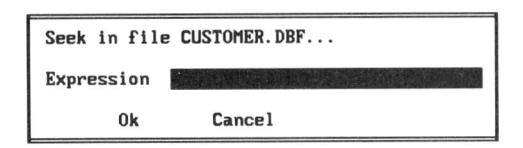

```
Seek in file CUSTOMER.DBF...
Expression  [                    ]
        Ok      Cancel
```

Figure 13-13: The Seek Dialog Box

Selecting *Ok* locates the first record in the file with a matching index key and highlights it for editing in the window. If not found, an error message is displayed in the message and prompt area, and the dialog box remains open on the screen. You can either *Cancel* or enter a new key value.

Goto

Goto jumps to a particular record number in the Browse window. When you select this menu item, a dialog box opens in which you enter a number, as shown in the figure below.

The DBU Menus

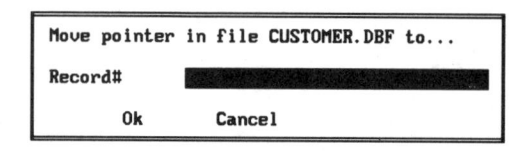

```
┌─────────────────────────────────────────────┐
│ Move pointer in file CUSTOMER.DBF to...     │
│ Record#        ███████████████████████████  │
│          Ok         Cancel                  │
└─────────────────────────────────────────────┘
```

Figure 13-14: The Goto Dialog Box

Selecting *Ok* locates the record with the indicated record number and highlights it for editing in the window. If there is no such record number, an error message is displayed in the message and prompt area, and the dialog box remains open on the screen. You can either *Cancel* or enter a new number.

Locate

Locate lets you search forward from the current record position for the first record matching a logical search condition. When you select this menu item, a dialog box opens and allows you to enter a logical expression into a fill-in field, as shown in the figure below.

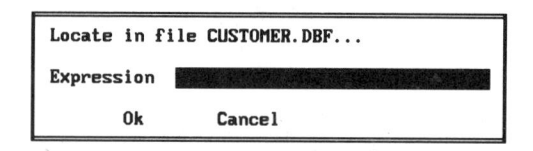

```
┌─────────────────────────────────────────────┐
│ Locate in file CUSTOMER.DBF...              │
│ Expression     ███████████████████████████  │
│          Ok         Cancel                  │
└─────────────────────────────────────────────┘
```

Figure 13-15: The Locate Dialog Box

Selecting *Ok* locates the next record in the file that meets the search condition and highlights it for editing in the window. If not found, an error message is displayed in the message and prompt area, and the dialog box remains open on the screen. You can either *Cancel* or enter a new search condition.

Skip

Skip allows you to move the record pointer forward or backward in the current work area. When you select this menu item, a dialog box (shown in the figure below) opens in which you enter the number of records to skip. A positive number is used to move the record pointer forward and a negative number to move it backward.

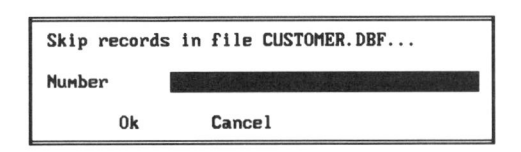

```
┌─────────────────────────────────────────────┐
│ Skip records in file CUSTOMER.DBF...        │
│ Number     ▓▓▓▓▓▓▓▓▓▓▓▓▓▓▓▓▓▓▓▓▓▓▓▓▓▓▓▓▓▓▓  │
│                                             │
│          Ok        Cancel                   │
└─────────────────────────────────────────────┘
```

Figure 13-16: The Skip Dialog Box

Selecting *Ok* updates the record pointer by skipping the specified number of records and highlights the new record for editing in the Browse window. Specifying a number which would advance the record pointer beyond the beginning or end of file selects the first or last record—no error message is displayed.

F8 Set

F8 Set allows you to establish relationships between database files and to set up a field list and filter condition to define a view. The menu items are as follows:

- Relation
- Filter
- Fields

All of these menu items require at least one open database file. Relation requires at least two in addition to the possible requirement of an index file.

To save the *Set* menu settings, you must select *F4 Save:View*. Otherwise, the settings will be lost when you exit DBU.

Relation

Relation allows you to model relationships between the open database files in the Files area, just as you would with the SET RELATION command, but from an interactive window. To use this item, make sure that all the necessary database and index files are open in an order that reflects the hierarchy of the relationship you want to establish—this is important since you will only be able to establish relations from left to right.

Selecting *Relation* opens the Set Relation window in which you select files and enter expressions to relate them. If there are already relations established between the files on the main screen, they will show up in the Set Relation window for you to edit. Otherwise, the window will be empty and you will have to start from scratch. The figure below shows a window in which a relation is being established.

```
┌─────────────────────────────Relations──────────────────────────────┐
│  ┌────────┐   ┌──────┐                                             │
│  │CUSTOMER├──▶│ORDERS│                                              │
│  └────────┘   └──────┘                                             │
│                                                                     │
│                                                                     │
│                                                                     │
│                                                                     │
│                                                                     │
└─────────────────────────────────────────────────────────────────────┘
```

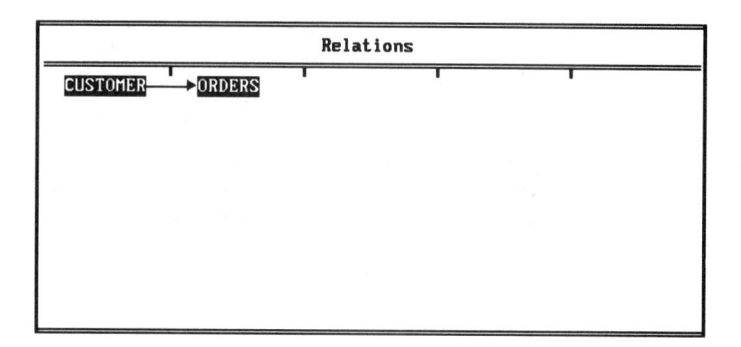

Figure 13-17: The Set Relation Window

To start, move the highlight in the window to the work area column containing the parent file and press *Return*. The filename will appear in the window.

To select a child file, press *Return* again. An arrow appears to visually indicate that a connection is being made between the two files.

Now move the highlight to the work area column containing the child file and press *Return*. A fill-in field area opens up just beneath the filenames where you type the relation expression.

Depending on whether or not the child file is indexed, the value that you enter here takes on a different significance. If the file is indexed, the two files will be related based on the controlling index key of the child; if not, they will be related based on record number. In the former (indexed) case, enter the field (or expression) in the parent that you want compared to the index key in the child. In the latter (unindexed) case, enter a numeric value (usually the RECNO() function) that you want compared to the record number in the child. When you are finished entering the expression, press *Return*.

At this point, *Dnarrow* can be used to establish another relation in the same manner. If a particular database file is the parent of more than one relation, a multiple parent-child relation is established. *Ins* inserts a relation definition and pushes existing ones down in the window, and *Del* deletes the current relation. You can establish up to 15 relations per view in DBU.

The following table summarizes the keys that are available for use when a Set Relation window is active.

Set Relation Window Keys

Key	Action
Return	Enter select mode Select parent or child Exit select mode
Esc	Exit select mode without saving Exit Set Relation
Uparrow	Go to previous relation
Dnarrow	Go to next relation
Leftarrow	In input mode, go to previous file
Rightarrow	In input mode, go to next file
Ins	Insert new relation
Del	Delete current relation
PgUp	Scroll relations back
PgDn	Scroll relations forward

After a relation is established, selecting *F5 Browse:View* reflects the relation by showing how the records in the various database files are connected. For example if two files are related using an index key, records with matching key values in the parent and child file will be on the same line, appearing as one record in the Browse window.

Filter

Filter allows you to impose a logical condition that each record in the database file must meet, just as you would with the SET FILTER command. This menu item operates on the currently highlighted database file, allowing you to set a filter condition for each open file.

When you select *Filter*, a dialog box (shown in the figure below) opens and allows you to enter a logical expression into a fill-in field.

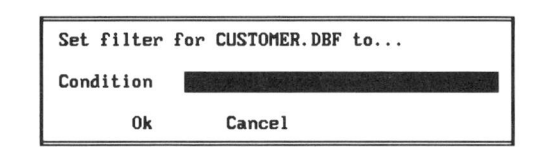

```
Set filter for CUSTOMER.DBF to...
Condition  ██████████████████████████
     Ok      Cancel
```

Figure 13-18: The Set Filter Dialog Box

The DBU Menus

Selecting *Ok* activates the filter for the current file, so that only those records meeting the condition are available for processing. If there is an error in the condition, an error message is displayed in the message and prompt area, and the dialog box remains open on the screen. You can either *Cancel* or enter a new condition.

After a filter is established, selecting *F5 Browse:Database* reflects the filter condition by showing only those records that meet the criteria. The filter condition is also reflected in the operation of certain *F6 Utility* menu items such as *Copy* and *Replace*. The filter remains in effect until you remove it by selecting *F8 Set:Filter* and deleting the condition.

Fields

Fields allows you to establish a field list for the current database file. This menu item operates on the currently highlighted database file, allowing you to set a field list for each open file.

Selecting *Fields* activates the Set Fields dialog box where you may select a single field name to add to the current field list. When you select *Ok*, the field is inserted in the Fields section for the current work area, pushing all other fields down one slot. The Set Fields dialog box is shown in the figure below.

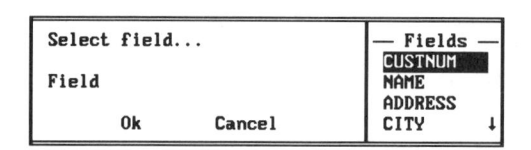

```
┌─────────────────────────────┬─── Fields ───┐
│ Select field...             │ CUSTNUM      │
│                             │ NAME         │
│ Field                       │ ADDRESS      │
│         Ok      Cancel      │ CITY       ↓ │
└─────────────────────────────┴──────────────┘
```

Figure 13-19: The Set Fields Dialog Box

The Set Fields dialog box may be activated by pressing *Return* or *Ins* instead of using the menu. If you know the name of the field that you want to include, you may also type it directly on the main screen.

The fields list that you see on the main screen and that you establish with the *F8 Set:Fields* dialog box is used by *F5 Browse* to determine which fields to show in the Browse window and in what order. All other operations in DBU use all fields defined in the database file structure.

Summary

The DBU system is provided as a utility for you to design database and index files for use in your applications. Additionally, it allows you to add and edit data and perform various batch utilities with the database files to give you a complete interactive database utility.

You can also design views that you can open and browse within the confines of DBU. The view file is not a *CA-Clipper* data structure, however, and cannot be used directly by a *CA-Clipper* program. If you want to have access to the view files that you create in DBU, take a look at the source code that was mentioned in the previous section. This way, you can understand the view file structure and incorporate it in your own programs.

Chapter 14
Report and Label Utility
RL.EXE

CA-Clipper gives you the ability to create and modify standard dBASE III PLUS report and label definitions by supplying a report and label utility, RL.EXE. RL.EXE is a stand-alone application written in *CA-Clipper* that emulates the dBASE CREATE | MODIFY REPORT and LABEL commands.

RL creates a binary report (.frm) or label (.lbl) file, which can later be used by the REPORT FORM or LABEL FORM commands to format and print data on a record by record basis from one or more open database files.

In This Chapter

This chapter explains how to use RL to create and modify report and label definitions and also provides information on how to use the resulting report and label definitions in your *CA-Clipper* applications.

The following general topics are covered in this chapter:

- Loading RL
- Creating and modifying reports
- Creating and modifying labels
- Leaving RL
- The RL system architecture
- Notes

Loading the Report and Label Utility

To load the Report and Label Utility, enter:

RL ↵

at the DOS prompt. Like all other *CA-Clipper* utilities, RL.EXE can be executed from any drive or directory since the appropriate directory was added to the PATH by the *CA-Clipper* installation program.

Creating and Modifying Reports

The Report and Label main menu contains a *Report* item that activates the report editor. From there, you can create and modify report form (.frm) files that are executable with the REPORT FORM command. The report definitions you can design are standard columnar reports with report and column headings. When totaling numeric columns you can have up to two levels of subtotals and generate a grand total for each column. For more information on the REPORT FORM command refer to the *Language Reference* chapter of the *Reference* guide.

Creating or Modifying a Report

To create or modify a report form, select *Report* from the main menu. A file dialog box opens like the one in the figure below.

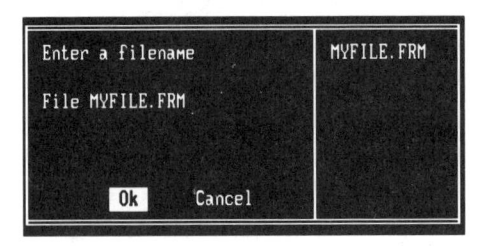

| Report | Label | Quit |

```
Enter a filename          MYFILE.FRM
File MYFILE.FRM

        Ok    Cancel
```

Figure 14-1: Create or open a (.frm) file

Creating and Modifying Reports

Here you can either create a new file by entering a new filename or select an existing file to modify from a picklist displayed on the right side of the dialog box. To create or open the (.frm) file you specified or highlighted, select the *OK* button and press *Return*. To cancel and return to the main RL menu, select *Cancel* and press *Return*.

Defining Report Columns

Having selected an (.frm) to edit, you are presented with the Column Definition screen where you can design a new report or make changes to an existing one. The Column Definition screen corresponds to the *F4 Columns* menu item. It is here that you define the columns of your report by adding, changing, or deleting columns. The maximum number of columns per report is 24.

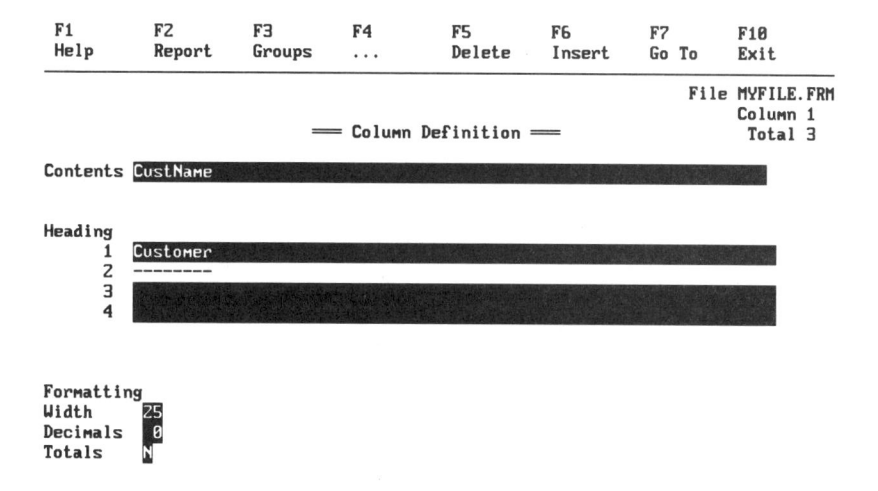

```
F1        F2        F3        F4        F5        F6        F7        F10
Help      Report    Groups    ...       Delete    Insert    Go To     Exit

                                                            File MYFILE.FRM
                                                                 Column 1
                        == Column Definition ==                  Total  3

Contents CustName


Heading
      1  Customer
      2  --------
      3
      4


Formatting
Width      25
Decimals   0
Totals     N
```

Figure 14-2: The Column Definition Screen

Columns appear on the resulting report in the order you define them. A detail line is displayed for each record in the corresponding database file.

A report column is defined by entering each of its attributes on the Column Definition screen. To do this, move the cursor to the setting that you want to define and type in the corresponding information. To add subsequent columns, press *PgDn* to navigate to a new Column Definition screen. Once you have defined several column definitions in this way, *PgUp* and *PgDn* can be used to navigate to the next or previous column definition for editing.

Each column definition has a contents expression, a heading, and a number of formatting attributes defined in the following sections.

Contents

Each column has a contents expression. It is here that you enter the name of a field variable or a more complicated expression—most data types including memo fields can be used. Values of each data type display in the default format and alignment settings, unless explicitly formatted. As a general rule, you can use the TRANSFORM() function to apply picture formatting to each data type. Refer to the *Language Reference* chapter of the *Reference* guide for the syntax and arguments. Specific formatting considerations for each data type are discussed below.

- ***Character strings*** display left-justified and can be formatted using TRANSFORM(). Character strings also wrap within the column boundaries if longer than the specified Width setting. To force a line break and wrap the rest of the column to a new line, embed a semicolon (;) using an expression like <cString1> + ";" + <cString2>.

- ***Memo fields*** are treated in the same way as character strings, but there are some special considerations since they may contain embedded soft or hard carriage returns which adversely affect word wrapping. For a memo field to display correctly, these characters must be replaced using either HARDCR() or MEMOTRAN().

- ***Date values*** display left-justified and are formatted according the current DATE SETting. To format the date differently, create a user-defined function.

- ***Logical values*** are left-justified and display as either (.T.) or (.F.). To define a new format, you can use either a user-defined function or the IF() function.

- ***Numeric values*** are right-justified and display according to the current DECIMALS and FIXED SETtings. To format numeric values to business or other common numeric formats, use TRANSFORM(). Note, however, that TRANSFORM() converts a numeric value to a character value which prevents totaling of the column. Unfortunately, this is a limitation of the REPORT FORM architecture.

Heading

Heading is a literal string of up to four lines that displays above the current column. The Heading displays left-justified regardless of the data type of the current column.

Formatting

- **Width** defines the display width of the current column. If the column type is numeric and the Decimals setting is greater than zero, the Width includes the decimal point and the decimal digits. The default width is 10. When the report is run, if the column contents exceed the specified Width and the type is character, it is wrapped to the next line in the column. If the type is numeric, a numeric overflow occurs and the column is filled with a row of asterisk (*) characters.

For numeric columns, additional attributes may also be defined. These attributes are listed below:

- **Decimals** defines the number of decimal digits to display. The default is zero.

- **Totals** determines whether or not a numeric column is totaled. A value of *Y* causes the column to be totaled. The default is *N*, indicating that no totaling takes place. Three levels of totaling are available when the Totals setting is *Y*. A grand total prints after all other report lines print and, if there are Groups or Subgroups defined, a subtotal prints for each level of grouping as well.

Deleting a Column

After you have added one or more column definitions, you can delete any column you do not want in the report definition. To delete the current column definition, delete the Contents expression by pressing *Ctrl-Y* and then use *F5 Delete*.

Note: *F5 Delete* will not delete a column definition unless the Contents area is empty. This is implemented as a precautionary measure to prevent you from deleting a column by accident.

Inserting a New Column

Since the format of the report directly relates to the order of the column definitions, you may need to insert new columns within current column definitions. To do this, position the cursor at the desired location and press *F6 Insert* to insert a new blank column definition. This new column becomes the current column. You can then enter the column definition attributes.

Note: *F6 Insert* will not insert a column before the first column definition in the report.

Creating and Modifying Reports

Locating a Column

To navigate to an existing column definition, press *F7 Goto*. A message appears prompting you to enter the number of the desired column. Pressing *Return* confirms your entry.

Defining Report Options

Columns are only a part, although a major one, of the entire report definition you can create. There are also options you can specify which apply at the report level. For example, headings and margins apply to the entire report rather than an individual column. To change the default report layout, press *F2 Report*. The following screen displays:

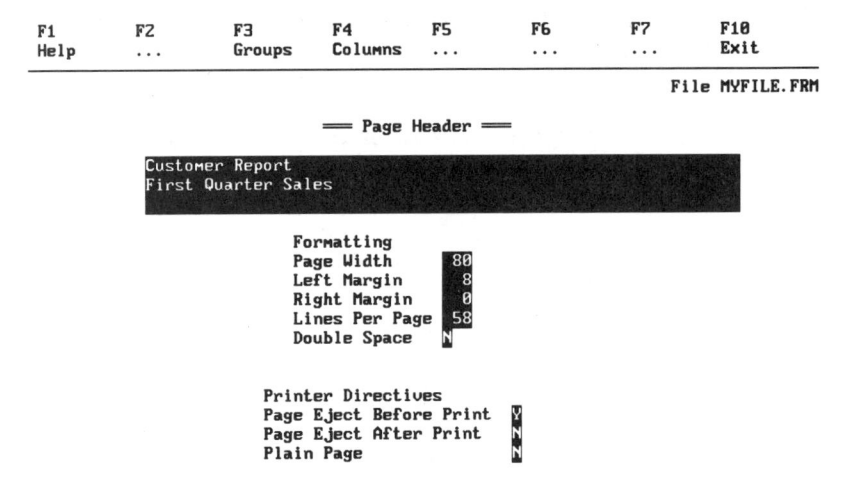

```
F1        F2        F3        F4        F5        F6        F7        F10
Help      ...       Groups    Columns   ...       ...       ...       Exit

                                                            File MYFILE.FRM

                          === Page Header ===
Customer Report
First Quarter Sales

                        Formatting
                        Page Width     80
                        Left Margin     8
                        Right Margin    0
                        Lines Per Page 58
                        Double Space   N

                        Printer Directives
                        Page Eject Before Print  Y
                        Page Eject After Print   N
                        Plain Page               N
```

Figure 14-3: The Report Options Screen

The report options screen allows you to define the following report attributes:

- **Page header** prints a four line literal string on the top and center of each page if the Plain Page option in the report definition is *N* (the default setting). The same effect can be achieved using the REPORT FORM command with the HEADING clause.

- **Page width** determines the number of characters allowed for the combined widths of all report columns. The default is 80.

14-6 CA-Clipper

- **Left margin** determines the number of characters the report is indented. If the report is directed to the printer, this value is added to the SET MARGIN value. The default is eight.

- **Right margin** performs no useful function and is supplied for dBASE compatibility purposes only. The default is zero.

- **Lines per page** determines the total number of lines to print on each page of the report. This number includes the page heading, column titles, detail lines, and total lines. The default is 58.

- **Double space** determines whether detail lines are double spaced when printed. The default is *N*.

- **Page eject before print** determines whether CHR(12), the formfeed character, is sent to the printer prior to the first report line. This setting can be overridden by specifying the NOEJECT clause on the REPORT FORM command line. The default is *Y*.

- **Page eject after print** determines whether a formfeed is sent to the printer after the last report line is printed. This setting cannot be overridden from the REPORT FORM command line. The default setting is *N*.

- **Plain page** corresponds to the PLAIN clause of the REPORT FORM command. If set to *Y*, the page heading, page number, and report date are not printed and the REPORT FORM HEADING is printed only on the first page. In addition, there are no page breaks. The default is *N*.

Defining Groups

RL allows you to define two levels of grouping in order to summarize information from one or more related database files. A group defines a set of consecutive records having the same key value.

- Each time there is a new key value, a new group begins and a group header line prints.

- Each time the current group ends, a summary line prints. If grouping criteria is defined and the Totals attribute on any Column Definition screen is set to *Y*, subtotals print for numeric report columns. Within a group, subgroups behave in the same manner.

To create groups for the current report, press *F3 Groups* and the following Group Definition screen displays:

```
F1        F2        F3        F4        F5        F6        F7        F10
Help      Report    ...       Columns   ...       ...       ...       Exit
```
                                                                    File MYFILE.FRM

=== Group Specifications ===

Group On Expression   DeptNo
                       Department

Summary Report Only   N
Page Eject After Group N

=== Sub-Group Specifications ===

Sub-Group On Expression   SalesGroup
Sub-Group Heading   Sales Group

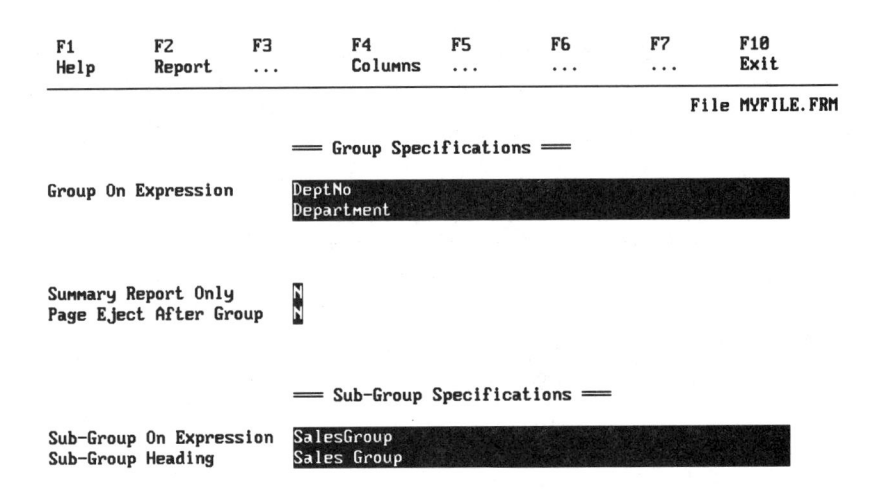

Figure 14-4: The Groups Definition Screen

This screen allows you to define both group and subgroup definitions for the current database file. If you do not define grouping criteria, only a grand total will be displayed for numeric columns with the Totals attribute on the Column Definition screen set to *Y*.

The options that you can define on this screen are described below:

- **Group on expression** specifies the primary key expression to group on. This may be a single field or any valid expression involving at least one field. As an example, the group expression might be *DeptNo* so you can generate column totals for each department.

- **Group heading** specifies the title of the group as a literal string. This heading prints before the first report line for each group and just to the left of the group expression.

- **Summary report only** determines whether detail lines are printed in a grouped report. If you specify *Y*, detail lines are suppressed and only group header and summary lines print. The default is *N*. Specifying *Y* is equivalent to specifying the SUMMARY clause on the REPORT FORM command line.

- **Page eject after group** determines whether there is a page eject at the end of each group of records. The default value is *N*. Specifying *Y* causes each group to begin on a new page.

- **Subgroup on expression** specifies a secondary key expression defining a second level of grouping. As an example, if the group expression is *DeptNo* the subgroup expression might be *SalesGroup* to obtain a breakdown of each sales group within a department.

- **Subgroup heading** specifies a literal string to print at the beginning of each subgroup.

In order for grouping to work properly in a report, the corresponding database file must be in order according to the group and subgroup expressions. The most convenient method for ordering a database file is with an index. In keeping with the example given above, the index key would be *DeptNo + SalesGroup*. See INDEX, in the *Language Reference* chapter of the *Reference* guide for more information.

Saving the Report Definition

From any screen, you can save the Report Definition by pressing *F10 Exit*. If you have not made any changes, selecting *F10* returns you directly to the main menu. Otherwise, you are presented with a dialog box in which you have three choices:

- Selecting *OK* and pressing *Return* saves all changes you have made to the report form (.frm) file and returns you to the main menu.

- Selecting *No* and pressing *Return* discards all changes, leaving the original report form file intact, and returns you to the main menu. In the case of a new file, the file is simply not created.

- Selecting *Cancel* and pressing *Return* returns you to the current report screen where you may continue editing the current report definition.

Printing a Report

To print a report form within an application, you must have a corresponding database file as well as an open index file if grouping was specified. Substitute the appropriate filenames in the following *CA-Clipper* commands and include the commands in a program to print the report:

```
USE <xcDatabase> INDEX <xcIndex list> NEW
REPORT FORM <xcReport> TO PRINTER
```

There are other options available for the REPORT FORM command. See the REPORT FORM entry in the *Language Reference* chapter of the *Reference* guide for more information on printing reports that you create with RL.

Reporting from Related Work Areas

Using a report form, you can print field information from more than one work area by relating work areas using the SET RELATION command. There are, however, several requirements to make this work.

In RL, *all* references to field variables must be specified using an alias, like this:

```
<idAlias>-><idField>
```

This guarantees that a field will always be available when the report is printed.

Before you execute the REPORT FORM command, you must have all the required database files in USE and RELATIONs SET in the proper order. In one-to-one relations, report from the main work area and not from the look up.

For example, suppose you have two databases files, Customer and Zipcode. Customer is the main and Zipcode is the lookup linked by a common field, Zip. The code to execute a report form that prints a Customer list would look something like this:

```
USE Zipcode INDEX Zipcode NEW          // Open the lookup first
USE Customer NEW                       // Report from the main
SET RELATION TO Zip INTO Zipcode       // Establish the link
REPORT FORM CustList TO PRINTER
```

If the relationship between the work areas is one-to-many, the order is reversed. Report from the lookup work area instead of the main work area. In addition, the main work area must have an index for the linking key. For example, this time you are reporting Invoice amounts for each Customer. The code to execute this kind of a report would look something like this:

```
USE Customer INDEX InvNum NEW          // Open the main first
USE Invoices INDEX CustNum NEW         // Report from the lookup
SET RELATION TO CustNum INTO Customer  // Establish the link
REPORT FORM CustInv TO PRINTER
```

Creating and Modifying Labels

The Report and Label main menu contains a *Label* item that activates the label editor. From there, you can create and modify label (.lbl) files executable with the LABEL FORM command. The label definitions you can design are quite versatile, allowing you to define label attributes such as lines per label and number of labels across. The (.lbl) files created by RL are fully compatible with dBASE III PLUS label files. For more information on LABEL FORM refer to the *Language Reference* chapter of the *Reference* guide.

Creating or Modifying a Label

To create or modify a label form, select *Label* from the main menu. A file dialog box opens like the one in the figure below.

Report **Label** Quit

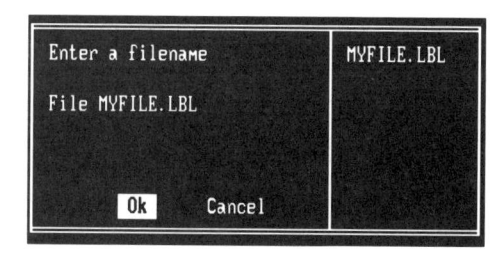

```
┌─────────────────────────┬──────────┐
│ Enter a filename        │ MYFILE.LBL│
│                         │          │
│ File MYFILE.LBL         │          │
│                         │          │
│                         │          │
│      Ok    Cancel       │          │
└─────────────────────────┴──────────┘
```

Figure 14-5: Create or open a (.lbl) file

Here you can create a new file by entering a new filename or select an existing file to modify from a picklist displayed on the right side of the dialog box. After a new filename has been entered or an existing file has been chosen, select the *OK* button and press *Return*. To cancel and return to the main RL menu, select *Cancel* and press *Return*.

The Label Editor Screen

After you create a new or select an existing (.lbl) file to edit, the Label editor screen displays. The editor screen is divided into two sections as indicated in the figure below. Within the top section you define the Dimensions and Formatting attributes of the label. In the bottom section, you define the Contents expressions for each label row.

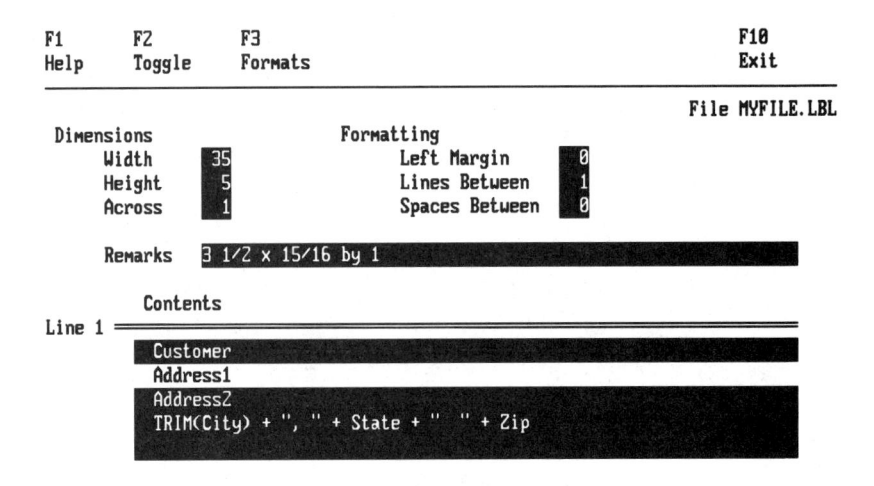

```
F1      F2       F3                                           F10
Help    Toggle   Formats                                      Exit
                                                     File MYFILE.LBL
   Dimensions              Formatting
       Width    35              Left Margin    0
       Height   5               Lines Between  1
       Across   1               Spaces Between 0

       Remarks  3 1/2 x 15/16 by 1

              Contents
   Line 1
              Customer
              Address1
              Address2
              TRIM(City) + ", " + State + " " + Zip
```

Figure 14-6: The Label Editor Screen

When you first access the editor, the Dimensions and Formatting section is active. Pressing *F2 Toggle* moves the cursor between sections.

Defining the Label Dimensions and Formatting

The following label attributes can be defined in the Dimensions and Formatting section of the Label Editor screen:

- **Width** determines the horizontal width of an individual label. This value can range from one to 255. The default is 35 characters.

- **Height** determines the vertical number of lines in an individual label. This value can range from 1 to 16 lines. The default is 5 lines.

- **Across** determines the number of labels printed across the page. For multiple across labels, the last label contents printed on a line is trimmed automatically. This value can range from one to 255. The default is 1 label across.

- **Left Margin** specifies the left margin and determines the first print position of the leftmost label. When the labels are printed, this value is added to the SET MARGIN value. This value can range from zero to 255. The default is zero.

- **Lines Between** determines the number of blank horizontal lines printed between labels. This value can range from zero to 255. The default is one blank line at the bottom of each label.

- **Spaces Between** determines the amount of vertical space printed between labels if the number of labels across is greater than one. This setting can also be used to set the left margin for labels after the first label, and its value can range from zero to 255. The default is zero.

- **Remarks** is a comment field that contains standard label format size definitions. Selecting one of the label formats described in the next section fills this area with the appropriate format definition. The default label definition is 3 1/2 x 15/16 by 1. The Remarks comment field can also be customized with your own label format size definitions.

Standard Label Formats

Instead of specifying the dimension and formatting criteria directly, you can select a standard label format from the *F3 Formats* menu. The table below gives the available standard formats and the resulting label dimensions and formatting.

Standard Label Formats

Remarks	Width	Height	Across	Margin	Lines	Spaces
3 1/2 x 15/16 by 1	35	5	1	0	1	0
3 1/2 x 15/16 by 2	35	5	2	0	1	2
3 1/2 x 15/16 by 3	35	5	3	0	1	2
4 x 17/16 by 1	40	8	1	0	1	0
3 2/10 x 11/12 by 3 (Cheshire)	32	5	3	0	1	2

To select one of these standard label formats, press *F3 Formats*. When the menu appears on the screen, move the highlight to the desired label format and press *Return*. The label dimension and formatting information will automatically be updated to your new selection.

Defining the Label Contents

After you have defined the label dimension and formatting attributes, *F2 Toggle* moves the cursor to the Contents section in the bottom half of the Label editor screen. It is here that you define the actual contents of each label row.

Depending on the Height you have chosen for your label definition, the number of available lines in the Contents area is defined. On each line,

Creating and Modifying Labels

you can enter an expression of any valid data type, and the expression itself can be up to 60 characters in length. The result of a contents expression can be any length, but a result longer than the specified label width is truncated when printed by LABEL FORM.

For each label line, enter the expression whose result you want to print on that line. For example, a familiar label content might be:

```
TRIM(FirstName) + " " + LastName
Address1
Address2
TRIM(City) + ", " + State + " " + Zip
```

To enter the contents for a line, move the highlight to that line and begin typing the expression. Pressing *Return* moves the highlight to the next row.

Note: In *CA-Clipper*, the label content does not support a list of expressions delimited by commas as in dBASE III PLUS. If more than one expression is specified, a runtime error occurs.

Blank Lines

In *CA-Clipper*, a blank contents expression prints a blank line when the label is printed. For example, if you want a blank line between the name and address, the label contents might look like this:

```
TRIM(FirstName) + " " + LastName

Address1
Address2
TRIM(City) + ", " + State + " " + Zip
```

If, however, a label contents expression that returns a null string ("") is specified, the resulting blank line is suppressed when the label is printed. A good example of this is the second address line in a mailing list which is often empty.

To display a blank line when the contents value is a null ("") string, the contents expression must return a nonprintable character. You can generally enforce this by specifying the following IF() expression on the label contents line:

```
IF(!EMPTY(<cString>), <cString>, CHR(255))
```

or create a user-defined function that performs the same action:

```
FUNCTION NoSkip(exp)
    RETURN IF(!EMPTY(exp), exp, CHR(255))
```

and use it as the label content (e.g., NoSkip(Address2)).

Saving the Label Design

When you have finished specifying the label, you can save the definition by pressing *F10 Exit*. If you have not made any changes to the current label definition, selecting *F10* returns you directly to the main menu. Otherwise, you are presented with a dialog box offering the following three choices:

- Selecting *OK* and pressing *Return* saves all changes made to the label (.lbl) file and returns you to the main menu.

- Selecting *No* and pressing *Return* discards all changes, leaving the original label file intact and returning you the main menu.

- Selecting *Cancel* and pressing *Return* returns you to the current Label screen where you may continue editing the current label definition.

Printing the Labels

To print the label using the label definition you have designed, you must have a corresponding database file in USE and then invoke the LABEL FORM command specifying the appropriate label definition, like this:

```
USE <xcDatabase> INDEX <xcIndex list>
LABEL FORM <xcLabel> TO PRINTER
```

The LABEL FORM command has a number of options to determine which records will be printed and whether a sample set of labels will be printed in place of the actual labels. Refer to the LABEL FORM entry in the *Language Reference* chapter of the *Reference* guide for more information on printing labels.

Leaving RL

When you have finished designing report and label definitions, you can exit RL by selecting *Quit* or by pressing *Esc* from the main menu.

The RL System Architecture

The RL system is written in *CA-Clipper* and provided as source code in the \CLIPPER5\SOURCE\RL directory where *CA-Clipper* is installed. It is also compiled and linked in the \CLIPPER5\BIN directory. The source files are:

- Rlfront.prg
- Rlback.prg
- Rldialg.prg

The system is provided as source code so you can modify it to suit your specific needs. You may want to make enhancements specific to your development environment or simply add the RL system to your application, thereby giving your users the power to create their own reports and labels.

To facilitate modification, the RL system is broken up into two primary subsystems: front and back. The RL front subsystem (found in Rlfront.prg) is the user interface of the entire RL system. The back subsystem (found in Rlback.prg) contains all of the routines for reading and writing (.frm) and (.lbl) files. The system is organized in this way so that you can create your own user interface and easily access the lower-level routines to read and write the form files.

If you decide to tailor this system by making changes to the source code, a make file, RL.RMK, is located in the same directory as the RL source code. Provided that your changes do not involve additional source modules or changes in the existing file dependencies, issuing the following command at the DOS prompt:

```
RMAKE RL ↵
```

will compile and link the system to create a new RL.EXE file. Refer to the *Program Maintenance—RMAKE.EXE* chapter in this guide for more information on make files.

Notes

- ***External functions:*** If you specify functions located in libraries other than CLIPPER.LIB in a REPORT or LABEL FORM definition and you don't explicitly reference these functions elsewhere in your program, you must declare them to the linker using the REQUEST statement.

The REQUEST statement forces the linker to bring the code for specified routines into your program even though there are no references to the routines in any of the compiled object modules. You can declare these functions external in any source file using the REQUEST statement as long it is compiled and linked into the current executable file. Refer to the *Language Reference* chapter of the *Reference* guide for more information on declaring external routines.

Summary

The RL system is provided as a utility for building report and label form files that can be used with the REPORT FORM and LABEL FORM commands in your *CA-Clipper* programs. The report and label designs you create with RL require an open database file when you invoke them. Although not required, it is recommended that you create the database files before attempting to create report and label definitions.

Chapter 15
Online Documentation
NG.EXE

The *CA-Clipper* reference documentation is provided in the form of several databases referred to collectively as *The Guide To CA-Clipper*. It contains the most timely reference information on *CA-Clipper* commands, functions, and utilities with new items denoted by the Ω symbol and changed or updated items denoted by the Σ symbol.

The Guide To CA-Clipper is accessible via the Norton Instant Access Engine™ (NG.EXE), a memory-resident program also included with *CA-Clipper*. In the default configuration, all of the documentation database files and the Instant Access Engine are installed in the \NG directory, and the PATH is updated to include this directory.

In This Chapter

In this chapter, the following general topics are discussed:

- Loading the Instant Access Engine
- How the Instant Access Engine searches for files
- Using the Access Window
- Viewing a documentation database
- Instant Access Engine navigation keys
- Configuring the Instant Access Engine
- Leaving the Instant Access Engine
- Creating your own documentation databases

Loading the Instant Access Engine

The Instant Access Engine (NG.EXE) can be loaded in two ways:

Loading the Instant Access Engine

- *Memory-resident mode* where the Instant Access Engine loads as a TSR (terminate and stay resident) program and remains in memory until explicitly removed.

- *Pass through mode* where the Instant Access Engine loads and then runs a specified application program. The Instant Access Engine remains in memory until the application program terminates.

In either mode, the Instant Access Engine occupies approximately 65K of RAM, no matter how large the documentation database you access.

NG Syntax and Arguments

To use *The Guide To CA-Clipper* or any other documentation database, you must first load NG.EXE, the Instant Access Engine. The general syntax is as follows:

```
NG [<command line>]
```

<command line> is any valid DOS command, including arguments. If <command line> is specified, NG is loaded in pass through mode. Otherwise, it is loaded in memory-resident mode.

If you installed the default configuration of the *CA-Clipper* development system, NG.EXE is located in the \NG directory and was added to the PATH statement in your AUTOEXEC.BAT. This allows you to easily access the Instant Access Engine from any drive and/or directory.

Note: If you are using DOS 5.0, you must press *Esc* immediately after *Shift-F1* to activate the Instant Access Engine. If you place the statement SWITCHES=/k in your CONFIG.SYS file, you will not have to press *Esc*.

Using Memory-Resident Mode

In memory-resident mode, the Instant Access Engine remains in memory until you explicitly remove it from memory using *Options:Uninstall* or until you reboot your computer. This means you have access to the current documentation database from within any program, including your editor or from the DOS prompt.

Using Pass Through Mode

In pass through mode, the Instant Access Engine is loaded into memory and then the command line program is executed. Terminating the command line program automatically removes the Instant Access

Engine from memory. This means that you only have access to a documentation database as long as the command line program is running. For example, entering:

```
NG PE Sample.prg ↵
```

at the DOS prompt loads the Instant Access Engine and then the *CA-Clipper* program editor. While operating in the editor, you can activate the Instant Access Engine at any time to view the current documentation database. When you save the program file you are editing and leave the editor, the Instant Access Engine is automatically removed from memory.

Accessing the Instant Access Engine

Once the Instant Access Engine is loaded, you can access the current documentation database by pressing the activation *hot key*. The default hot key is *Shift-F1*. If this conflicts with a key used by another program, you can change it to a new value using the *Options:Hot key* menu item described below.

Note: The Instant Access Engine does not display if your screen is currently in graphics mode. If the screen is in either EGA 43-line or VGA 50-line mode, the Instant Access Engine displays but treats the screen as if it only has 24 lines.

How the Instant Access Engine Searches for Files

The Instant Access Engine can display documentation databases from any drive or directory as long as they are located in the same directory as NG.EXE or the current directory. When you access the *Options:Database* to select a new documentation database to view, the Instant Access Engine presents a picklist of documentation databases from the NG.EXE directory followed by documentation databases found in the current directory.

If you installed the default configuration of the *CA-Clipper* development system, all documentation databases included were copied into the \NG directory and the PATH was then updated with this directory. In this configuration, you can invoke the Instant Access Engine from any drive or directory.

Note: If you purchase a third-party product that includes a compatible documentation database, copy the database into the \NG directory.

This makes the database accessible from the Options:Database picklist independent of your current directory location.

Using the Access Window

When the Instant Access Engine is first activated, a window appears, which is similar to the one in the figure below:

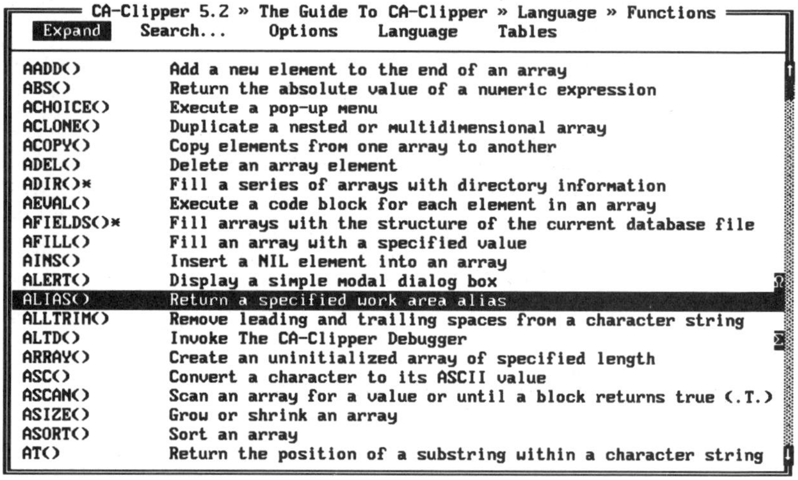

```
═══ CA-Clipper 5.2 » The Guide To CA-Clipper » Language » Functions ═══
│ Expand    Search...    Options    Language    Tables                │
│ AADD()      Add a new element to the end of an array               │
│ ABS()       Return the absolute value of a numeric expression      │
│ ACHOICE()   Execute a pop-up menu                                  │
│ ACLONE()    Duplicate a nested or multidimensional array           │
│ ACOPY()     Copy elements from one array to another                │
│ ADEL()      Delete an array element                                │
│ ADIR()*     Fill a series of arrays with directory information     │
│ AEVAL()     Execute a code block for each element in an array      │
│ AFIELDS()*  Fill arrays with the structure of the current database file │
│ AFILL()     Fill an array with a specified value                   │
│ AINS()      Insert a NIL element into an array                     │
│ ALERT()     Display a simple modal dialog box                      │
│ ALIAS()     Return a specified work area alias                     │
│ ALLTRIM()   Remove leading and trailing spaces from a character string │
│ ALTD()      Invoke The CA-Clipper Debugger                         │
│ ARRAY()     Create an uninitialized array of specified length      │
│ ASC()       Convert a character to its ASCII value                 │
│ ASCAN()     Scan an array for a value or until a block returns true (.T.) │
│ ASIZE()     Grow or shrink an array                                │
│ ASORT()     Sort an array                                          │
│ AT()        Return the position of a substring within a character string │
```

Figure 15-1: The Instant Access Engine Window

At the top of the screen is the name of the current documentation database followed by the major and minor category names. Just below this is the menu bar.

The Menu Bar

The menu bar provides you, via pulldown menus, with options to select different documentation databases to view, search for keywords in the current list of short entries, and change the default configuration of the Instant Access Engine. The first three menus displayed are Instant Access Engine system menus and are always displayed. The remaining menus are placed on the menu bar by the current documentation database.

A menu can execute a function directly, prompt you for input, or pull down a menu. Of the system menus, *Expand* executes a function directly, *Search* displays an input field, and *Options* pulls down a menu of configuration items.

Selecting Menus and Menu Items

When the menu bar is active, you can select a menu either by pressing the first letter of the menu name or by moving the highlight using the *Rightarrow* or *Leftarrow* keys. If you use the first letter method to select either the *Expand* or the *Search* menu, pressing the key executes the menu. If you use the navigation method, you must press *Return* to execute the menu.

When you access a pulldown menu, a list of *menu items* appears. You can then select a menu item the same way you select a menu: either by using the first letter method or by moving the highlight to the desired item with *Uparrow* or *Dnarrow* and pressing *Return*.

To cancel a pulldown menu, press *Esc* and the highlight moves back to the *Expand* menu.

Sizing and Moving the Access Window

By default, the Access window takes up the entire screen. If you need to see the application program screen without exiting the Instant Access Engine, the *Options:Full-screen* menu item or *F9* toggles the size of the Access window between full-screen and half-screen modes. A check mark indicates the current mode is full-screen.

When the Access window is activated in half-screen mode, it pops up either at the top or the bottom of the screen, whichever is furthest away from the current cursor position. If the Access window obscures important information on the application program screen, you can scroll the window up or down on the screen.

To do this, press *Scroll-Lock* and then use *Uparrow* and *Dnarrow* to move the window to a new location. When you have finished, press *Scroll-Lock* again to freeze the window position. Note, however, that the Instant Access Engine does not remember the new window location. Each time you exit and reactivate in half-screen mode, the Access window always pops up away from the cursor.

Getting Help

While viewing in the Access window, you can pop up a help window like the one in the figure below by pressing *F1*.

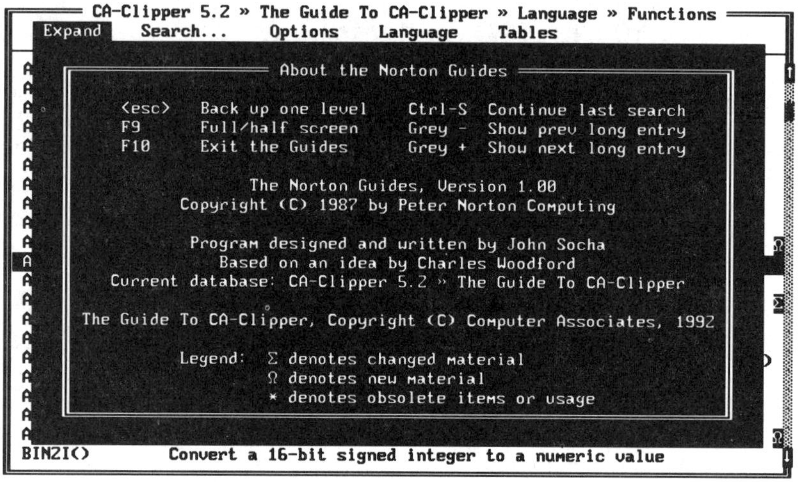

```
╔══ CA-Clipper 5.2 » The Guide To CA-Clipper » Language » Functions ══
║ Expand    Search...    Options    Language    Tables
║ ┌──────────────── About the Norton Guides ────────────────┐
║ │                                                          │
║ │   <esc>   Back up one level    Ctrl-S  Continue last search │
║ │   F9      Full/half screen     Grey -  Show prev long entry │
║ │   F10     Exit the Guides      Grey +  Show next long entry │
║ │                                                          │
║ │         The Norton Guides, Version 1.00                  │
║ │      Copyright (C) 1987 by Peter Norton Computing        │
║ │                                                          │
║ │      Program designed and written by John Socha          │
║ │        Based on an idea by Charles Woodford              │
║ │   Current database: CA-Clipper 5.2 » The Guide To CA-Clipper │
║ │                                                          │
║ │  The Guide To CA-Clipper, Copyright (C) Computer Associates, 1992 │
║ │                                                          │
║ │      Legend:  Σ denotes changed material                 │
║ │              Ω denotes new material                      │
║ │              * denotes obsolete items or usage           │
║ │                                                          │
║ └──────────────────────────────────────────────────────────┘
║ BIN2I()      Convert a 16-bit signed integer to a numeric value
```

Figure 15-2: Getting Help with F1:Help

This window gives you information about the Instant Access Engine and the current documentation database. If the current documentation database is one of *The Guide To CA-Clipper* databases, there is a legend of symbols used in the short item list.

Viewing a Documentation Database

Basically, a documentation database is a hierarchical structure consisting of the following items:

- Menus
- Menu items
- Short entries
- Long entries
- See also references

In a typical documentation database, a menu lists a series of categories as its items. Each menu item, in turn, refers to a list of short entries or a single long entry. Each short entry can also refer to another list of short entries or a long entry. If you consider a documentation database as a tree structure, long entries are nodes of the tree and contain the actual topical information. Since long entries are the nodes of the information tree, they can only refer to other long entries by way of the *see also* list.

The Short Entry List

Having selected a subject area of the current documentation database using the menu system, you are generally presented with a list of short entries. Short entries usually consist of a keyword such as a command or function name followed by a short description. For example, the figure below shows the short entry list of *CA-Clipper* functions.

```
┌──── CA-Clipper 5.2 » The Guide To CA-Clipper » Language » Functions ────┐
│ [Expand]  Search...   Options    Language    Tables                     │
├─────────────────────────────────────────────────────────────────────────┤
│ AADD()         Add a new element to the end of an array                 │
│ ABS()          Return the absolute value of a numeric expression        │
│ ACHOICE()      Execute a pop-up menu                                    │
│ ACLONE()       Duplicate a nested or multidimensional array             │
│ ACOPY()        Copy elements from one array to another                  │
│ ADEL()         Delete an array element                                  │
│ ADIR()*        Fill a series of arrays with directory information       │
│ AEVAL()        Execute a code block for each element in an array        │
│ AFIELDS()*     Fill arrays with the structure of the current database file │
│ AFILL()        Fill an array with a specified value                     │
│ AINS()         Insert a NIL element into an array                       │
│ ALERT()        Display a simple modal dialog box                        │
│ ALIAS()        Return a specified work area alias                       │
│ ALLTRIM()      Remove leading and trailing spaces from a character string │
│ ALTD()         Invoke The CA-Clipper Debugger                           │
│ ARRAY()        Create an uninitialized array of specified length        │
│ ASC()          Convert a character to its ASCII value                   │
│ ASCAN()        Scan an array for a value or until a block returns true (.T.) │
│ ASIZE()        Grow or shrink an array                                  │
│ ASORT()        Sort an array                                            │
│ AT()           Return the position of a substring within a character string │
└─────────────────────────────────────────────────────────────────────────┘
```

Figure 15-3: A Typical Short Entry List

To navigate a short entry list, you can use the typical cursor navigation keys (refer to the Instant Access Engine Navigation Keys table later in this chapter) or the *Search* menu.

Searching a Short Entry List

Instead of scrolling or paging through a list of short entries, you can search for a short entry containing a specified text string using the *Search* menu. When you select *Search*, a dialog box pops up into which you can enter a string to locate. If you have already performed a search, the previous search string reappears. To enter a new search string, begin typing and the previous entry is automatically erased. If you wish to edit the existing text, press *Home* (or any other cursor movement key) before typing any new text.

Once you have entered the search string, pressing *Return* performs the search operation. Searching begins with the next entry and continues to the end of the list. If there is no match, the search continues from the top of the list.

The search operation is not case-sensitive, allowing you to use any combination of upper and lowercase letters. In addition, each search scans short entries for a matching substring, allowing you to specify a partial search string. For example, if *Functions* is the current short entry list, searching for "ST" locates the short entry for STR().

To continue a search for the next occurrence of the search string, reexecute the *Search* menu or press *Ctrl-S*.

Expanding an Entry—Moving Down a Level

Once you have located a short entry of interest, you can move to a lower-level in the database hierarchy by executing the *Expand* menu. The referenced entry can be another short entry list or a long entry.

Note that *Expand* is always the default menu unless you explicitly access another menu. Even then, after you make a selection the highlight automatically moves back to *Expand*. This makes moving down a level very easy since pressing *Return* almost invariably executes *Expand*.

If the referenced entry is a long entry, the Access window display area is replaced with the long entry text and the menu bar with the associated see also list. Once you are located within a long entry, you can navigate using the cursor keys. For example, *Uparrow*, *Dnarrow*, *PgUp*, and *PgDn* move within the window, while *Leftarrow* and *Rightarrow* navigate the see also references (if there are any).

See Also References

Within a long entry, the menu bar is replaced with a list of *see also* references. A see also reference allows you to navigate directly to another long entry without navigating through the menus and short entry lists. To select a see also reference, move the highlight using *Leftarrow* or *Rightarrow* and press *Return*. You can also select a see also reference using the first letter method and pressing *Return*. The following figure demonstrates a typical long entry and see also list.

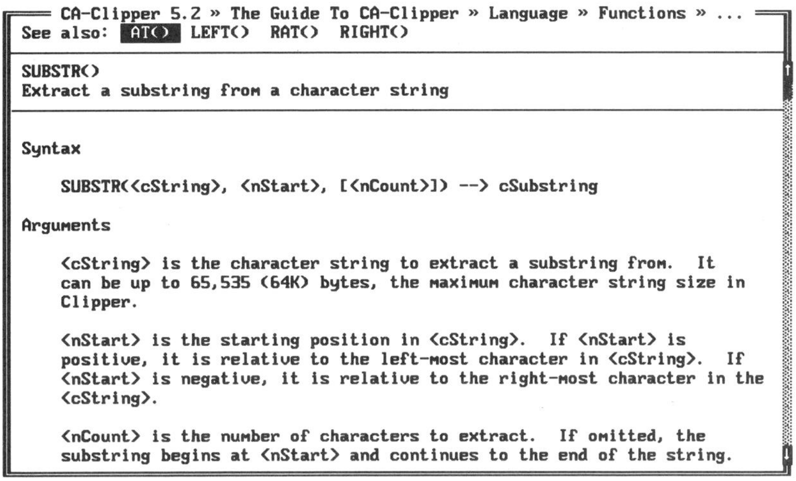

Viewing a Documentation Database

```
┌──── CA-Clipper 5.2 » The Guide To CA-Clipper » Language » Functions » ... ────┐
│ See also:  AT()  LEFT()  RAT()  RIGHT()                                       │
├───────────────────────────────────────────────────────────────────────────────┤
│ SUBSTR()                                                                      │
│ Extract a substring from a character string                                   │
│                                                                               │
│ Syntax                                                                        │
│                                                                               │
│     SUBSTR(<cString>, <nStart>, [<nCount>]) --> cSubstring                    │
│                                                                               │
│ Arguments                                                                     │
│                                                                               │
│     <cString> is the character string to extract a substring from.  It        │
│     can be up to 65,535 (64K) bytes, the maximum character string size in     │
│     Clipper.                                                                  │
│                                                                               │
│     <nStart> is the starting position in <cString>.  If <nStart> is           │
│     positive, it is relative to the left-most character in <cString>.  If     │
│     <nStart> is negative, it is relative to the right-most character in the   │
│     <cString>.                                                                │
│                                                                               │
│     <nCount> is the number of characters to extract.  If omitted, the         │
│     substring begins at <nStart> and continues to the end of the string.      │
└───────────────────────────────────────────────────────────────────────────────┘
```

Figure 15-4: A Typical Long Entry

If you navigated to the current long entry by way of a see also reference, the Instant Access Engine provides two ways to return to the referring long entry.

- If the current long entry has the referring long entry name as a see also reference, that reference is highlighted. To return to the referring long entry, press *Return*. As an example, if you begin with the SUBSTR() function long entry and execute the AT() see also reference, the SUBSTR() see also reference in the AT() long entry is highlighted. By pressing *Return* you return to the SUBSTR() long entry, and the AT() see also reference is again highlighted. In this way, you can move quickly between two related long entries by pressing *Return* consecutively.

- If there is no see also reference to the referring long entry in the current see also list, the Instant Access Engine puts a special see also reference, *Previous,* at the end of the current list. As before, you can press *Return* to return to the referring long entry. Note, however, the Instant Access Engine remembers only one level of reference with *Previous.*

Moving Up a Level

From any level in the database hierarchy, you can move up to the previous level by pressing *Esc*. If you are currently located in a long entry, this moves you up to the referring short entry list. If you are at

Viewing a Documentation Database

the top level of the database hierarchy, *Esc* causes you to exit the Instant Access Engine altogether.

Selecting a New Documentation Database

To select another documentation database to view, use the *Options:Database* menu item. This allows you to select a documentation database from a picklist as shown in the following screen:

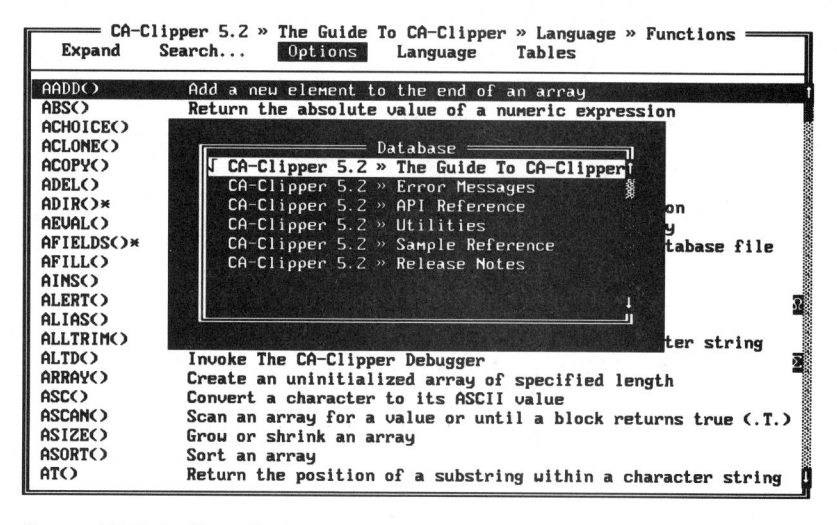

```
═══ CA-Clipper 5.2 » The Guide To CA-Clipper » Language » Functions ═══
     Expand    Search...   Options     Language    Tables
   ┌──────────┬─────────────────────────────────────────────────────────┐
   │ AADD()   │ Add a new element to the end of an array                │
   │ ABS()    │ Return the absolute value of a numeric expression       │
   │ ACHOICE()│                                                         │
   │ ACLONE() │        ┌──────────── Database ────────────┐             │
   │ ACOPY()  │      √ │ CA-Clipper 5.2 » The Guide To CA-Clipper│      │
   │ ADEL()   │        │ CA-Clipper 5.2 » Error Messages  │             │
   │ ADIR()*  │        │ CA-Clipper 5.2 » API Reference   │   on        │
   │ AEVAL()  │        │ CA-Clipper 5.2 » Utilities       │   y         │
   │ AFIELDS()│        │ CA-Clipper 5.2 » Sample Reference│   tabase file│
   │ AFILL()  │        │ CA-Clipper 5.2 » Release Notes   │             │
   │ AINS()   │        │                                  │             │
   │ ALERT()  │        │                                  │             │
   │ ALIAS()  │        │                                  │             │
   │ ALLTRIM()│        └──────────────────────────────────┘   ter string│
   │ ALTD()   │ Invoke The CA-Clipper Debugger                          │
   │ ARRAY()  │ Create an uninitialized array of specified length       │
   │ ASC()    │ Convert a character to its ASCII value                  │
   │ ASCAN()  │ Scan an array for a value or until a block returns true (.T.)│
   │ ASIZE()  │ Grow or shrink an array                                 │
   │ ASORT()  │ Sort an array                                           │
   │ AT()     │ Return the position of a substring within a character string│
   └──────────┴─────────────────────────────────────────────────────────┘
```

Figure 15-5: Options:Database Picklist

Documentation databases found in the NG.EXE directory are listed first, followed by those found in the current directory. Documentation databases are arranged in the order the actual database (.NG) files are encountered. To change the order, you must physically sort the \NG directory with a disk utility.

To select a new documentation database to view, navigate the picklist using *Uparrow*, *Dnarrow*, *Home*, or *End*. Pressing *Return* selects the currently highlighted documentation database. *Esc* cancels the selection.

Instant Access Engine Navigation Keys

The following is a list of all the navigation keys available while you are operating within the Instant Access Engine:

Instant Access Engine Navigation Keys

Key	Function	Mode
F1	Help	All
F9	Full/half-screen toggle	All
F10, *Shift-F1*	Exit	All
Esc	Move up a level	Short, Long
	Exit	Top-level Short
Return	Move down a level	Short
	Select item	Menu bar, Menu
	Select See also	Long
Gray -	Show previous long item	Long
Gray +	Show next long item	Long
Ctrl-S	Continue last Search	Long
Scroll-Lock	Enable/disable moving	All
Uparrow	Go to previous item	Menu, Short
	Go to previous line	Long
	Move window up one line	Move
Dnarrow	Go to next item	Menu, Short
	Go to next line	Long
	Move window down one line	Move
Leftarrow	Go to previous item	Menu bar, See Also
Rightarrow	Go to next item	Menu bar, See Also
PgUp	Go to last item previous screen	Short
	Go to last line previous screen	Long
	Move window screen top	Move

Instant Access Engine Navigation Keys (cont.)

Key	Function	Mode
PgDn	Go to first item next screen	Short
	Go to first line next screen	Long
	Move window screen bottom	Move
Home	Go to first short item	Short
	Go to first line	Long
	Move window screen top	Move
End	Go to last short item	Short
	Go to last line	Long
	Move window screen bottom	Move

Configuring the Instant Access Engine

The *Options* menu allows you to change certain parameters affecting the display and operation of the Instant Access Engine. The menu contains the following items:

- Database
- Color
- Full-screen F9
- Auto look up
- Hot key
- Uninstall
- Save options

Toggling Color

The *Options:Color* menu item toggles color on and off and is only useful if you have a color monitor. Color is on by default, indicated by a check mark next to the menu item.

Toggling Auto Lookup

The *Options:Auto look up* menu item toggles the automatic search mode on or off. A check mark indicates the mode is turned on.

Configuring the Instant Access Engine

When this option is on, activating the Instant Access Engine automatically searches the current short entry list for the word at the application program cursor position. If found, the matching entry is highlighted in the short entry list.

If Auto lookup is off, the entry highlighted when you last left the Instant Access Engine is still highlighted when you reactivate it.

This menu item is very useful if you are accessing the Instant Access Engine from your editor while writing or editing a program. For instance, if you do not know the correct syntax for a function you can type the function name and press *Shift-F1*. The function will be highlighted automatically and pressing *Return* zooms into the long entry for that function allowing you to view the function's syntax.

Changing the Hot key

The *Options:Hot key* menu item allows you to change the key pressed to activate or exit the Instant Access Engine. The default hot key is *Shift-F1*, but this can be changed to any valid key including the *Shift*, *Alt*, and *Ctrl* key combinations. When you select *Options:Hot key*, the dialog box shown in the figure below displays.

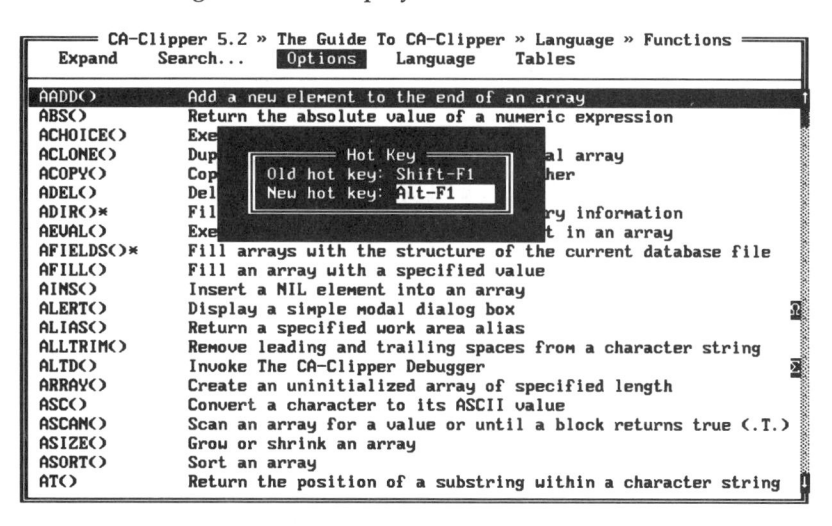

```
═══ CA-Clipper 5.2 » The Guide To CA-Clipper » Language » Functions ═══
     Expand    Search...    Options   Language    Tables

    AADD()           Add a new element to the end of an array
    ABS()            Return the absolute value of a numeric expression
    ACHOICE()        Exe
    ACLONE()         Dup ══════ Hot Key ══════  al array
    ACOPY()          Cop  Old hot key: Shift-F1  her
    ADEL()           Del  New hot key: Alt-F1
    ADIR()*          Fil ═══════════════════════ ry information
    AEVAL()          Exe                         t in an array
    AFIELDS()*       Fill arrays with the structure of the current database file
    AFILL()          Fill an array with a specified value
    AINS()           Insert a NIL element into an array
    ALERT()          Display a simple modal dialog box
    ALIAS()          Return a specified work area alias
    ALLTRIM()        Remove leading and trailing spaces from a character string
    ALTD()           Invoke The CA-Clipper Debugger
    ARRAY()          Create an uninitialized array of specified length
    ASC()            Convert a character to its ASCII value
    ASCAN()          Scan an array for a value or until a block returns true (.T.)
    ASIZE()          Grow or shrink an array
    ASORT()          Sort an array
    AT()             Return the position of a substring within a character string
```

Figure 15-6: Options:Hot Key Dialog Box

To change the hot key, press the new hot key or key combination that you want to use. Pressing *Return* confirms your choice, and pressing *Esc* cancels any changes you have made.

Note: Since the Instant Access Engine hot key takes precedence over your application program key definition for the same key, be sure to choose a hot key that does not conflict with other software that you are using.

Saving the New Configuration

While you are using the Instant Access Engine, items you change in the *Options* menu remain in effect until you remove the Instant Access Engine from memory. If you want to save your changes as the new default settings, execute the *Options:Save options* menu item. Then, the next time you load the Instant Access Engine, these settings will be the default settings.

When you execute the *Options:Save options* menu item, the Instant Access Engine saves the current options in NG.INI, located in the same directory as NG.EXE.

Leaving the Instant Access Engine

To leave the Instant Access Engine, you can either return to where you were in your application program or remove the Instant Access Engine from memory.

Exiting the Instant Access Engine

When you have finished viewing in the Instant Access Engine, you can return to the original position in your application program by pressing the hot key or *F10* from any level or view. The next time you activate the Instant Access Engine, you will return to the same screen. This is useful for quickly moving between the current view of information and your application program screen.

You can also return to your application program screen by pressing *Esc* from the top-level short entry list. Note that this may cause you to exit accidentally from time to time, since *Esc* also moves you up a level from a lower-level short entry list. If this happens, reactivate the Instant Access Engine using the hot key.

Uninstalling the Instant Access Engine

If you loaded the Instant Access Engine in memory-resident mode, you can remove the Instant Access Engine from memory using the *Options:Uninstall* menu item. If you loaded the Instant Access Engine in pass through mode, the Engine is automatically unloaded from memory when the launched application is terminated, and the *Options:Uninstall* is not operational.

In some cases, you may not be able to remove the Instant Access Engine from memory. This may occur if you load another memory-resident program after the Instant Access Engine, and it is still memory-resident. If this happens, uninstall the memory-resident program loaded after the Instant Access Engine, and then uninstall the Instant Access Engine. Failing this, you may have to reboot your computer using *Ctrl-Alt-Del*.

Summary

In this chapter, you learned about the Instant Access Engine and *The Guide To CA-Clipper*. A brief summary is given below:

- To load the Instant Access Engine as a memory-resident program, type *NG* from the DOS prompt.
- To load the Instant Access Engine in pass through mode, use the *NG <command line>* form.
- Press *Shift-F1* (or current hot key) from wherever you are located to access *The Guide To CA-Clipper* after the Instant Access Engine is loaded.
- Use the arrow keys and *Return* to select the menu item you want.
- Press *F10*, the current hot key, or *Esc* from the top level to exit.
- To remove the Instant Access Engine from memory, use the *Options:Uninstall* menu item.

For a tutorial session on using the Instant Access Engine and *The Guide To CA-Clipper*, refer the *Online Documentation* chapter in the *Getting Started* guide.

Glossary

Abbreviation
(preprocessor) A source token whose leftmost characters exactly match the leftmost characters of a keyword in a translation directive. Abbreviations must be at least four characters in length.

Activation
(procedure and function) The commencement of execution of a procedure. Each call to a procedure is referred to as an *activation* of that procedure. If the procedure in turn calls another procedure (or calls itself recursively), a *new activation* is said to have occurred. The earlier activation is then referred to as a *pending activation*.

Pending activations are often referred to as *higher level activations* or *higher level procedures*. When one procedure calls another (i.e., creates a new activation), the latter is often referred to as a *lower level activation* or a *lower level procedure*. (*See also:* Activation Record, Activation Stack, Function, Procedure)

Activation Record
(procedure and function) An internal data structure that contains information pertaining to an activation. (*See also:* Activation, Activation Stack)

Activation Stack
(procedure and function) An internal data structure that contains an activation record for the current activation and all pending activations. (*See also:* Activation, Activation Record, Stack)

Active Window
(debugger) The window to which all keystrokes (except those valid in the Command Window) apply. An active window is indicated by a highlighted border. The *Tab* and *Shift-Tab* keys are used to select the next and previous window, respectively.

Algorithm
(algorithm) A set of rules and/or a finite series of steps that will accomplish a particular task. (*See also:* Sequence, Selection, Iteration)

Alias
(database) The name of a work area; an alternate name given to a database file. Aliases are often used to give database files descriptive names and are assigned when the database file is opened. If no alias is specified when the database file is USEd, the name of the database file becomes the alias. (*See also:* Work Area)

An alias can be used to reference both fields and expressions (including user-defined functions). In order to alias an expression, the expression must be enclosed in parentheses.

Animate Mode
(debugger) The mode of execution in which an application runs one line at a time until a Breakpoint or Tracepoint is

Application

reached, with the Execution Bar moving to each line as it is executed.

Application
(configuration) A program designed to execute a set of interrelated tasks. Typically referring to a system designed to address a particular business purpose (e.g., Order Entry/Inventory/Invoicing, a document tracking database, or an insurance claims calculator).

Argument
(variable) Generally, a value or variable supplied in a function or procedure call, or an operand supplied to an operator. In function and procedure calls, arguments are often referred to as *actual parameters*. (*See also:* Parameter)

Array
(array) A data structure that contains an ordered series of values called *elements*. The elements of an array are referred to by ordinal number; the first element is number 1, the second is number 2, etc. A numeric expression used to specify an element of an array is referred to as a *subscript* or *index*. In *CA-Clipper*, the elements of an array may be values of any type, including references to other arrays. (*See also:* Array Reference, Nested Array, Subarray, Subscript)

Array Functions
(array) Those functions that specifically perform their tasks on arrays. (*See also:* Array, Function, Element, Subscript)

Array Iterator
(array) A function that traverses an array, performing an operation on each element. (*See also:* Array)

Array Reference
(array) A special data value that allows access to an array. In *CA-Clipper*, program variables and array elements cannot directly contain arrays; they may, however, contain array references. A variable that contains a reference to a particular array is said to *refer to* that array, and the array's elements may be accessed by applying a subscript to the variable. If the value of a variable containing an array reference is assigned to a second variable, the second variable will contain a copy of the array reference; both variables then refer to the same array, and the array's elements may be accessed by applying a subscript to either variable. (*See also:* Array, Multi-dimensional Array, Nested Array, Subarray, Subscript)

ASCII
(general) An acronym for the American Standard Code for Information Interchange, the agreed upon standard for representing characters (alphabetic, symbolic, etc.) in the memory of the computer.

Assignment
(expression) The act of copying a new value into a variable. In *CA-Clipper* this is done with the simple assignment operators (=) and (:=), or the compound operators (+=, -=, *=, **=).

Attribute
(database) As a formal DBMS term, refers to a column or field in a table or database file. (*See also:* Column, Field)

Background Color
(user-interface) The color that appears behind displayed text of another color (the foreground color). (*See also:* Foreground color)

Beginning-of-file
(database) The top of the database file. In *CA-Clipper* there is no beginning-of-file area or record. Instead, it is indicated by BOF() returning true (.T.) if an attempt is made to move the record pointer above the first record in the database file or the database file is empty.

Binary
(general) The numerical base upon which computer programs are modeled. Also referred to as base 2, it is a mathematical way of representing the electrical circuitry upon which computers operate. A 0 represents an *off* state, and a 1 indicates an *on* state. A combination of eight ons and offs are used to express bytes of program or data in memory. (*See also:* Byte)

Binary File
(file) A file that contains an unformatted sequence of bytes. Carriage-return, line-feed, or end-of-file characters have no special meaning in a binary file. Binary files include executable files, graphics files, or data files. (*See also:* Text File)

Binary Operator
(expression) An operator that operates on two operands. For example, the addition operator. (*See also:* Operator)

Blockify
(preprocessor) To change an expression in the source text into a code block definition. Blockifying is accomplished by surrounding the source text with braces and placing an empty block parameter list (a pair of vertical bars) just inside the braces. When the resulting code block is evaluated at runtime, the original expression will be evaluated.

Blockify Result-marker
(preprocessor) A result-marker of the form <{id}>. *id* must correspond to the name of a match-marker. A blockify result-marker specifies that the corresponding source text is to be blockified. If the matched source text is a list of expressions, each expression in the list is individually blockified. If no source text matched the corresponding match-marker, an empty result is produced.

Branching
(language) Changing the sequence of execution in a program. Execution normally proceeds in sequence from the top of a function or procedure to the bottom. When control is transferred to a statement that is not in sequence, execution is said to have *branched*.

Breakpoint
(debugger) A point at which an application pauses execution and returns control to the debugger.

Buffer
(general) A temporary data storage location in memory. As an example, a *disk input-output buffer* is an area of memory that stores data read from the disk to temporary locations while processing it.

Byte
(general) Eight bits of data, the smallest unit of information stored in the computer's memory. As an example, one byte is required to represent one ASCII character.

Calling Program
(procedure and function) The procedure, or user-defined function that transferred control to the currently executing procedure or function. When the current

procedure or user-defined function terminates with a RETURN statement, this is where control will return.

Callstack
(debugger) A list containing the names of all pending activations at the current point in an application.

Callstack Window
(debugger) The window in which the Callstack is displayed.

Cell
(database) In a table, a cell is the intersection of a Row and a Column. (*See also:* Column, Row, Table)

Character
(general) A letter, digit, punctuation mark, or special symbol stored and processed by the computer. (*See also:* ASCII)

(data type) A special data type consisting of one or more values in the IBM extended character set. Characters can be grouped together to form strings. The maximum size of a character string in *CA-Clipper* is 65,534 bytes. (*See also:* String)

Character Functions
(expression) Those functions that act upon individual characters or strings of ASCII characters in the performance of their tasks. (*See also:* ASCII, Function, String)

Classes
(object-oriented) A class defines the variables contained in an object, and the operations applied when the object receives a message. Every object is an instance of a class and responds to messages of that class. Each object, however, has its own copy of the variables specified in the class definition. New objects are created in *CA-Clipper* by a calling a special function that begins with the class name followed by the *New* suffix. (*See also:* Object, Instance Variables, Messages, Methods)

Clause
(command) An optional or required section of a *CA-Clipper* command beginning with a keyword that modifies or enhances the command.

Code Block
(data type) A special data type that refers to a piece of compiled program code. In a program, the source code that specifies the creation of a code block.

Code Window
(debugger) The window in which source code is displayed.

Collision
(network) An attempt by more than one user (typically on a networked, multi-user system) to update a database simultaneously, usually resulting in data corruption. Generally due to poor implementation of the code to support multi-user operations. (*See also:* LAN, Local Area Network, Deadly Embrace)

Column
(database) A database term used to describe a field in a table or database file. (*See also:* Field)

(user-interface) A numeric value that represents a position on the display screen or on the printed page.

Command
(command) A statement to be translated by the *CA-Clipper* preprocessor into

source code that will perform a particular operation. All *CA-Clipper* commands are defined in the standard header file, STD.CH, located in \CLIPPER5\INCLUDE. Also, the preprocessor directives that define a command. (*See also:* Statement, Header files, STD.CH)

Command Window
(debugger) The window in which commands are displayed and entered.

Comment
(language) Text in a source program that is ignored by the compiler. Usually used to make descriptive comments about the surrounding source code.

Compiler
(program) A program that translates source code output from the preprocessor into object code. The resulting object file can then be linked to produce an executable program using the linker. (*See also:* Linker, Object File, Program File)

Concatenate
(expression) To combine two groups of character data together by placing them in a sequence to form a new string of characters. (*See also:* Data Type)

Concurrency
(database) The degree to which data can be accessed by more than one user at the same time.

Condition
(command, database, expression) A logical expression that determines whether an operation will take place. With database commands, a logical expression that determines what records are included in an operation. Conditions are specified as arguments of the FOR or WHILE clause. (*See also:* Scope)

Conditional Compilation
(preprocessor) Selective exclusion by the preprocessor of certain source code. The affected source code is bracketed by the #ifdef and #endif directives. It is excluded if the argument in the #ifdef directive is not a #defined identifier.

Console Input/Output
(user-interface) A style of operation of the keyboard and display that emulates a simple typewriter-like interface. Console input echoes each key typed and provides processing for the backspace and return keys. Console output wraps to the next line when the output reaches the right edge of the visible display, and scrolls the display when the output reaches the bottom of the visible display. (*See also:* Full-screen Input/Output)

Constant
(variable) The representation of an actual value. For example, .T. is a logical constant, *string* is a character constant, 21 is a numeric constant. There are no date and memo constants.

Constant Array
(array) *See* Literal Array.

Control Structure
(language) Any program structure that alters the flow of program control. In *CA-Clipper*, these include:

- BEGIN SEQUENCE...END
- DO WHILE...ENDDO
- DO CASE...ENDCASE
- FOR...NEXT
- IF...ENDIF

Controlling/Master Index
(database) The index currently being used to refer to records by key value or sequential record movement commands. (*See also:* Index, Natural Order)

Conversion Functions
(expression) Generally referring to a category of functions whose purpose is to change one data type to another (e.g., to change a number or a date to a character string).

Cursor
(user-interface) An on-screen indicator used to show the current keyboard input focus and is displayed as a block or underline character. The cursor moves in response to characters or control keys typed by the user. (*See also:* Highlight, Input Focus)

Cursor
(debugger) The cursor indicates the current line and/or column position in the active window or dialog box. Note that some windows, such as the Monitor Window, do not utilize the cursor. When a window that does not utilize the cursor is active, the cursor appears in the Code Window.

Data Independence
(modular programming) The technique of writing a program, function, or procedure in such a way that it will be able to perform its operation on data supplied to it without a *built-in* description of the data format.

Data Type
(data type) The category of a data value. A data type is distinguished by the set of allowable values for that type, the set of operators that can be applied, and the storage format used to represent these values. In *CA-Clipper*, the following data types are defined: character, numeric, date, logical, array, object, code block, and NIL. Program variables may contain values of any type. Database field variables are limited to character, numeric, date, logical, and a special type called memo which is treated the same as character.

Database
(database) An aggregation of related operational data used by an application system. A database can contain one or more data files or tables. (*See also:* Field, Record, Tuple, View)

Date Functions
(expression) Functions that operate on date values (as opposed to character, numeric or other values). (*See also:* Function)

Date Type
(data type) A special data type consisting of digits to store year, month, and day values. Operations on date values are based on chronological values.

DBMS
(database) An acronym for the term *database management system*. A DBMS is a software system that mediates access to a database through a data manipulation language.

Deadly Embrace
(network) In a multi-user database management system, the result of an unforeseen flaw that has allowed a situation wherein two or more simultaneous operations cannot be completed because each has instigated locks on files or records that the other needs to complete its task.

Debugger
(debugger) A tool used to track down errors in a program.

Debugging
(error handling) A phase of software development where errors are identified and fixed.

Declaration
(variable) A statement used by the compiler to define a variable, procedure, or function identifier. The scope of the declaration is determined by the position of the declaration statement in the source file. (*See also:* Identifier, Scope)

Decrement
(expression) To decrease a value by a fixed amount, usually one. In *CA-Clipper*, the decrement operator (--) can be used to decrement a numeric value in a variable. (*See also:* Assignment, Increment)

Define
(preprocessor) To #define an identifier to the preprocessor and optionally specify text to be substituted for occurrences of the identifier.

Delimited File
(file, database) A text file that contains variable-length database records with each record separated by a carriage-return/line feed pair (CHR(13) + CHR(10)) and terminated with an end-of-file mark (CHR(26)). Each field within a delimited file is variable length, not padded with either leading or trailing spaces, and separated by a comma. Character strings are optionally delimited to allow for embedded commas. (*See also:* Database File, Text File, SDF File)

Delimiter
(general) A character or other symbol that marks a boundary.

Destination
(expression) The variable or array element to receive data in an assignment. (*See also:* Assignment)

(general) The work area, file, or device to which data is sent.

Device
(general) Either an actual physical component of the computer system such as printer or a DOS *handle* that refers to it (e.g., PRN:), or a *logical device* that behaves and is addressed the same way as a *physical device* (e.g., a print spooler).

Dialog box
(debugger) A box displayed from within the debugger whenever further input is required.

Dimension
(array) The maximum number of subscripts required to specify an array element. For example, a two-dimensional array must have two subscripts, a three-dimensional array must have three subscripts and so on. (*See also:* Subscript)

Directive
(preprocessor) An instruction to the preprocessor. Preprocessor directives must appear on a separate line, and must begin with a hash mark (#). Their scope or effect extends from the point where they are encountered to the end of the source file in which they appear.

Directory
(file) The major operating system facility for cataloging files. A directory contains a list of files and references to child

directories (subdirectories), and is identified by name. Directories can be nested forming a hierarchical tree structure. The operating system provides a number of facilities that allow users to create and delete directories. (*See also:* Disk, File, Path, Volume)

Disk

A magnetic storage medium designed for long-term storage. Disks come in two varieties: hard disks (fast but fixed) or floppy disks (slow but removable). A disk can be partitioned into multiple volumes, each containing a tree-structured directory system that holds files accessible by programs. (*See also:* Volume, Directory, File)

Drive

(file) A disk drive or a letter (normally followed by a colon) that designates a disk drive. On most computers, the letters A and B refer to floppy disk drives; other letters refer to fixed disk drives or *logical* drives (e.g., fixed disk partitions or network drives).

Dumb Stringify Result-marker

(preprocessor) A result-marker of the form #<{id}>. *id* must correspond to the name of a match-marker. A dumb stringify result-marker specifies that the corresponding source text is to be enclosed in quotes. If the matched source text constitutes a list of expressions, each expression in the list is individually stringified. If no source text was matched, an empty pair of quotes is produced.

Dynamic

(general) Used generically to refer to data or algorithms that change with time. Often used specifically to describe algorithms that automatically adjust to prevailing conditions.

Dynamic Overlay

(linker) Allows a module's code to be divided into pages to be brought into and out of memory on a least recently used basis.

Dynamic Scoping

(variable) A method of determining an item's existence or visibility based on the state of a program during execution. Example: A *CA-Clipper* public variable may or may not be visible within a particular function, depending on whether the variable has been created and whether a previously called function has obscured it by creating a private variable with the same name. (*See also:* Lexical Scoping, Scope of a Variable)

Element

(array) A component unit of an array, usually referred to by a numeric subscript or index. (*See also:* Array, Subscript)

Empty Result

(preprocessor) An absence of result text; the effect of certain result-markers when the corresponding match-marker did not match any source text (but when the translation directive as a whole was matched). An empty result simply implies that no result text is written to output.

Encapsulation

(modular programming) Generically, the design of a function or program that obeys the principle of information hiding. A function or program is said to be encapsulated if other programs or functions have no knowledge of its inner workings. A data structure is said to be encapsulated if knowledge of its internal organization is limited to a single function or module. (*See also:* Information Hiding, Lexical Scoping, Modularity, Side Effect)

End-of-file
(database) The bottom of a database file. In *CA-Clipper*, this is LASTREC() + 1 and is indicated by EOF() returning true (.T.).

Enhanced Color
(user-interface) The color used to display GETs or PROMPTs (if INTENSITY is ON). (*See also:* Standard Color)

Environment Variables
(configuration) Operating system variables that can be used to communicate configuration information to executable programs. Environment variables are manipulated using the DOS SET command. The *CA-Clipper* compiler and linker respond to certain environment variables. *CA-Clipper* programs can inspect the settings of environment variables using the GETENV() function.

Error
(error handling) The presence of some element of an operation that does not satisfy the requirements of the operation. An error when encountered causes failure, which in turn raises an exception. (*See also:* Exception, Failure, Runtime Error)

Error Handling
(error handling) The concept of including code in a program so that exceptions to normal operational states that occur during the program execution can be anticipated and dispatched with the least possible detrimental consequences to the use of the program and the data being worked on. (*See also:* Exceptions, Runtime Errors)

Evaluate
(expression) To execute part of a program in order to produce a value. For an expression, to execute the program code associated with the expression and return the resulting value. For the macro operator, to compile the macro string, execute the resulting program code, and return the resulting value. (*See also:* Expression)

Exception
(error handling) An occurrence of an abnormal condition during the execution of an operation. An exception is said to be raised when an operation fails. (*See also:* Error, Failure, Runtime Error)

Exclusive
(network) In a network, to assure that no other user will write data to a file, it may be opened in an Exclusive mode. Only the user opening the file exclusively may then access it until exclusive use of the file (by closing it, or opening it SHARED) is relinquished.

Executable File
(configuration) A file output from the linker directly executable from the operating system command-line. Executable files have an .EXE extension. (*See also:* Linker)

Execution Bar
(debugger) The highlight bar which is positioned on the line of code to be executed next.

Exported Instance Variables
(object-oriented) *See* Instance Variables.

Expression
(expression) A combination of constants, identifiers, operators, and functions that yield a single value when evaluated.

Extension
(file) A filename extension normally used for identifying the type or originating program of a file. (*See also:* Drive, Filename, Path)

Failure
(error handling) The inability of an operation to satisfy its purpose. When a failure occurs an exception is raised. Failures are due in large part to errors. (*See also:* Exception, Error, Retry, Runtime Error)

Field
(database) The basic column unit of a database file. A field has four attributes: name, type, length, and decimals if the type is numeric. (*See also:* Database, Record, Tuple, Vector, View)

Field Variable
(variable, database) A variable that refers to data in a database field, as opposed to data in memory. (*See also:* Local Variable, Memory Variable, Private Variable, Public Variable, Static Variable, Variable)

File
(file) A file is an organized collection of bytes stored on disk, maintained by the operating system, and referenced by name. Its internal structure is solely determined by its creator. (See also: Binary File, Database File, Text File)

File Handle
(file) An integer numeric value returned from FOPEN() or FCREATE() when a file is opened or created. This value is used to identify the file for other operations until it is closed.

File Locking
(network) The process by which a user is guaranteed exclusive access to a database file. The file is only available to the user that applied the lock. (*See also:* Record Locking)

File Server
(network) A computer on a network dedicated to providing data storage to other computers (i.e., workstations) for the purpose of sharing information among multiple users. File servers tend to provide other services such as E-Mail service and shared printer support as well.

File-wide Declaration
(variable) A variable declaration statement that has the scope of the entire source file. File-wide declarations are specified before the first procedure or function declaration in a program file and the program file must be compiled with the /N option. (*See also:* Scope, Storage Class)

Filename
(file) The name of a disk file that may optionally include a drive designator, path, and extension. (*See also:* Drive, Extension, Path)

Foreground Color
(user-interface) The color of text appearing on the screen, usually on a different colored background. (*See also:* Background Color)

Form-feed
(general) A special character (CHR(12)) that by convention causes most printers to move the printhead to the top of the next page. (*See also:* Hard Carriage Return, Line-feed)

Freeformat
(linker) The suggested command interface for *.RTLink*, allowing you to specify linker

commands in any order on the command-line. It is easier to create, examine and change FREEFORMAT command-lines. It is compatible with the Plink86-Plus syntax.

Full-screen Input/Output
(user-interface) A style of operation of the keyboard and display used for complex data entry and display tasks. Full-screen input and output are generally performed using the @..SAY, @..GET and READ commands. Full-screen output is distinguished from console-style output by the fact that control characters (e.g., backspace, carriage-return) are not processed, and wrapping and scrolling do not occur at the boundaries of the visible display area. (*See also:* Console Input/Output)

Function
(procedure and function) An executable block of code with an assigned name. Alternately, the collection of source code statements that define a function. Certain functions are supplied as part of *CA-Clipper*; others are defined by the programmer using the FUNCTION or PROCEDURE declaration statements. The latter are referred to as *user-defined* functions.

The terms *procedure* and *function* are generally interchangeable. By convention, a function returns a value, while a procedure does not. (*See also:* Activation, Parameter, Procedure)

Group
(linker) An Intel 8086 addressing classification defining a collection of segments to be addressed using the same segment register. Note that a group is not a section but rather a logical concept used only for addressing.

Hard Carriage Return
(general) An explicit carriage-return character at the end of a line in a text file, as opposed to a soft carriage return that might be inserted into text by a program designed to handle word-wrapping. A hard carriage-return character is generated by the expression (CHR(13)) where a soft carriage-return character is generated with the expression (CHR(141)).

Header File
(configuration, pre-processor) A source file containing manifest constant definitions; command or pseudo-functions; and/or program statements merged into another source file using the #include pre-processor directive. (*See also:* Program File, Source File, STD.CH)

Help Window
(debugger) The window in which on-line help is displayed.

Hexadecimal
(general) A representation of a value in base-16 rather than decimal which is base-10. Hexadecimal values are easily converted to and from binary (base-2) which is the form of data the computer actually uses. Hexadecimal values are represented by digits zero through nine and A through F for values between 10 and 15.

Hidden
(modular programming) The resulting state of a module of a program written to conform to the principals of information hiding. (*See also:* Information Hiding)

Highlight
(user-interface) Indicates input focus for menus, browsers, or GETs. With menus and browsers, the currently selected item

or cell has input focus and is displayed in the current enhanced color or inverse video. With GETs, the current GET is highlighted in the current enhanced color or inverse video while the other GETs are displayed in the current standard color if an unselected color setting is active. (*See also:* Cell, Standard Color, Enhanced Color, Input focus, Unselected Color)

IBM Extended Character Set
(general) The character set built into the ROM of the IBM-PC. This character set is a superset of ASCII, containing additional special characters (such as a line drawing character set) that may be used to enhance your program screens.

Identifier
(pre-processor, procedure and function, variable) A name that identifies a function, procedure, variable, constant or other named entity in a source program. In *CA-Clipper*, identifiers must begin with an alphabetic character and may contain alphabetic characters, numeric characters, and the underscore character.

Include File
(pre-processor) *See* Header File.

Increment
(expression) To increase a value by a fixed amount, usually one. In *CA-Clipper* the increment operator (++) can be used to increment a numeric value in a variable.

Incremental Linking
(linker) The ability to link only the modules of an application that have been changed, greatly increasing the speed in which the link occurs. (*See also:* Linking, Module)

Index
(database) An ordered set of key values that provides a logical ordering of the records in an associated database file. Each key in an index is associated with a particular record in the database file. The records can be processed sequentially in key order, and any record can be located by performing a SEEK operation with the associated key value. (*See also:* Controlling/Master Index, Key Value, Natural Order)

Information Hiding
(procedure and function) A fundamental programming principle that states that functions and programs should conceal their inner workings from other functions and programs. Stated simply: a function should possess only the knowledge necessary for it to accomplish its task. When one function calls another, the calling function should possess only the knowledge explicitly required to call the other function. (*See also:* Encapsulation, Lexical Scoping, Modularity, Side Effect)

Initialize
(variable) To assign a starting value to a variable. If initialization is specified as part of a declaration or variable creation statement, the value to be assigned is called an *initializer*. (*See also:* Assignment)

Input Focus
(user-interface) The GET, browse cell, or menu item where user interaction can take place is said to have input focus. The item with input focus usually is displayed in enhanced color or inverse video.

Insert Mode
(user-interface) A data entry mode entered when the user presses the insert key. When this mode is active, characters are inserted at the cursor position. Text

to the right of the cursor is shifted right.
(*See also*: Overstrike Mode)

Inspecting
(debugger) The process of examining work areas, variables, expressions and activations inside the debugger.

Instance Variables
(object-oriented) Instance variables are the attributes or the data portion of an object as defined by the object's class. Each object, when created, is given its own unique set of instance variables initialized to their default values. Instance variables that are accessible are called *exported* instance variables. Exported instance variables can be inspected and—in some cases—assigned using the send operator. The instance variables of an object persist as long as the object it belongs to. (*See also*: Class, Object)

Integer
(data type) A number with no decimal digits. Note that *CA-Clipper* does not provide a separate data type for integer values.

Iteration
(algorithm) One of the three basic building blocks of algorithm development (the others are sequence and selection). Iteration refers to operations that are performed repeatedly, usually until some condition is satisfied. (*See also*: Selection, Sequence)

Join
(database) An operation that takes two tables as operands and produces one table as a result. It is, in fact, a combination of other operations including selection and projection. (*See also*: Selection, Projection)

Key Expression
(database) An expression, typically based on one or more database fields, that when evaluated, yields a key value for a database record. Key expressions are most often used to create indexes or for summarization operations. (*See also*: Index, Key Value)

Key Value
(database) The value produced by evaluating a key expression. When placed in an index, a key value identifies the logical position of the associated record in its database file. (*See also*: Index, Key Expression)

Keyboard Buffer
(user-interface) An area of memory dedicated to storing input from the keyboard while a program is unable to process the input. When the program is able to accept the input, the keyboard buffer is emptied.

Keyword
(command, language) A word that has a special meaning to a compiler or other utility program. Commands, directives, or options are often recognized by examining supplied text to see if it contains keywords.

LAN
(network) An acronym for Local Area Network. Generally used to describe a system by which microcomputers are connected together to perform such functions as file and peripheral sharing, electronic mail, and centralized backup of data.

Lexical Scoping
(variable) A method of determining an item's existence, visibility, or applicability (i.e., the item's *scope*) by it's position

within the text of a program. (*See also:* Local, Variable, Scope of a Variable)

Lexically Scoped Variable
(variable) A variable that is only accessible in a particular section of a program, where that section is defined using simple textual rules. For example, a local variable is only accessible within the procedure that declares it. (*See also:* Dynamic Scoping, Private Variable, Public Variable, Static Variable)

Library
(linker) A file containing one or more object modules. Modules are extracted by linker and combined with object files to form an executable (.EXE) file or a pre-link library (.PLL) file.

Library File
(configuration) A file containing one or more object modules. The linker searches specified libraries to resolve references to functions or procedures that were not defined in the object files being linked. (*See also:* Linker, Module, Object File)

Lifetime of a Variable
(variable) The period of time during which a variable retains its assigned value. The lifetime of a variable depends on its storage class. (*See also:* Scope of a Variable, Visibility)

Line-feed
(general) A special character (CHR(10)) that by convention causes the cursor or printhead to move to the next line or to terminate a line in a text file. It is usually used in combination with a hard carriage return. (*See also:* Form-feed, Hard Carriage Return)

Linker
(program) A program that combines object files created by a compiler to produce an executable program. The linker examines the supplied object files to resolve symbol references between modules. If a module refers to a symbol that is not defined by any of the modules, the linker searches one or more libraries to resolve the reference. (*See also:* Library File, Object File)

Linking
(linker) The process in which object files and libraries are combined and references are resolved to produce a relocatable memory image (generally, an executable).

List
(expression, command) A list of expressions, field names, or filenames, separated by commas specified generally as command, procedure, or function arguments. Code blocks can also execute a list of expressions.

List Match-marker
(pre-processor) A match-marker indicating a position that will successfully match a list of one or more arbitrarily complex expressions. A list match-marker marks a part of a command that is expected to consist of a list of programmer-supplied expressions. A list match-marker has the form <*id*,...>. *id* associates a name with the match-marker. The name can be used in a result-marker to specify how the matching source text is to be handled.

Literal
(data type) A source code element interpreted literally (as encountered), and assumed to have no abstract meaning. Generally a constant. (*See also:* Constant).

Literal Array
(array) In *CA-Clipper*, an array specified by enclosing a series of expressions in curly ({}) braces. A literal array is an expression that evaluates to an array reference. (*See also:* Array, Array Reference)

Local Area Network
(network) A system by which microcomputers are connected (via coaxial cable, optical fiber, twisted pair phone wire, or other media), relying on sophisticated operating software to perform such functions as file and peripheral sharing, electronic mail, and centralized backup of data. (*See also:* LAN).

Local Variable
(variable) A variable that exists and retains its value only as long as the procedure in which it is declared is active (i.e., until the procedure returns control to a higher level procedure). Local variables are lexically scoped; they are accessible by name only within the procedure where they are declared. (*See also:* Dynamic Scoping, Lexical Scoping, Private Variable, Public Variable, Static Variable)

Logical Type
(data type) A special data type consisting of true (.T.) or false (.F.) values. (*See also:* Condition)

Logify
(pre-processor) To change an expression in the source text into a logical value. Logifying is accomplished by surrounding the expression with periods.

Logify Result-marker
(pre-processor) A result-marker of the form #<.id.>. *id* must correspond to the name of a match-marker. This result-marker writes true (.T.) to the result text if any input text is matched; otherwise, it writes false (.F.) to the result text. The input text itself is not written to the result text.

Macro
(expression) In *CA-Clipper*, an operation that allows source code to be compiled and executed at runtime. In *CA-Clipper*, the macro symbol (&) does not perform *text substitution* unless embedded within a character string. Instead, it is generally treated as a unary operator that operates on a character string. The text in the character string is compiled *on the fly* using a special runtime compiler. The resulting code is then executed, and the value obtained is returned as the result of the macro operation. (*See also:* Code Blocks, Unary Operator)

Make
(program) A program used to maintain multi-file program systems. A make program takes as its input a file (make file) specifying the relationships between files. When executed the make program compares the date and time stamps of specified target files to the specified dependent files. If any of the dependent files have a more recent date and time stamp than the associated target files, a series of actions are performed. (*See also:* Make File)

Make File
(configuration) A text file used as input to a make utility containing the specifications and actions required to build a program or a system of programs. This file is often referred to as a description file. (*See also:* Make)

Manifest Constant
(pre-processor) An identifier specified in a #define directive. The pre-processor substitutes the specified result text whenever it encounters the identifier in the source text.

Map File (.MAP)
(linker) The map file (.MAP) contains information about symbol and segment addresses within the memory image created by *.RTLink*. It is generated when requested through the use of the appropriate command-line switch. A map file generated during a link will have much more information than one produced during the creation of a pre-linked library. During the .PLL creation, only symbols, names, and some relative addresses are known. During a link, the final memory layout is known, and a more detailed map can be created.

Master Index
(database) *See* Controlling/Master Index.

Match
(pre-processor) A successful comparison of source text with a match-pattern (or part of a match-pattern).

Match-marker
(pre-processor) A construct used in a match-pattern to indicate a position that will successfully match a particular type of source text. There are several types of match-markers, each of which will successfully match a particular type of source text.

Match-pattern
(pre-processor) The part of a translation directive that specifies the format of source text to be affected by the directive. A match-pattern generally consists of words and match-markers.

Memo Type
(data type, database) A special database field type consisting of one or more characters in the IBM extended character set. The maximum size of a memo field in *CA-Clipper* is 65,534 bytes. A memo field differs only from a character string by the fact it is stored in a separate memo (.DBT file) and the field length is variable-length (*See also:* Character, String)

Memory Variable
(variable) In general, a variable that resides in memory, as opposed to a database field variable. Sometimes used specifically to refer to variables of the MEMVAR storage class (private and public variables), as opposed to static or local variables. (*See also:* Field Variable, Local Variable, Private Variable, Public Variable, Static Variable, Variable)

Menu
(user-interface) An on-screen list of choices from which the user selects. Menus range from simple to elaborate forms. Two examples are menus that *pull-down* from the top of the screen (an elaborate type requiring more programming), or a simple list of numbered items from which the user selects by entering the appropriate number.

Menu Bar
(debugger) The bar at the top of the debugger screen, on which the available menu choices are displayed.

Messages
(object-oriented) A message is the way an object is requested to perform some action. Messages are sent to an object and composed of the object name, the send operator, and the selector name followed by arguments enclosed in parentheses.

The selector has the same name as the method it is calling. Sending a message produces a return value, much like a function call with the return value varying depending on the operation performed. (*See also:* Objects, Methods, Instance Variables)

Metasymbol
(language) Descriptive symbols used in syntax to represent information that must be supplied as part of a source code statement. A metasymbol is constructed using two information components: a data type prefix and a logical descriptor.*See also:*

Methods
(object-oriented) A method is the operation performed in response to a message sent to an object. (*See also:* Classes, Messages, Objects)

Modularity
(modular programming) Roughly, a measure of a system's adherence to the principles of modular programming. The principles of modular programming are not precisely defined, but may be said to comprise these basic ideas: programs should be organized as well-defined *modules*; modules should correspond with syntactic units of the programming language (such as functions or source files); a module should accomplish a well-defined task; a module should interact with as few other modules as possible; interactions between modules should be explicitly specified in the source code for the modules; modules should obey the principle of information hiding. (*See also:* Encapsulation, Information Hiding, Lexical Scoping, Side Effect)

Module
(modular programming) Generically, a procedure or function (or a set of related procedures and functions) that can be treated as a unit. Sometimes used to refer specifically to the code in a single object file, normally the result of compiling a single source file. (*See also:* Object File, Source File)

Module
(linker) A portion of the object code that is a discrete unit. If any part of a module is linked, the entire module must be linked.

Monitor Window
(debugger) The window in which monitored variables are displayed.

Monitored variable
(debugger) A variable which is selected by the options on the Monitor Menu and displayed in the Monitor Window.

Multi-dimensional Array
(array) In *CA-Clipper*, an array whose elements consist entirely of references to other arrays (called *subarrays*). The elements of the subarrays may, in turn, contain references to other arrays. Arrays organized in this fashion are said to be *nested*. Each level of nesting may be viewed as a *dimension* of the main array, and the elements of the subarrays may be accessed by applying multiple subscripts to the main array. (*See also:* Array, Array Reference, Nested Array, Subscript)

Natural Order
(database) For a database file, the order determined by the sequence in which records were originally entered into the file. Also called unindexed order. (*See also:* Index)

Nested Array
(array) In *CA-Clipper*, two arrays are said to be *nested* if one of them contains a reference to the other. When an array contains a reference to a second array, the second array is sometimes called a *subarray* of the first array. (*See also:* Array, Array Reference, Multi-dimensional Array, Subscript)

NIL
(data type) A special data type that has only one allowable value. The special value (NIL) is automatically assigned to all uninitialized variables except publics, and is also passed as a substitute when arguments are omitted in a procedure or function call.

Normal Stringify Result-marker
(pre-processor) A result-marker of the form <*"id"*>. *id* must correspond to the name of a match-marker. A normal stringify result-marker specifies that the corresponding source text is to be enclosed in quotes. If the matched source text is a list, each element of the list is individually stringified. If no source text was matched, an empty result is produced.

Normalization
(database) The process of elimination and consolidation of redundant data elements in a database system.

Numeric Type
(data type) A special data type consisting of values that indicate magnitude. Numeric values consist of digits between zero and nine, a sign, and a decimal point.

Object File
(configuration) A file that contains the output of a compiler or other language translator, generally the result of compiling a single source file. Object files are linked to create an executable program. (*See also:* Linking, Source File)

Objects
(object-oriented, data type) An object is an instance of a class. Each object has one or more attributes (called instance variables) and a series of operations (methods) that execute when a message is sent to the object. The object's instance variables can only be accessed or assigned by sending messages to the object. Objects are created by calling a special function associated with a class. (*See also:* Classes, Instance Variables, Messages, Methods)

Operand
(expression) A value that is operated on by an operator, or the term in an expression that specifies such a value. For example, in the expression $x + 5$, x and 5 are operands. (*See also:* Operator)

Operating System
The basic software program that organizes and services the computer and its peripheral devices. The operating system supported by *CA-Clipper*, MS/PC-DOS is organized into several layers as follows:

- *Loader* is the layer which brings the operating system software into memory.

- *BIOS* is the basic hardware interface layer that provides services to the kernel and consists of initialization code and device drivers.

- *Kernel* is the application interface layer and provides services for process control, memory management, peripheral support, and a file system.

- *User interface shell (COMMAND.COM)* provides basic services to the user including an interactive mode, directory management, and a service for loading and executing application programs.
- *Support programs* provide extended operating services not resident in the user interface shell.

Operator
(expression) A symbol that identifies a basic operation. For example, the multiplication operator (*) denotes that two values are to be multiplied. Operators are categorized as either unary or binary, depending on whether they require one or two operands, respectively. (*See also:* Binary Operator, Operand, Unary Operator)

Optional Clause
(command, pre-processor) A portion of a match-pattern that is enclosed in square ([ ]) brackets. An optional clause specifies part of a match-pattern that need not be present for source text to match the pattern. An optional clause may contain any of the components legal within a match-pattern, including other optional clauses. When a match-pattern contains a series of optional clauses that are immediately adjacent to each other, the matching portions of the source text are not required to appear in the same order as the clauses in the match-pattern. If an optional clause is matched by more than one part of the source text, the multiple matches may be handled using a repeating clause in the result-pattern.

Overlay
(linker) A section of an executable program that shares memory with other sections of the same program. An overlay is read into memory when the code residing in it is requested by the root (non-overlayed) section or another overlay. (*See also:* Dynamic overlay)

Overstrike Mode
(user-interface) A data entry mode entered when the user presses the insert key. When this mode is active, characters are entered at the cursor position and text to the right of the cursor remains stationary. (*See also:* Insert Mode)

Parameter
(variable) A identifier that receives a value or reference passed to a procedure or user-defined function. A parameter is sometimes referred to as a *formal parameter*. (*See also:* Activation, Argument, Function, Procedure, Reference)

Path
(file) A literal string that specifies the location of a disk directory in the tree structured directory system. A path specification consists of the following elements: an optional disk drive letter followed by a colon, an optional backslash indicating that the path starts at the root directory of the specified drive, the names of all the directories from the root directory to the target directory, separated by backslash (\) characters. Example: C:\CLIPPER5\INCLUDE. A path list is a series of path specifications separated by semicolons.

Picture
(user-interface) A string that defines the format for data entry or display in a GET, SAY, or the return value of TRANSFORM(). Picture strings are comprised of functions which affect the formatting as a whole and a series of template characters that affect formatting

on a character by character basis. (*See also:* Template)

Port
(general) A designation for the hardware that allows the processor to communicate with peripheral devices.

Positional
(linker) The POSITIONAL command interface requires that certain items appear on the input line in a specific order. This syntax is similar to Microsoft LINK interface. Because this syntax limits the use of *.RTLink* overlays to one overlay area, it is recommended that the FREEFORMAT syntax be used.

Pre-linked Library
(linker) Part of the executable program that is stored external to the .EXE file. Pre-linked libraries are created before producing an executable in a multi-step link. This allows you to create a run-time library with code that you access from different programs, considerably speeding up the linking process.

Preprocessor
(program, preprocessor) A translation program that prepares source code for compilation by applying selective text replacements. The replacements to be made are specified by directives in the source file. In *CA-Clipper*, the pre-processor operates transparently as a part of the compiler program (CLIPPER.EXE). (*See also:* Compiler)

Precedence
(expression) The stature of an operator in the hierarchy that determines the order in which expressions are evaluated. For example, the expression 5 + 2 * 3 is interpreted as 5 + (2 * 3) because the multiply operator (*) has a higher precedence than the addition operator (+). (*See also:* Expression)

Primitive
(procedure and function) A simple, low-level function used by other high-level functions or programs to perform a more complex task.

Print Spooler
A program running either on a local workstation or on the file server that captures print jobs to a file and then queues them for printing later. Print spoolers generally operate as background tasks in order to facilitate printing while other tasks are operating in the foreground.

Private Variable
(variable) A variable of the MEMVAR storage class. Private variables are created dynamically at runtime using the PRIVATE statement, and accessible within the creating procedure and any lower level procedures unless obscured by another private variable with the same name. (*See also:* Activation, Dynamic Scoping, Function, Lexical Scoping, Local Variable, Procedure, Public Variable, Static Variable)

Procedure
(procedure and function) An executable block of code with an assigned name. Alternately, the collection of source code statements that define a procedure. In *CA-Clipper*, this can be a source file (.prg), a format file (.fmt), an explicitly declared procedure (PROCEDURE), or an explicitly declared function (FUNCTION).

The terms *procedure* and *function* are generally interchangeable. By convention, a function returns a value, while a procedure does not. (*See also:* Activation, Function, Parameter)

Procedure File
(configuration) An ASCII text file containing *CA-Clipper* procedure and function definitions usually ending with a (.prg) extension; a program file. (*See* Program File)

Program Editor
(program) A program that creates and edits text files or programs.

Program File
(configuration) An ASCII text file containing *CA-Clipper* source code. Program files usually end with a (.prg) extension. The compiler reads the program file, translates the source code, and produces an object file, that is then linked to produce an executable program. (*See also:* Linking, Object File, Source Code)

Projection
(database) A DBMS term specifying a subset of fields. In *CA-Clipper*, the analogy is the FIELDS clause. (*See also:* Join, Selection)

Prompt
(user-interface) A series of characters displayed on the screen indicating that input from the keyboard is expected.

Pseudo-function
(pre-processor) A function-like construct that is replaced with another expression via the #define directive, rather than compiled into a conventional function call. Pseudo-functions may contain parenthesized arguments that may be included in the substituted text.

Public Variable
(variable) A variable of the MEMVAR storage class. Public variables are created dynamically at runtime using the PUBLIC statement, and are accessible from any procedure at any level unless obscured by a private variable with the same name. (*See also:* Activation, Dynamic Scoping, Function, Lexical Scoping, Local Variable, Private Variable, Procedure, Static Variable)

Query
(database) A request for information to be retrieved from a database. Alternately, a data structure in which such a request is encoded.

(general) A query is also a general term used when you want to interrogate a setting or an exported instance variable for its current value.

Queue
(general) A data structure of variable length where elements are added to one end and retrieved from the other. A queue is often described as *first in, first out*. (*See also:* Stack, Print Spooler)

Record Locking
(network) The process by which one user obtains exclusive access to a record in a database to prevent another user from attempting to write data to it concurrently. A record lock must be applied prior to writing to a database in use by more than one user.

Recovery
(error handling) The process of attempting to handle an exception or runtime error. Generally, recovery consists of three possible actions: terminate processing, retry the failed operation, or resume processing with the

next operation. In all cases, the environment of the program must be restored to a stable state. (*See also:* Exception, Error, Failure, Retry, Runtime Error)

Recursion
(expression) The calling of a procedure by a statement in that same procedure. When a procedure calls itself it is said to *recurse*. A recursive call causes a new activation of the procedure. If the source code for the procedure includes a declaration of local variables, a new set of local variables is created for each activation. A private variable created by the procedure is associated with the activation in which it is created, and is visible in that activation and any lower level activations, unless obscured by a private variable created in a lower level activation. (*See also:* Activation, Function, Private Variable, Procedure)

Reference
(array, variable) A special value that refers indirectly to a variable or array. If one variable contains a reference to a second variable (achieved by passing the second variable by reference in a function or procedure call), operations on the first variable (including assignment) are *passed through* to the second variable. If a variable contains a reference to an array, the elements of the array can be accessed by applying a subscript to the variable. (*See also:* Array Reference, Parameter)

Regular Match-marker
(pre-processor) A match-marker indicating a position that will successfully match an arbitrarily complex expression in the source text. A regular match-marker generally marks a part of a command that is expected to consist of arbitrary programmer-supplied text, as opposed to a keyword or other restrictive component. In order for the source text to match, it must constitute a properly formed expression. A regular match-marker has the form <id>. id associates a name with the match-marker. The name can be used in a result-marker to specify how the matching source text is to be handled.

Regular Result-marker
(pre-processor) A result-marker of the form <id>. id must correspond to the name of a match-marker. This result-marker writes the matched input text to the result text, or nothing if no input text is matched.

Relation
(database) A link between database files that allows the record pointer to move in more than one database file based on the value of a common field or expression. This allows information to be accessed from more than one database file at a time.

Relational Database System
(database) A system that stores data in rows and columns, without system dependencies within the data. In other words, relationships between different databases are not stored in the actual database itself, as is the case in a system that uses record pointers.

Relative Addressing
(user-interface) To refer to a memory address, array element, screen location, or printer location with respect to another value, rather than referring to a specific address or element.

Repeating clause
(pre-processor) A portion of a result-pattern surrounded by square ([ ])

brackets. The text specified by the repeating clause is written to output once for each successfully matched match-marker in the corresponding match-pattern.

Restricted Match-marker
(pre-processor) A match-marker indicating a position that will successfully match one or more specified keywords. A restricted match-marker marks a part of a command that is expected to be a keyword. A restricted match-marker has the form <id: wordList> where *wordList* is a list of one or more keywords. Source text is successfully matched only if it matches one of the keywords (or is an acceptable abbreviation). *id* associates a name with the match-marker. The name can be used in a result-marker to specify how the matching source text is to be handled.

Result Text
(pre-processor) The text that results from formatting matched source text using a result-pattern. If the result text matches a match-pattern in another pre-processor directive, then it becomes the source text for that directive. Otherwise, the result text is passed as input to the compiler.

Result-pattern
(pre-processor) The part of a translation directive that specifies the text to be substituted for source text that matches the match-pattern. A result-pattern generally consists of operators and result-markers.

Retry
(network) Upon failing to lock a record in a database that is opened in shared mode, retry refers to attempting the write again based upon certain programmatic parameters. (*See also:* Shared, Exclusive)

(error handling) After an exception has been raised and the conditions of a failure corrected, an attempt is made to re-execute the failed operation. (*See also:* Exception, Error, Failure, Runtime Error)

Return Value
(procedure and function) The value or reference returned by a function or method from a function call or message send.

Root
(linker) A special section of the program that has the lowest address of all sections. This is the first section of the program loaded into memory by DOS or the RTLINKST.COM start-up code. Other sections are loaded by the overlay manager.

Row
(database) A group of related column or field values that are treated as a single entity. It is the same as a *CA-Clipper* record. (*See also:* Column, Field, Record)

(user-interface) A numeric expression that evaluates to an integer identifying a screen or printer row position.

Run Mode
(debugger) The mode of execution in which an application executes without pausing, until a Breakpoint or Tracepoint is reached.

Runtime Error
(error handling) An error that halts a program while it is executing.

Scope
(command, database) In a database command, a clause that specifies a range of database records to be addressed by the command. The scope clause uses the

qualifiers ALL, NEXT, RECORD, and REST to define the record scope. (*See also:* Condition)

Scoreboard
(user-interface) An area of the display on line 0 beginning at column 60 that displays status information during certain data entry operations.

Script File
(configuration) A text file that contains command input to a compiler, linker, or other utility program. A script file is often used in lieu of equivalent keyboard input. For the *CA-Clipper* compiler, script files contain a list of source files to be compiled into a single object file.

Script File
(debugger) A file in which frequently used debugger commands are stored and from which those commands can be executed.

Scrolling
(user-interface) The action that takes place when the user attempts to move the cursor or highlight beyond the window boundary to access information not currently displayed. (*See also:* Window)

SDF File
(file, database) A text file that contains fixed-length database records with each record separated by a carriage return/line feed pair (CHR(13) + CHR(10)) and terminated with an end-of-file mark (CHR(26)). Each field within an SDF file is fixed-length with character strings padded with trailing spaces and numeric values padded with leading spaces. There are no field separators. (*See also:* Database, Text File, Delimited File)

Search Condition
(database) *See* Condition, Scope

Section
(linker) Load module portion of an .EXE or .OVL file loaded into memory as a single unit. In a program with overlays, the root section containing the main program module loads when the program is executed. Other sections are loaded as overlays when modules within them are invoked. (*See also:* Dynamic Overlay, Linking)

SeeRecord
(database) The basic row unit of a database file consisting of one or more field elements. (*See also:* Database, Field, Table, Tuple)

Segment
(linker) Code or data handled by the linker as a indivisible unit.

Selection
- (algorithm) One of the three basic building blocks of algorithm development (the others are sequence and iteration). Selection allows control to flow along a number of possible paths, depending on the circumstance encountered.

- (database) A DBMS term that specifies a subset of records meeting a condition. The selection itself is obtained with a selection operator. In *CA-Clipper*, the analogy is the FOR clause.

(*See also:* Sequence, Iteration, Join, Projection)

Self
(object-oriented) An object-oriented term describing a reference to the object that

received the current message. In *CA-Clipper*, this reference is often the return value of a message send.

Send Operator
(object-oriented) A new operator (:) in *CA-Clipper* used to send messages to user interface objects.

Separator
(file, database) The character or set of characters that differentiate fields or records from one another. In *CA-Clipper*, the DELIMITED and SDF file types have separators. The DELIMITED file uses a comma as the field separator and a carriage return/line feed pair as the record separator. The SDF file type has no field separator, but also uses a carriage return/line feed pair as the record separator. (*See also:* Delimiter)

Sequence
- (algorithm) One of the three basic building blocks of algorithm development (the others are selection and iteration). A sequence is a series of discrete steps that must be performed in a particular order.
- (language) In *CA-Clipper*, a series of statements enclosed in a BEGIN SEQUENCE control structure. (*See also:* Algorithm, Iteration, Selection)

Set Colors Window
(debugger) The window in which the Debugger color settings can be inspected.

Shared
(network) A mode in which a file is opened that allows it to be accessed by more than one user at the same time. The inverse of Exclusive.

Shortcutting
(expression) A compiler optimization that causes expressions to be evaluated only to the extent required to determine their outcome. For example, in the expression $f()$ .OR. $g()$ function g need not be executed if function f returns true (.T.). *CA-Clipper* performs shortcutting on all logical operators (.OR. .AND. .NOT.).

Side Effect
(modular programming) An *unexpected effect* of executing a function or program. When a function changes the state of a system in a way that is not explicitly specified by the function's name or calling protocol, the change is called a *side effect*. Reliance on side effects is contrary to the principles of modular programming. (*See also:* Encapsulation, Information Hiding, Lexical Scoping, Modularity)

Single Step Mode
(debugger) The mode of execution in which only the line of code highlighted by the Execution Bar is executed, and its output displayed.

Single-dimensional Array
(array) In *CA-Clipper*, an array whose elements do not contain references to other arrays. (*See also:* Array, Array Reference, Multi-dimensional Array, Nested Array, Subarray, Subscript)

Skeleton
(command) A wildcard mask used to specify a group of filenames or memory variables. The * is used to specify one or more characters and the ? to specify a single character.

Smart Stringify Result-marker
(pre-processor) A result-marker of the form <*(id)*>. *id* must correspond to the name of a match-marker. A smart

stringify result-marker specifies that the corresponding source text is to be enclosed in quotes unless the source text was enclosed in parentheses. If the matched source text is a list, each element of the list is individually processed. If no source text was matched, an empty result is produced. The smart stringify result-marker is used to implement commands that allow extended expressions (a part of a command that may be either an unquoted literal or a character expression).

Soft Carriage Return
(general) A carriage return that is introduced into text usually in order to implement some sort of wrap operation, as opposed to a Hard Carriage Return that was specifically entered into the text when it was created.

Sort Order
(array, database) Describes the various ways database files and arrays are ordered.

- Ascending

    Causes the order of data in a sort to be from lowest value to highest value.

- Descending

    Causes the order of data in a sort to be from highest value to lowest value.

- Chronological

    Causes data in a sort to be ordered based on a date value, from earliest to most recent.

- ASCII

    Causes data in a sort to be ordered according to the ASCII Code values of the data to be sorted.

- Dictionary

    The data in a sort is ordered in the way it would appear if the items sorted were entries in a dictionary of the English language.

- Collating Sequence

    Data in a sort will be placed in sequence following the order of characters in the IBM Extended Character Set.

- Natural

    The order in which data was entered into the database.

Source Code
(procedure and function) The textual representation of a program or procedure. (*See also:* Source File, Object File)

Source File
(configuration) *See* Program File, Header File.

Source Text
(pre-processor) Text from a source file, processed by the pre-processor. Source text is examined to see if it matches a previously specified match-pattern. If so, the corresponding result-pattern is substituted for the matching source text.

Spooler
(network) *See* Print Spooler.

Stack
(general) A data structure of variable length whose elements are added and retrieved from the same end. A stack is often described as *first in, last out*. (*See also:* Queue)

Standard Color
(user-interface) The color pair definition that is used by all output options (such as SAY and ?), with the exception of GETs and PROMPTs, that use the enhanced color pair. (*See also:* Enhanced Color).

Statement
(language) In *CA-Clipper*, the basic unit of source code. A statement is normally a single line of text. Multiple statements can be placed on the same line by separating them with semicolons. A statement may be continued to another line by placing a semicolon at the end of the line to be continued. If the text of a statement matches a command definition (defined with a pre-processor directive), it is translated into the form specified by the command definition. (*See also:* Command)

Static Overlay
(linker) A section of the program that is not always resident in RAM, and shares memory with other sections. The section that is currently in use is loaded into memory, allowing a larger program to execute in less available RAM.

Static Variable
(variable) A variable that exists and retains its value for the duration of execution. Static variables are lexically scoped; they are only accessible within the procedure that declares them, unless they are declared as *file-wide*, in which case they are accessible to any procedure in the source file that contains the declaration. (*See also:* Dynamic Scoping, Lexical Scoping, Local Variable, Private Variable, Public Variable)

STD.CH
(pre-processor) The *standard header file* containing definitions for all *CA-Clipper* commands.

Storage Class
(variable) Defines the two characteristics of variables: lifetime and visibility. (*See also:* Lifetime, Scope, Visibility)

String
(data type) Generically, a value of type character. In source code, a series of characters enclosed in single or double quotes. (*See also:* Character)

Stringify
(pre-processor) To change source text into a literal character string by surrounding the text with quotes.

Stub
(error handling) A procedure used for debugging purposes that only simulates the intended actions of the real procedure. It may display an indicating message, return a constant value, or do nothing.

Subarray
(array) In *CA-Clipper*, an array that is referred to by an element of another array. (*See also:* Array, Array Reference, Multi-dimensional Array, Nested Array, Subscript)

Subdirectory
(file) *See* Directory.

Subscript
(array) A numeric value used to designate a particular element of an array. Applying a subscript to an array is called *subscripting* the array. In *CA-Clipper* programs, subscripting is specified by

enclosing a numeric expression in square ([ ]) brackets after the name of a program variable. The variable is then said to be *subscripted*. (*See also:* Array, Array Reference, Multi-dimensional Array, Nested Array, Subarray)

Substring
(data type) A string within a string, usually to be specified as an argument of a function or command.

Swapfile
(linker) Also known as the *workfile*, used by *.RTLink* to swap data and code in and out of memory during the linking process.

Symbol
(linker) An assigned name for a value representing a constant or the address of code or data. There are four types of symbols used by the linker defined as follows:

- *Absolute symbol:* a constant

- *Relative symbol:* address of code or data

- *Public symbol:* accessed by modules other than the module in which they are defined. Public symbols are used to share procedures and variables between modules. As such, the relative address of a public symbol is assigned by the compiler during compilation.

- *External symbol:* a public symbol not defined in the current module. Generally, these are references into CLIPPER.LIB or EXTEND.LIB, but the compiler generates them whenever there is a procedure or user-defined function referenced but not compiled into the current module.

Syntax
(language) The rules that dictate the form of statements or commands as defined by the implementors of the language. Also, a complete description of the forms that a statement or command can take.

Table
(database) A DBMS term defining a collection of column definitions and row values. In *CA-Clipper*, it is represented and referred to as a database file.

Template
(user-interface) A mask that specifies the format in which data should be displayed. For example, you might want to store phone numbers as "9999999999" to save space, but use a template to display the number to the user as "(999) 999-9999."

Text File
(file) A file consisting entirely of ASCII characters. Each line is separated by a carriage return/line feed pair (CHR(13) + CHR(10)) and the file is terminated with a end-of-file mark (CHR(26)). (*See also:* Delimited File, Program File, SDF File)

Text Replacement
(pre-processor) The process of removing portions of input text and substituting different text in its place.

Toggle
(command) As a verb, to choose between an *on* or *off* state. As a noun, a value or setting that can be either on or off. A toggle is often represented using a logical value, with true (.T.) representing on, and false (.F.) representing off.

Token
(pre-processor) An elemental sequence of characters having a collective meaning.

The pre-processor groups characters into tokens as it reads the input text stream. Tokens include identifiers, keywords, constants, and operators. Whitespace, and certain special characters, serve to mark the end of a token to the pre-processor.

Trace Mode
(debugger) A mode of execution similar to Single Step Mode, the difference being that Trace Mode traces over function and procedure calls.

Tracepoint
(debugger) A variable or expression whose value is displayed in the Watch Window, and which causes an application to pause whenever that value changes.

Translation Directive
(pre-processor) A pre-processor instruction containing a translation rule. The two translation directives are #command and #translate.

Translation Rule
(pre-processor) The portion of a translation directive containing a match-pattern followed by the special symbol (=>) followed by a result-pattern.

Translation Rule
(pre-processor) The portion of a translation directive containing a match-pattern followed by the special symbol (=>) followed by a result-pattern.

Truncate
(expression) To remove insignificant information from the end of an item of data. With numerics, to ignore any part of the number that falls outside of the specified precision.

Tuple
(database) A formal DBMS term that refers to a row in a table or a record in a database file. In DIF files, tuple also refers to the equivalent of a *CA-Clipper* record. (*See also:* Database, Field, Record)

Two-dimensional Array
(array) An array that has two dimensions. In *CA-Clipper*, an array whose elements contain references to other arrays, all of which have the same length and do not refer to other arrays. (*See also:* Array, Array Reference, Nested Array, Subscript)

Typeahead Buffer
(user-interface) *See* Keyboard Buffer.

Unary Operator
(expression) An operator that operates on a single operand. For example, the .NOT. operator. (*See also:* Binary Operator, Operator)

Undefine
(pre-processor) To remove an identifier from the pre-processor's list of defined identifiers via the #undefine directive.

Undefined Symbol
(linker) An *unresolved symbol*) that was never declared public by a module, but which is referenced by another module. After the public symbol definition is encountered, the symbol becomes defined (resolved). When a symbol is referenced, but not defined, it is said to be undefined.

Unselected Color
(user-interface) The color pair definition used to display all but the current GET or the GET that has input focus. If this color setting is specified, the current GET is displayed using the current enhanced color. (*See also:* Enhanced Color).

Update
(database) The process of changing the value of fields in one or more records. Database fields are updated by various commands and the assignment operator.

User Function
(user-interface) A user-defined function called by ACHOICE(), DBEDIT(), or MEMOEDIT() to handle key exceptions. A user function is supplied to one of these functions by passing a parameter consisting of a string containing the function's name.

User Interface
(user-interface) The way a program interacts with its user (i.e., menu operation and selection, data input methods, etc.)

User-defined Function
(procedure and function) *See* Function.

Variable
(variable) An area of memory that contains a stored value. Also, the source code identifier that names a variable. (*See also:* Local Variable, Private Variable, Public Variable, Static Variable)

Vector
(database) In a DIF file, vector refers to the equivalent of a *CA-Clipper* field. (*See also:* Database, Field, Record, Tuple)

Verb
(command) The first word of a command that describes the action to perform. (*See also:* Command)

View
(database) A DBMS term that defines a virtual table. A virtual table does not actually exist but is derived from existing tables and maintained as a definition. The definition in turn is maintained in a separate file or as an entry in a system dictionary file. In *CA-Clipper*, views are supported only by DBU.EXE and are maintained in (.vew) files. (*See also:* Database, Field, Record)

View Sets Window
(debugger) The window in which *CA-Clipper* status settings can be inspected.

View Workareas Window
(debugger) The window in which work area information is displayed.

Visibility
(variable) The set of conditions under which a variable is accessible by name. A variable's visibility depends on its storage class. (*See also:* Dynamic Scoping, Lexical Scoping)

Volume
(file) A unit of disk storage uniquely identified by a *label* and of fixed size. A hard disk can be partitioned into one or more volumes by an operation system utility. Volumes are subdivided into one or more directories organized in tree structure. (*See also:* Directory, Disk)

Wait State
(user-interface) A wait state is any mode that extracts keys from the keyboard except for INKEY(). These modes include ACHOICE(), DBEDIT(), MEMOEDIT(), ACCEPT, INPUT, READ and WAIT.

Watch Window
(debugger) The window in which Watchpoints and Tracepoints are displayed.

Watchpoint
(debugger) A variable or expression whose value is displayed in the Watch Window and updated as an application executes.

Wild Match-marker
(pre-processor) A match-marker indicating a position that will successfully match any source text. A wild match-marker matches all source text from the current position to the end of the source line. A wild match-marker has the form <*id*>. *id* associates a name with the match-marker. The name can be used in a result-marker to specify how the matching source text is to be handled.

Window
(user-interface) A rectangular screen region used for display. A window may be the same size or smaller than the physical screen. Attempting to display information that extends beyond the specified boundaries of the window clips the output at the window edge.

Word
(pre-processor) A series of characters in a match-pattern or result-pattern. Source text matches a word in a match-pattern if the text is identical to the word or is an acceptable abbreviation of it. A word that appears in a result-pattern is copied unmodified into the result text.

Word-wrapping
(user-interface) The process of continuing the current text on the next line of a display when a boundary is reached and breaking the text on a word boundary.

Work Area
(database) The basic containment area of a database file and its associated indexes. Work areas can be referred to by alias name, number, or a letter designator. (*See also:* Alias)

Workfile
(linker) *See* Swapfile.

Workstation
(network) A personal computer connected to a network used to run applications and front end processes. (*See also:* File Server)

Index

! (negate), 2–39, 2–45
!= (not equal), 2–30, 2–34, 2–37, 2–39, 2–40, 2–44
(not equal), 2–30, 2–34, 2–37, 2–39, 2–40, 2–44
#include, 2–6, *See also* Header files, 8–6
$ (substring), 2–30, 2–44
% (modulus), 2–37, 2–44
%= (modulus and assign), 2–46, 2–48
& (compile and run), 2–50, 2–55, 2–65, 2–77
&& (comment), 2–6
() (group), 2–42, 2–50, 2–54
* (comment), 2–6
* (multiplication), 2–37, 2–44
** (exponentiation), 2–37, 2–44
**= (exponentiation and assign), 2–46, 2–48
*/ (comment), 2–6
*= (multiplication and assign), 2–46, 2–48
+ (addition), 2–34, 2–37, 2–43, 2–44
+ (concatenation), 2–30, 2–43
++ (increment), 2–34, 2–37, 2–49
+= (addition and assign), 2–46, 2–48
+= (concatenate and assign), 2–46, 2–48
- (concatenation), 2–30, 2–43
- (subtraction), 2–34, 2–37, 2–43, 2–44
-- (decrement), 2–34, 2–37, 2–49
-= (concatenate and assign), 2–46, 2–48
-= (subtraction and assign), 2–46, 2–48
-> (alias), 2–24, 2–50

.AND., 2–39, 2–45
    truth table, 2–45
.clp files, *See* Script files
.dbf files, 2–84, *See also* Database files
.dbt files, 2–33, *See also* Memo fields, Memo files, 2–85

.EXE files, *See* Executable files
.fmt files, *See* Format files
.INF files, *See* Information files
.LIB files, *See* Library files
.lnk files, *See* Script files
.MAP files, *See* Map files
.ndx files, 2–87, *See also* Index files
.NOT., 2–39, 2–45
    truth table, 2–45
.ntx files, 2–87, *See also* Index files
.OBJ files, *See* Object files
.OR., 2–39, 2–45
    truth table, 2–45
.OVL files, *See* Overlay files, Static overlays
.PLL files, *See* Prelinked library files
.PLT files, *See* Prelinked transfer files
.ppo files, 8–4, 8–8, 8–12
.prg files, *See* Procedure files, Program files
.RMK file, *See* Make file
.RTLink, *See* Linker
.vew file, 13–10, 13–14, 13–19

/ (division), 2–37, 2–44
/* (comment), 2–6
// (comment), 2–6
/= (division and assign), 2–46, 2–48
/B compiler option, *See* Compiler options, *See* Compiler options
: (send), 2–78, 2–79, 2–80
:= (inline assign), 2–46, 2–47, 2–48, 2–63, 2–80
; (continuation), 2–7
< (less than), 2–30, 2–34, 2–37, 2–39, 2–40, 2–44
<= (less than or equal), 2–30, 2–34, 2–37, 2–39, 2–40, 2–44

<> (not equal), 2–30, 2–34, 2–37, 2–39, 2–40, 2–44
= (assign), 2–46, 2–47, 2–48, 2–63
= (equal), 2–30, 2–34, 2–37, 2–39, 2–40, 2–44, 2–47
== (array equivalence), 2–68
== (equal), 2–34, 2–37, 2–39, 2–40, 2–44
== (exactly equal), 2–30, 2–44
> (greater than), 2–30, 2–34, 2–37, 2–39, 2–40, 2–44
>= (greater than or equal), 2–30, 2–34, 2–37, 2–39, 2–40, 2–44
?, 10–3, 10–4, 10–36, 10–37, 10–39, 10–40, 10–47, 10–74, 10–75, 10–76, 10–77, 10–79
@ (pass by reference), 2–13
@ (pass-by-reference), 2–50
@...GET
    in a network environment, 4–11
@...SAY, 2–91
[] (array element), 2–50
^ (exponentiation), 2–37, 2–44
^= (exponentiation and assign), 2–46, 2–48
{} (array delimiter), 2–50, 2–65
{} (code block delimiter), 2–50, 2–72
| | (code block argument delimiter), 2–50, 2–72

A

AADD(), 2–67, 2–69
Accessing DOS, 10–45, 10–56
ACLONE(), 2–70
ACOPY(), 2–70
Adding records
    in a network environment, 4–12
AddRec(), 4–12
ADEL(), 2–69
AEVAL(), 2–73
    example, 2–67
    general discussion, 2–48
AFILL(), 2–63
AINS(), 2–69
Alias
    expressions, 2–50
    field, 2–47
    operator, 2–50, 2–82
ALTD(), 2–95, 10–7
Animate, *See* Run Animate
APPEND BLANK
    in a network environment, 4–12
APPEND FROM
    in a network environment, 4–8
ARRAY(), 2–61, 2–63
Arrays
    adding new elements, 2–69
    addressing elements, 2–62
    and the macro operator, 2–58
    as return values, 2–16, 2–65
    assigning values to, 2–63
    changing the size of, 2–67, 2–69
    comparison, 2–68
    constant, 2–50, 2–65
    copying elements, 2–70
    creating, 2–61
    definition, 2–61
    deleting elements, 2–69
    determining number of elements in, 2–68
    determining size of, 2–68
    duplicating, 2–70
    duplicating subarrays, 2–70
    dynamic, 2–61
    empty, 2–67, 2–68
    equivalence, 2–68
    general discussion, 2–61
    growing, 2–67, 2–69
    initializing, 2–63
    inserting elements, 2–69
    lexically scoped, 2–61
    literal, 2–65
    local, 2–61
    multidimensional, 2–61, 2–64, 2–70
    passing as parameters, 2–12, 2–13, 2–65
    private, 2–61
    public, 2–61
    scanning for a value, 2–71
    shrinking, 2–69
    sorting, 2–70
    static, 2–61
    subscript numbering, 2–62
    syntax for addressing, 2–62
    syntax for creating, 2–61

traversing, 2–66
used with macro operator, 2–65
ASCAN(), 2–71
ASCII files, *See also* Output to text file
ASIZE(), 2–67, 2–69
ASORT(), 2–70
Assignment
   array, 2–63
   operators, 2–46, 2–47, 2–48, 2–63
   precedence of operators, 2–52, 2–54
   table of compound operators, 2–48
   table of operators, 2–46
AUTOEXEC.BAT, 3–1

B

BADCACHE, 3–8, 3–10
Beginning-of-file, 2–98
BIN directory, 8–6, 9–13, 12–1, 13–1
Binary file
   closing a, 2–99
   creating a, 2–97
   moving pointer in a, 2–98
   opening a, 2–97
   reading a, 2–97, 2–98
   writing to a, 2–98
Boolean algebra, 2–39, 2–45
BP, 10–8, 10–32, 10–41, 10–50, *See also* Point Breakpoint
Branching
   conditional, 2–19, 2–20
Breakpoints, 10–3, 10–33, 10–40, 10–43, 10–58, 10–60, 10–104, 10–114
   definition, 10–40, 10–50, 10–95
   deleting, 10–32, 10–41, 10–50, 10–53, 10–95
   listing, 10–41, 10–42, 10–67
   setting, 10–8, 10–32, 10–36, 10–41, 10–50, 10–95
Buffer
   management in a network environment, 4–11

C

Calling conventions
   command, 2–4, 2–13
   function, 2–4, 2–13
Callstack, 10–22, 10–52, *See also* View Callstack
Callstack Window, *See* Windows
CASE, *See also* DO CASE
CGACURS, 3–8
Character functions
   table, 2–30
Character string
   comparison, 2–30, 2–33
   constants, 2–30
   delimiters, 2–30
   fixed length, 2–30
   literals, 2–30
   maximum length of, 2–30
   operations, 2–30
   operators, 2–30, 2–43, 2–46, 2–48
   precedence of operators, 2–52
   table of operators, 2–43
   variable length, 2–33
CHR(), 2–94
CL.BAT, 8–6, 9–13
Class, *See also* Objects
   definition, 2–78
CLD.EXE
   command line syntax, 10–5
CLD.LIB, 10–6, 10–7
CLEAR TYPEAHEAD, 2–96
CLIPPER
   BADCACHE setting, 3–8, 3–10
   CGACURS setting, 3–8
   DYNF setting, 3–8
   E setting, 3–9
   F setting, 3–9
   INFO setting, 3–10
   NOIDLE setting, 3–10
   SWAPK setting, 3–11
   SWAPPATH setting, 3–11
   TEMPPATH setting, 3–11
   X setting, 3–11

CLIPPER.EXE, *See also* Compiler, 8–1
    command line syntax, 8–2
CLIPPER.LIB, 8–12, 9–27, 9–35
    search paths, 9–18
CLIPPERCMD, 8–2, 8–9
Closing files
    database, 2–89
    index, 2–89
    low-level, 2–99
Code blocks, 10–3, 10–4, 10–42
    and the macro operator, 2–59
    as return values, 2–16
    compared to macro expressions, 2–59, 2–72
    compiling with macro operator, 2–56, 2–77
    creating, 2–50, 2–72
    evaluating, 2–73
    evaluating for arrays, 2–67
    evaluating for database records, 2–87
    example using, 2–75, 2–76
    general discussion, 2–72
    operations, 2–73
    passing as parameters, 2–12
    syntax for defining, 2–72
    table of operations, 2–73
    used to sort arrays, 2–70
    variable scoping within a, 2–22, 2–73
Code Window, *See* Windows
Command Window, *See* Windows
Commands
    user-defined, 2–72, 2–87
Comments, 2–6
COMMIT, 2–89
Compatibility, *See also* dBASE III PLUS compatibility
    previous releases, 2–96, 3–4
Compilation
    conditional, 8–10
Compile and link batch file, 8–6, 9–13
Compiler
    checking syntax with, 8–12
    configuration, 8–2
    DOS return code, 8–4, 8–6
    error handling, 8–4, 8–6
    general discussion, 8–1
    include file directory, 8–6, 8–10
    options, 8–9
    output files, 8–7
    script file, 8–2, 8–3, 8–5
    switches, 8–9
    syntax for, 8–2
    temporary file directory, 8–7, 8–12
Compiler assumptions, 2–82
Compiler options
    /B, 10–4, 10–5
    /P, 10–4
Compiler switches
    /A, 8–9
    /B, 8–9, 8–11
    /CREDIT, 8–10
    /D, 8–10
    /ES, 8–10
    /ES0, 8–10
    /ES1, 8–10
    /ES2, 8–10
    /I, 8–6, 8–10
    /L, 8–2, 8–9, 8–11, 8–12
    /M, 8–3, 8–5, 8–11, 8–12
    /N, 2–3, 2–15, 2–22, 2–26, 8–11
    /O, 8–2, 8–7, 8–12
    /P, 8–4, 8–8, 8–12
    /Q, 8–12
    /R, 8–12
    /S, 8–12
    /T, 8–7, 8–12
    /U, 8–2, 8–6, 8–13
    /V, 2–24, 8–13
    /W, 2–24, 8–13
    /Z, 8–13
Compiling program code, 10–4
Console commands
    directing output to a file, 2–93
    general discussion, 2–90
    printing, 2–92
Control structures
    decision-making, 2–19
    DO CASE, 2–20
    DO WHILE, 2–18
    FOR, 2–18, 2–66
    general discussion, 2–17
    IF, 2–19

looping, 2–17, 2–66
nesting, 2–17
Conversion functions
   BIN2I(), 2–97
   BIN2L(), 2–97
   BIN2W(), 2–97
   I2BIN(), 2–98
   L2BIN(), 2–98
COPY FILE
   in a network environment, 4–8
COPY STRUCTURE
   in a network environment, 4–8
COPY STRUCTURE EXTENDED
   in a network environment, 4–8
COPY TO
   in a network environment, 4–8
COUNT
   in a network environment, 4–11
CREATE
   full-screen, 13–14
   in a network environment, 4–8
CREATE FROM
   in a network environment, 4–8
Creating files
   in DBU, 13–14
CTOD(), 2–34
Cursor
   controlling use of, 3–8

D

Data structures, 2–30
Data type
   array, 2–30, 2–61
   character, 2–30, 2–43
   code block, 2–30, 2–72
   date, 2–34, 2–43
   general discussion, 2–30
   logical, 2–39, 2–45
   memo, 2–30, 2–33, 2–43
   NIL, 2–40, 2–44, 2–46
   numeric, 2–37, 2–43, 2–44
   object, 2–78
Database commands
   syntax for specifying scope, 2–86
   table, 2–85
Database files
   attributes of, 2–85
   closing, 2–89
   creating in DBU, 13–14
   definition, 2–84
   general discussion, 2–81
   maximum number of open in DBU, 13–8
   modifying records, 5–11
   opening, 2–82
   opening in DBU, 13–13
   operations, 2–85
   processing with DBEVAL(), 2–87
   saving structure in DBU, 13–19
   table of attributes, 2–85
Database functions
   table, 2–85
Database utility, 2–81, *See* DBU
Date functions
   table, 2–34
Dates
   blank, 2–34
   comparison, 2–34
   constants, 2–34
   literal, 2–34
   operations, 2–34
   operators, 2–34, 2–43, 2–46, 2–48
   precedence of operators, 2–52
   range of, 2–34
   separators, 2–34
   validation, 2–34
dBASE III PLUS
   database driver, 2–87
dBASE III PLUS compatibility
   and network programming, 4–3
   compiling dBASE source code, 8–5
   CREATE LABEL, 14–1
   CREATE REPORT, 14–1
   database file format, 2–84
   indexing, 2–87
   LABEL FORM, 14–13
DBEVAL(), 2–73, 2–87
DBFNTX.LIB, 8–12, 9–27, 9–35
DBSTRUCT(), 2–61, 2–63
DBU
   appending to a file in, 13–24

Browse menu, 13–1, 13–6, 13–9, 13–20, 13–27, 13–29, 13–32
Browse window, 13–20, 13–27
browsing a database file, 13–20
browsing a view, 13–21
color settings, 13–1
copying a file in, 13–23
Create menu, 13–6, 13–14, 13–19
creating database files in, 13–14
creating files, 13–14
creating index files in, 13–18
data types, 13–16
decimal restrictions, 13–17
dialog boxes, 13–4
dialog controls, 13–4
editing a memo field in, 13–22
error messages, 13–4
executing DOS commands from, 13–26
field naming conventions, 13–16
field width restrictions, 13–17
Help menu, 13–4, 13–6, 13–11
Help window, 13–11
leaving, 13–10
main screen navigation keys, 13–9
main screen usages, 13–7
marking deleted records in, 13–20
menu bar, 13–3, 13–11
menu bar navigation keys, 13–11
menu navigation, 13–3
Move menu, 13–20, 13–27
Open menu, 13–8, 13–10, 13–12
opening database files in, 13–8, 13–13
opening files, 13–12
opening index files in, 13–8, 13–13
opening view files in, 13–14
performing logical searches in, 13–28
prompts, 13–4
removing deleted records in, 13–26
replacing fields in, 13–25
Save menu, 13–10, 13–17, 13–19, 13–29
saving a database file structure, 13–17, 13–19
saving a view file, 13–10, 13–19
searching indexes in, 13–27
Set menu, 13–9, 13–29
Set Relation window, 13–29
setting a field list in, 13–9, 13–32
setting a filter in, 13–31
setting relations in, 13–29
skipping records in, 13–29
Structure window, 13–6, 13–14
Utility menu, 13–11, 13–22
windows, 13–6
DBU.EXE, 2–81, 13–1, *See also* DBU
command line syntax, 13–1
DBU.HLP, 13–1
Debugger, 2–95
Debugging, 8–9, 8–11, 9–24, 11–2
Declaration statements
general discussion, 2–25
Declarations
compile time, 2–5, 2–21
runtime, 2–21
Decrement
operators, 2–49
DELETE
in a network environment, 4–11
Delete, 10–32, 10–38, 10–41, 10–50, 10–53, *See also* Point Breakpoint, Point Delete
Dependency file, 11–5, 11–9
searching for, 11–7
Dependency rule, 11–5, 11–8
syntax for, 11–9
Dependency statement, 11–9, 11–11
Dialog boxes, 10–26, 10–38, 10–39
Directing output to a file, 2–93, *See also* Output to text file
DIRECTORY(), 2–61, 2–63
DISPLAY
in a network environment, 4–8
Displaying
variables and expressions, 10–47
DO
compiler implications, 8–5
DO CASE, 2–20
DO WHILE, 2–18
Documentation
online, 15–1
DOS, *See* File DOS
640K memory limitation, 9–33
access from a make file, 11–16
BUFFERS setting, 3–2

command line length limitation, 9–6, 11–19
COMMAND.COM, 3–5, 8–1, 9–1
COMSPEC setting, 3–5
CONFIG.SYS, 3–1, 3–9, 8–1, 9–1
default directory, 2–97
error number, 2–99
ERRORLEVEL, 8–4, 8–6, 9–13, 9–17, 11–3, 11–5
executing commands from DBU, 13–26
file attributes, 2–97
file pointer, 2–97, 2–98
FILES setting, 3–2, 3–9
flushing buffers, 2–99
open mode, 2–97, 4–8
PATH, 9–15, 12–1, 12–3, 13–1, 15–1, 15–2
redirection, 8–4
RENAME command, 9–15
SHELL directive, 3–3, 8–1, 9–1
version requirements for networking, 4–2
version specific information, 3–2, 9–15
Dynamic overlays
  creating, 9–23, 9–32
  specifying file handles at runtime, 3–8
DYNF, 3–8

E

E, 3–9
ELSE, *See also* IF
ELSEIF, *See also* IF
EMM, 3–8, 3–10
EMPTY(), 2–67
End-of-file, 2–98
ENDCASE, *See also* DO CASE
ENDDO, *See also* DO WHILE
ENDIF, *See also* IF
ENDTEXT, *See also* TEXT
Environment functions
  GETACTIVE(), 6–33
Environment variables, 8–1, 11–15
  CLIPPER, 3–2, 3–8, 3–9, 3–10, 3–11
  CLIPPERCMD, 8–2, 8–9
  defining in make files, 11–11
  INCLUDE, 8–6, 8–10, 11–15

LIB, 3–3, 9–14, 9–18, 9–19
OBJ, 9–18
PATH, 9–15, 13–1
PLL, 3–3, 9–14, 9–19
RMAKE, 11–2, 11–3, 11–4, 11–6, 11–7, 11–17
RTLINKCMD, 9–7, 9–10, 9–11, 9–20
TMP, 3–4, 8–7, 8–12, 9–16
Error block
  BREAK, 7–8, 7–13, 7–15
  ERRORBLOCK(), 7–13, 7–15
  ERRORNEW(), 7–8, 7–13, 7–15
  EVAL(), 7–8, 7–13, 7–15
  RECOVER, 7–8, 7–13, 7–15
  SEQUENCE, 7–8
Error class
  canDefault, 7–19
  canRetry, 7–19
  canSubstitute, 7–19
  DefError(), 7–19, 7–21
  example, 7–19
Error handling
  ALERT(), 7–21
  compiler, 8–4, 8–6
  default handling, 7–6
  error block vs SEQUENCE, 7–15
  error objects, 7–16
  error processing, 7–3
  error scoping, 7–3
  example, 7–18, 7–23, 7–25, 7–29
  exception handling concepts, 7–2
  exception mechanisms, 7–8
  general, 7–1
  handling errors, 7–6
  in a network environment, 4–3, 4–4, 4–7, 4–10
  linker, 9–13
  localization, 7–3
  low-level, 2–99
  modular programming, 7–2
  NETERR(), 7–29
  on a network environment, 7–29
  posted error block, 7–13
  raising errors, 7–3
  RMAKE, 11–3, 11–5
  runtime, 2–43, 7–1, 8–11

SEQUENCE construct, 7–8
strategies, 7–1, 7–21
substitution, 7–6
VAL(), 7–18, 7–19, 7–21
Error messages
.RTLink file containing, 9–3
Error object
DefError(), 7–29
Error.ch, 7–19
ERRORNEW(), 7–16, 7–18
RECOVER, 7–25
ERRORBLOCK(), 7–21
EVAL(), 2–73
Exclusive mode, 4–1, 4–2, 4–3, 4–5, 4–8
commands that require, 4–13
Executable files, *See also* runtime , 8–1, 8–5, 9–8
default extension, 9–5, 9–15
default name, 9–5, 9–15
discussion of runtime environment, 3–1
execution speed of, 9–41
generating, 9–4, 9–7, 9–15
relationship to .PLL files, 9–14, 9–15, 9–38
relationship to .PLT files, 9–38
relationship to prelinked library files, 3–3
searching for, 9–15
Executing
code blocks, 10–3
individual functions and procedures, 10–37, 10–47
program code, 10–33, 10–34, 10–35, 10–36, 10–103, 10–104, 10–105, 10–106
Execution modes, 10–33
Animate Mode, 10–33, 10–36, 10–37, 10–84, 10–92, 10–103, 10–107
Run Mode, 10–33, 10–35, 10–36, 10–104, 10–105, 10–109
Single Step Mode, 10–34, 10–36, 10–82, 10–108
Trace Mode, 10–35, 10–36, 10–110
EXIT, *See also* DO WHILE, FOR
Expanded Memory Mangager, 3–8, 3–10
Expanded memory usage, 3–8, 3–9, 3–10
Expressions
displaying, 10–47
evaluation of, 2–42
general discussion, 2–29

inspecting, 10–36, 10–39
passing as parameters, 2–12
precedence used to evaluate, 2–52
EXTEND.LIB, 8–12, 9–27, 9–35
search paths, 9–18
EXTERNAL
and macro expressions, 2–60
in LABEL FORMs, 14–16
in REPORT FORMs, 14–16

F

F, 3–9
FIELD alias, 2–24
Fields
declaring, 2–29
passing as parameters, 2–12
FILE
FREEFORMAT option, 9–4, 9–5, 9–6
File DOS, 10–45, 10–56
File Exit, 10–10, 10–57
File handle
accessing a, 2–97
obtaining a, 2–97
specifying number at runtime, 3–9
File menu, 10–31
DOS Access option, 10–56
Exit option, 10–57
Open option, 10–58
Resume option, 10–60
File Open, 10–3, 10–43, 10–58, *See also* View, *See also* View, 10–60
File Resume, 10–42, 10–58, 10–60, 10–114
File server, 4–2
FilLock(), 4–6, 4–10
Find, *See* Locate Find
FLOCK(), 4–3, 4–4, 4–6, 4–11
FOPEN()
in a network environment, 4–8
FOR, 2–18
FOR condition, 2–86
Format files
compiling, 8–5
FOUND(), 2–89

FREEFORMAT
    command line mode, 9–6
    prompt mode, 9–6
    script file mode, 9–7
    syntax, 9–4, 9–37
FREEFORMAT options, *See also* Linker
        options
    BATCH, 9–21
    BEGINAREA, 9–30
    DEBUG, 9–24
    DEFAULTLIBRARYSEARCH, 9–27
    DYNAMIC, 9–23
    ENDAREA, 9–30
    EXCLUDE, 9–28
    EXTDICTIONARY, 9–27
    FILE, 9–4, 9–5, 9–6
    HELP, 9–24
    IGNORECASE, 9–22
    INCREMENTAL, 9–26
    LIBRARY, 9–4, 9–5
    MAP, 9–16, 9–24
    MODULE, 9–30
    NOBATCH, 9–21
    NODEFAULTLIBRARYSEARCH, 9–27
    NOEXTDICTIONARY, 9–27
    NOIGNORECASE, 9–22
    NOINCREMENTAL, 9–26
    OUTPUT, 9–4, 9–5
    PLL, 9–28
    PRELINK, 9–28
    PRELOAD, 9–30
    REFER, 9–28
    RESIDENT, 9–23
    SECTION, 9–30
    SILENT, 9–24
    STACK, 9–27
    VERBOSE, 9–24
Full-screen commands
    directing output to a file, 2–93
    printing, 2–92
Full-screen operations
    general discussion, 2–91
FUNCTION
    compiler implications, 8–13
Function keys, 10–32
    Ctrl-F5 Run Next, 10–33

F1 Help, 10–10, 10–23
F10 Trace, 10–35, 10–36
F4 Application Screen, 10–37
F5 Execute Application, 10–33, 10–35,
        10–36
F7 Run to Cursor, 10–33
F8 Step, 10–34, 10–36
F9 Set/Delete Breakpoint, 10–32, 10–36,
        10–41
programming, 2–96
Functions, *See also* Library functions, User-
        defined functions
    calling conventions, 2–4, 2–13
    character, 2–30
    date, 2–34
    defining, 2–5
    executing, 10–37, 10–47
    logical, 2–39
    memo, 2–30, 2–33
    numeric, 2–37

G

Get class
    assign method, 6–6
    block, 5–11, 6–2, 6–5
    buffer, 6–6
    cargo, 6–11, 6–25
    clear, 6–6, 6–30
    col, 6–2
    colorDisp method, 6–6
    colorSpec, 6–6
    data dictionary, 6–6
    decPos, 6–6
    display method, 6–6
    Exported Instance Variables, 6–2
    extending Read layer, 6–10
    Get
        block, 6–4
        cargo, 6–5
    GetActive method, 6–8, 6–10
    GetNew(), 6–2
    insert method, 6–6
    messages, 6–20

name, 6–2
original, 6–6
overStrike method, 6–6
pos, 6–6
ReadVar method, 6–11
row, 6–2
setFocus method, 6–6
setting and retrieving values, 6–5
undo method, 6–6
unTransform method, 6–30
varGet method, 6–5
varPut method, 6–5, 6–30
Get class exported methods
    Get
        delEnd(), 6–32
        delete(), 6–32
        unTransform(), 6–32
Get class instance variables
    Get
        clear, 6–32
        minus, 6–32
Get Function Layer, 6–28
    creating new, 6–30
Get system
    Code Blocks, 6–13
    Getsys.prg, 6–33, 6–34, 6–35, 6–36, 6–37, 6–38
    SET KEY, 6–13
    using code blocks for WHEN/VALID, 6–13
Get system commands
    @ GET, 6–7
    @...GET, 6–10, 6–11, 6–13, 6–16, 6–25
    CLEAR GETS, 6–16
    READ, 6–7
Get system functions
    GETACTIVE(), 6–33, 6–34
    GETAPPLYKEY(), 6–7, 6–23, 6–25, 6–28, 6–30
    GETDOSETKEY(), 6–7, 6–35
    GETPOSTVALIDATE(), 6–7, 6–36
    GETPREVALIDATE(), 6–7, 6–23, 6–37
    GETREADER(), 6–8, 6–23, 6–24, 6–25, 6–30, 6–38
    READFORMAT(), 6–39
    READKILL(), 6–40

READUPDATED(), 6–41
GETACTIVE(), 6–33
Go, *See* Run Go
Goto, *See* Locate Goto
Guide To Clipper, 15–1

H

HARDCR()
    in REPORT FORMs, 14–4
Header files, 10–3, 10–43, 10–58
    changing the standard, 8–13
    editing, 12–1
    general discussion, 2–6
    RMAKE, 11–15
    searching for, 8–6, 8–10, 8–13
    specifying a directory for, 8–6
Help, 2–96, 10–10, 10–23, 10–64
    .RTLink, 9–24, 9–42
    .RTLink file containing, 9–3
    in DBU, 13–11
    online, 1–1, 15–1
Help menu, 10–31
    Commands option, 10–64
    Keys option, 10–64
    Menus option, 10–64
    Windows option, 10–64
Help Window, *See* Windows

I

IF, 2–19
IF()
    in LABEL FORMs, 14–13
    in REPORT FORMs, 14–4
INCLUDE directory, 8–6, 8–10
Include files, *See* Header files
Increment
    operators, 2–49
Incremental linking, 9–26, 9–38
INDEX, 2–88
    in a network environment, 4–8
Index files
    closing, 2–89

controlling index, 2–88, 2–89
creating, 2–88
creating in DBU, 13–18
general discussion, 2–87
maximum number of open, 2–88
maximum number of open in DBU, 13–8
opening, 2–88
opening in DBU, 13–13
order of, 2–88
relative searching, 2–89
searching, 2–89
updating, 2–89
Index key
maximum length, 2–88
unique values, 2–88
Inference rule, 11–8
syntax for, 11–11
INFO, 3–10
Information files
default extension, 9–17, 9–38
default name, 9–38
generating, 9–38
Init.cld, 10–5, 10–8
Input, *See* Options Restore
Insert mode, 10–18
Inspecting
code blocks, 10–3
macros, 10–4
variables and expressions, 10–36, 10–39
Installation
default directory structure, 8–6, 9–3, 9–13, 9–18, 12–1, 12–3, 13–1, 15–1, 15–2
Instance variables, 2–79
accessing, 2–80
assignable, 2–80
exported, 2–79
syntax for assigning, 2–80
syntax for retrieving, 2–80
Instances
creating, 2–78
definition, 2–78
ISCOLOR(), 13–1

J

JOIN
in a network environment, 4–8

K

KEYBOARD, 2–94, 2–96
controlling the, 2–94, 2–95
number of programmable keys available, 2–96
polling the, 2–95
programming, 2–96
Keyboard buffer
and function keys, 2–95
changing the size of, 2–94
clearing the, 2–94, 2–96
disabling the, 2–94
extracting a key from, 2–95
general discussion, 2–94
maximum number of characters in, 2–94
minimum size of, 2–94
reading pending key, 2–95
returning last key extracted, 2–95
stuffing the, 2–94
KEYBOARD command, 6–11
Keyboard functions
INKEY(), 2–94, 2–95
LASTKEY(), 2–95
NEXTKEY(), 2–95

L

LABEL FORM
creating, 14–1
in a network environment, 4–8
printing, 14–15
LAN, 4–1
definition of, 4–2
requirements for using CA-Clipper with, 4–2
workstation requirements, 4–2

LEN(), 2–30, 2–68
LIB directory, 3–3, 9–14, 9–18, 9–19
LIBRARY
    FREEFORMAT option, 9–4, 9–5
Library files, 9–8, 9–35
    default extension, 9–5
    embedded name in object file, 8–12, 9–27
    search paths, 9–18, 9–21
    searching for, 9–18, 9–21
Linker
    command line mode, 9–6, 9–9
    command line syntax, 9–4, 9–7, 9–37
    configuration, 9–11
    DOS return code, 9–13, 9–17, 9–21, 9–24
    error handling, 9–13
    example, 9–20, 9–24, 9–30, 9–33
    extended dictionary, 9–27
    getting help, 9–24, 9–42
    input files, 9–17
    input modes, 9–4, 9–6, 9–7, 9–9, 9–10
    interfaces, 9–4
    NIL keyword, 9–21, 9–30
    output files, 9–14
    path searching, 9–17
    prelink mode syntax, 9–37
    prompt mode, 9–6, 9–9
    resolving external references, 8–12, 9–27, 9–28, 9–36
    script file, 9–4, 9–7, 9–10, 9–12, 9–19, 9–20, 9–22
    script file mode, 9–7, 9–10
    specifying stack size, 9–27
    temporary file directory, 9–16
Linker options, *See also* FREEFORMAT options, POSITIONAL options
    /FREEFORMAT, 9–12, 9–22
    /POSITIONAL, 9–7, 9–11, 9–12, 9–22
    abbreviations, 9–20
    BATCH, 9–17, 9–21
    batch mode, 9–21
    BEGINAREA, 9–30
    case sensitivity, 9–22
    configuration, 9–12, 9–22
    DEBUG, 9–24, 9–40
    DEFAULTLIBRARYSEARCH, 9–27
    DYNAMIC, 9–16, 9–23, 9–32, 9–33
    dynamic overlaying, 9–23, 9–24, 9–32, 9–33
    ENDAREA, 9–30
    entering numbers, 9–21
    EXCLUDE, 9–28, 9–34, 9–37
    EXTDICTIONARY, 9–27
    FILE, 9–4, 9–5, 9–6
    HELP, 9–3, 9–24
    IGNORECASE, 9–20, 9–22
    INCREMENTAL, 9–11, 9–17, 9–20, 9–26, 9–38
    incremental linking, 9–26, 9–38
    LIBRARY, 9–4, 9–5, 9–7
    list of, 9–19
    listing, 9–24
    MAP, 9–7, 9–8, 9–16, 9–24
    miscellaneous, 9–27
    MODULE, 9–21, 9–30
    NOBATCH, 9–21
    NODEFAULTLIBRARYSEARCH, 9–27, 9–36
    NOEXTDICTIONARY, 9–27
    NOIGNORECASE, 9–20, 9–22
    NOINCREMENTAL, 9–20, 9–26
    OUTPUT, 9–4, 9–5, 9–7
    PLL, 9–5, 9–15, 9–19, 9–28, 9–34, 9–38
    precedence of, 9–20
    PRELINK, 3–3, 9–3, 9–5, 9–14, 9–15, 9–19, 9–28, 9–34, 9–37
    prelink, 9–28, 9–34
    PRELOAD, 9–30
    REFER, 9–5, 9–28, 9–34, 9–37
    RESIDENT, 9–23, 9–32, 9–33
    SECTION, 9–16, 9–30, 9–41
    SILENT, 9–7, 9–10, 9–24
    specifying FREEFORMAT, 9–5
    specifying POSITIONAL, 9–8
    STACK, 9–27
    static overlaying, 9–30, 9–39
    VERBOSE, 9–24, 9–36, 9–38
LIST
    in a network environment, 4–8, 4–11
List, 10–19, 10–39, 10–41, 10–42, 10–53, 10–67, 10–95, 10–96
Local Area Network, *See* LAN
Locate Case, 10–68, 10–69, 10–72, 10–73

Locate Find, 10–68, 10–69, 10–72, 10–73
Locate Goto, 10–71, 10–81, 10–85
Locate menu, 10–31
    Case sensitive option, 10–68, 10–69, 10–72, 10–73
    Find option, 10–69, 10–72, 10–73
    Goto Line option, 10–71
    Next option, 10–72
    Previous option, 10–73
Locate Next, 10–68, 10–72
Locate Previous, 10–68, 10–73
Locking
    commands that require, 4–11
    file locking, 4–1, 4–3, 4–4, 4–6, 4–10, 4–11
    programming strategies, 4–5, 4–10
    record locking, 4–1, 4–3, 4–4, 4–7, 4–10, 4–11
    releasing locks, 4–3, 4–11
Locks.prg, 4–12, 4–20
    FilLock(), 4–5
    NetUse(), 4–8
    RecLock(), 4–5
Logical
    comparison, 2–39, 2–45
    constants, 2–39
    delimiters, 2–39
    literals, 2–39
    operations, 2–39
    operators, 2–39, 2–45
    precedence of operators, 2–52, 2–54
    table of operators, 2–45
    truth tables, 2–45
Logical functions
    table, 2–39
LOOP, *See also* DO WHILE, FOR
Looping structures
    conditional, 2–18
    counter, 2–18, 2–66
Low-level file functions
    error table, 2–99
    FCLOSE(), 2–99
    FCREATE(), 2–97
    FERROR(), 2–99
    FOPEN(), 2–97
    FREAD(), 2–97
    FREADSTR(), 2–97

    FSEEK(), 2–98
    FWRITE(), 2–98
    general discussion, 2–97

M

Macro expressions
    and external references, 2–60
    compared to code blocks, 2–72
    creating, 2–50
    general discussion, 2–55
Macro operator
    and arrays, 2–58
    and code blocks, 2–59
    compared to extended expressions, 2–57
    general discussion, 2–55
    limitations, 2–56, 2–57, 2–58, 2–59
    nesting, 2–60
    used for compiling on the fly, 2–56
    used for text substitution, 2–55
Macro variable
    definition, 2–55
    terminating, 2–55
Macros, 10–4
Main procedure, *See* Startup procedure
Make, *See* RMAKE
Make file, 11–2, 11–5, 11–8
    comments in, 11–9
    default extension, 11–6
    defining environment variables in, 11–9
    DOS access from, 11–16
    example, 11–19
    line continuation in, 11–9
    searching for, 11–6
    using quotation marks in, 11–7, 11–8, 11–9, 11–13, 11–17
Makepath
    RMAKE macro, 11–7, 11–9, 11–13, 11–15
Manifest constants, 10–4
    defining, 2–6, 8–10
MAP
    FREEFORMAT option, 9–16
    POSITIONAL option, 9–8
Map files
    default extension, 9–16

generating, 9–8, 9–16, 9–24
Mathematical operators, 2–43, 2–44
Memo fields
    and character strings, 2–33
    comparison, 2–30, 2–33
    maximum length of, 2–33
    operations, 2–30, 2–33
    operators, 2–43
    relationship to memo file, 2–85
Memo files, *See also* .dbt files
    general information, 2–85
Memo functions
    table, 2–30, 2–33
MEMOEDIT()
    example, 12–4
MEMOREAD()
    in a network environment, 4–8
MEMOTRAN()
    in REPORT FORMs, 14–4
MEMOWRIT()
    in a network environment, 4–8
MEMVAR, 8–9
MEMVAR alias, 2–24
Menus
    accelerator keys, 10–29
    accessing, 10–28
    example, 2–5
    File menu, 10–31
    Help menu, 10–23, 10–31
    hot keys, 10–29
    Locate menu, 10–31
    menu commands, 10–29
    Monitor menu, 10–21, 10–31, 10–36
    Options menu, 10–25, 10–31
    Point menu, 10–19, 10–31
    Run menu, 10–31
    selecting an option, 10–28
    the Menu Bar, 10–28, 10–30, 10–86
    View menu, 10–24, 10–25, 10–31
    Window menu, 10–31
Messages
    arguments, 2–79
    sending, 2–78, 2–79
    syntax for sending, 2–79
    use of parentheses, 2–79
Methods

definition, 2–79
Modes, *See also* Execution modes
    Insert, 10–18
    Overwrite, 10–18
MODIFY STRUCTURE
    full-screen, 13–14
Monitor Local, 10–74, 10–75
Monitor menu, 10–31, 10–36
    Local option, 10–74, 10–75
    Private option, 10–76
    Public option, 10–77
    Sort option, 10–21, 10–78
    Static option, 10–79
Monitor Private, 10–76
Monitor Public, 10–77
Monitor Sort, 10–21, 10–78
Monitor Static, 10–79
Monitor Window, *See* Windows
Multistatement command lines, 10–3

N

Nested READs, 6–10
NETERR(), 4–3, 4–4, 4–7
NETNAME(), 4–4
NetUse(), 4–7, 4–8
Network commands, 4–3
Network functions, 4–4
Network programming, 4–5
Network software, 4–2
Networking, 4–1
    assigning rights, 3–12
    hardware requirements, 3–12
    runtime considerations, 3–12
NEXT, *See also* FOR
Next, *See* Locate Next
NG directory, 15–1, 15–2
NG.EXE, 15–1, *See also* Norton Guides
    command line syntax, 15–2
NG.INI, 15–14
NIL
    as a return value, 2–4, 2–5, 2–16
    as an initializer, 2–5, 2–40, 2–63
    assignment, 2–40, 2–46

assignment to omitted parameters, 2–11, 2–15
comparison, 2–40, 2–44, 2–67
functions, 2–40
operations, 2–40, 2–44
operators, 2–46
table of operations, 2–40
NIL keyword
    with linker, 9–21, 9–30
NOIDLE, 3–10
Norton Guides, *See also* Guide To CA-Clipper
    access window, 15–4, 15–5
    accessing, 15–3
    configuration, 15–12
    Expand menu, 15–4, 15–5
    getting help, 15–6
    hot key, 15–3, 15–13, 15–14
    Instant Access Engine, 15–1
    leaving, 15–14
    loading, 15–2
    memory used by, 15–1
    memory-resident mode, 15–1, 15–2
    menu bar, 15–4
    menu bar navigation keys, 15–5
    menu navigation, 15–5
    navigation, 15–11
    NG.INI, 15–14
    operation, 15–3
    Options menu, 15–4, 15–5
    pass through mode, 15–1, 15–2
    Search menu, 15–4, 15–7
    syntax, 15–2
    uninstalling, 15–15
Norton guides
    navigation, 15–7
Null
    date, 2–34
    string, 2–30
    terminator, 2–30, 2–33
Num, 10–81, *See also* Options Line
Numeric
    comparison, 2–37
    constants, 2–37
    limitations, 2–37
    literals, 2–37
    operations, 2–37
    operators, 2–34, 2–37, 2–43, 2–44, 2–46, 2–48
    overflow in REPORT FORMs, 14–5
    precedence of operators, 2–52, 2–53
    precision, 2–37
    range of values, 2–37
    storage format, 2–37
    table of operators, 2–44
Numeric functions
    table, 2–37

O

OBJ directory, 9–18
Object files, 9–8, 9–35
    default extension, 9–5
    default name, 8–2, 8–3, 8–7, 8–12
    embedding library name in, 8–12, 9–27
    generating, 8–1, 8–2, 8–5
    linking, 9–4, 9–7
    reducing size of, 8–11
    relationship to .PLT files, 9–15
    search paths, 9–18, 9–21
    searching for, 9–18, 9–21
    specifying a directory for, 8–12
    specifying a name for, 8–7, 8–12
Objects
    as return values, 2–16, 2–78
    creating new objects, 2–78
    general discussion, 2–78
    passing as parameters, 2–12, 2–13, 2–78
    sending messages, 2–78, 2–79
Opening files
    database, 2–82
    in a network environment, 4–8, 4–10
    in DBU, 13–12
    in shared mode, 4–7
    index, 2–88
Operations
    array, 2–61
    character, 2–30
    code block, 2–72, 2–73
    database file, 2–85
    date, 2–34
    index, 2–87

logical, 2–39
memo, 2–30, 2–33
numeric, 2–37
Operators
    addition, 2–34, 2–37, 2–43, 2–44
    alias, 2–24, 2–50
    array element, 2–50
    array equivalence, 2–68
    assign, 2–46, 2–47, 2–63
    binary, 2–42
    character, 2–30, 2–43, 2–46, 2–48
    code block, 2–50
    code block arguments, 2–72
    compile and run, 2–50, 2–55, 2–56, 2–65, 2–77
    compound assign, 2–48
    concatenation, 2–30, 2–43
    constant array, 2–50, 2–65
    date, 2–34, 2–43, 2–46, 2–48, 2–49
    decrement, 2–34, 2–37, 2–49
    division, 2–37, 2–44
    equal, 2–30, 2–34, 2–37, 2–39, 2–40, 2–44
    exactly equal, 2–30, 2–44
    exponentiation, 2–37, 2–44
    general discussion, 2–30, 2–42
    greater than, 2–30, 2–34, 2–37, 2–39, 2–40, 2–44
    greater than or equal, 2–30, 2–34, 2–37, 2–39, 2–40, 2–44
    grouping, 2–50
    increment, 2–34, 2–37, 2–49
    infix, 2–42
    inline addition and assign, 2–46, 2–48
    inline assign, 2–46, 2–47, 2–48, 2–63, 2–80
    inline concatenation and assign, 2–46, 2–48
    inline division and assign, 2–46, 2–48
    inline exponentiation and assign, 2–46, 2–48
    inline modulus and assign, 2–46, 2–48
    inline multiplication and assign, 2–46, 2–48
    inline subtraction and assign, 2–46, 2–48
    less than, 2–30, 2–34, 2–37, 2–39, 2–40, 2–44
    less than or equal, 2–30, 2–34, 2–37, 2–39, 2–40, 2–44
    logical, 2–39, 2–45
    macro, 2–55
    mathematical, 2–49
    memo, 2–30, 2–43
    modulus, 2–37, 2–44
    multiple inline assign, 2–47
    multiplication, 2–37, 2–44
    not equal, 2–30, 2–34, 2–37, 2–39, 2–40, 2–44
    numeric, 2–34, 2–37, 2–43, 2–44, 2–46, 2–48
    overloading, 2–42, 2–47
    pass-by-reference, 2–13, 2–50
    postdecrement, 2–34, 2–49
    postfix, 2–34, 2–37, 2–42, 2–49
    postincrement, 2–34, 2–37, 2–49
    precedence of, 2–42, 2–43, 2–51, 2–52
    predecrement, 2–34, 2–37, 2–49
    prefix, 2–34, 2–37, 2–42, 2–49
    preincrement, 2–34, 2–37, 2–49
    relational, 2–30, 2–34, 2–37, 2–39, 2–40, 2–44, 2–68
    send, 2–78, 2–79, 2–80
    special, 2–50
    substring, 2–30, 2–44
    subtraction, 2–34, 2–37, 2–43, 2–44
    table of decrement, 2–49
    table of increment, 2–49
    table of special, 2–50
    unary, 2–42
Options Codeblock, 10–82
Options Color, 10–25, 10–83
Options Exchange, 10–33, 10–37, 10–82, 10–84, 10–92
Options Line, 10–85, *See also* Num
Options menu, 10–31, 10–86
    Codeblock Trace option, 10–82
    Color option, 10–25, 10–83
    Exchange Screens option, 10–84, 10–103
    Line Numbers option, 10–85
    Menu Bar option, 10–86
    Mono Display option, 10–87
    Path for Files option, 10–88
    Preprocessed Code option, 10–89

Restore Settings option, 10–90
Save Settings option, 10–91
Swap on Input option, 10–92
Tab Width option, 10–93
Options Mono, 10–87
Options Path, 10–88
Options Preprocessed, 10–4, 10–89
Options Restore, 10–8, 10–90
Options Save, 10–8, 10–91
Options Swap, 10–92
Options Tab, 10–93
Organization
    manual, 1–1
OTHERWISE, See also DO CASE
OUTPUT
    FREEFORMAT option, 9–4, 9–5
Output, See View App
Output to text file
    @...SAY, 2–93
    console commands, 2–93
Overlay files
    creating, 9–23, 9–30
    default extension, 9–16
    dynamic, 9–23, 9–33
    generating, 9–16
    searching for, 9–16
    static, 9–30, 9–41
Overlays
    dynamic, 9–23, 9–32, 9–33
    dynamic external, 9–23
    dynamic internal, 9–23
    in a network environment, 4–13
    static, 9–30, 9–39
    static external, 9–30, 9–41
    static internal, 9–30, 9–41
Overwrite mode, 10–18

P

PACK
    in a network environment, 4–13
Parameter
    declared, 2–11
    definition of actual, 2–11
    definition of formal, 2–11

methods of specifying, 2–11
number passed, 2–11, 2–15
omitting, 2–11, 2–15
optional, 2–11, 2–15
passing, 2–11, 2–15
passing arrays as, 2–12, 2–13, 2–65
passing by reference, 2–13
passing by value, 2–12, 2–13
passing code blocks as, 2–12
passing fields as, 2–12
passing from DOS command line, 2–15
passing objects as, 2–12, 2–13, 2–78
passing with DO...WITH, 2–13
PARAMETERS, 8–9
Parentheses
    and precedence, 2–42, 2–54
Pass-by-reference, 2–13, 2–50
PCOL(), 2–92
PCOUNT()
    effect of omitting parameters on, 2–11, 2–15
PE.EXE, 12–1, See also Program editor
PLL directory, 3–3, 9–14, 9–19
Point Breakpoint, 10–36, 10–41, 10–95, See also BP
Point Delete, 10–19, 10–38, 10–96, See also Delete
Point menu, 10–31
    Breakpoint option, 10–95
    Delete option, 10–38, 10–96
    Tracepoint option, 10–38, 10–97
    Watchpoint option, 10–38, 10–98
Point Tracepoint, 10–8, 10–19, 10–38, 10–97
Point Watchpoint, 10–8, 10–19, 10–38, 10–98
POSITIONAL
    command line mode, 9–9
    prompt mode, 9–9
    script file mode, 9–10
    specifying executable file, 9–8
    specifying library files, 9–8
    specifying map file, 9–8
    specifying object files, 9–8
    syntax, 9–7, 9–37
POSITIONAL options, See also Linker options
    /BATCH, 9–21
    /DEBUG, 9–24

Programming and Utilities Guide   Index–17

/DEFAULTLIBRARYSEARCH, 9–27
/DYNAMIC, 9–23
/EXCLUDE, 9–28
/EXTDICTIONARY, 9–27
/HELP, 9–24
/IGNORECASE, 9–22
/INCREMENTAL, 9–26
/MAP, 9–8, 9–24
/NOBATCH, 9–21
/NODEFAULTLIBRARYSEARCH, 9–27
/NOEXTDICTIONARY, 9–27
/NOIGNORECASE, 9–22
/NOINCREMENTAL, 9–26
/PLL, 9–28
/PRELINK, 9–28
/REFER, 9–28
/RESIDENT, 9–23
/SILENT, 9–24
/STACK, 9–27
/VERBOSE, 9–24
Postdecrement
    operators, 2–49
    precedence of operators, 2–52, 2–54
Postincrement
    operators, 2–49
    precedence of operators, 2–52, 2–54
Precedence, 2–42, 2–51
    and algebraic expression evaluation, 2–52
Predecrement
    operators, 2–49
    precedence of operators, 2–52, 2–53
Preincrement
    operators, 2–49
    precedence of operators, 2–52, 2–53
Prelinked libraries, *See also* Prelinked library files
    creating, 9–5
    files necessary for generating, 9–3
    nesting, 9–28
    steps to build, 9–35
    to save disk space, 9–34, 9–36
    to shorten link time, 9–34, 9–36
Prelinked library files, 9–28, 9–34
    default extension, 3–3, 9–5, 9–14
    default name, 3–3, 9–5, 9–14
    generating, 9–37

relationship to .EXE files, 9–14, 9–15, 9–38
relationship to executable files, 3–3
searching for, 3–3, 9–14
using, 9–38
Prelinked transfer files, *See also* Prelinked library files, 9–28, 9–34
    default extension, 9–15
    default name, 9–15
    relationship to .EXE files, 9–38
    relationship to object files, 9–15
    search paths, 9–19, 9–21
    searching for, 9–19, 9–21
    using, 9–38
Prelinking, 9–28, 9–34
Preprocessed output
    viewing, 10–4, 10–89
Preprocessor, 8–4
    defining an identifier, 8–10
    source listing file, 8–8, 8–12
    switches, 8–9
Preprocessor directives, 8–13
    #define, 2–6
    #include, 2–6, 8–6
    summary of, 8–4
Prev, *See* Locate Previous
Printing
    @...SAY, 2–92
    general discussion, 2–92
    in a network environment, 4–3
    redirecting output to a file, 2–93
    with console commands, 2–92
PRIVATE, 8–9
PROCEDURE
    compiler implications, 8–13
Procedure files
    compiling, 8–5
Procedures
    calling conventions, 2–4, 2–13
    declaring, 2–5
    defining, 2–5
    executing, 2–4, 10–37, 10–47
    passing parameters to, 2–11
    startup, 2–3
PROCLINE(), 2–96
PROCNAME(), 2–96
Program

comments, 2–6
multistatement line, 2–7
termination, 2–95
Program editor, 12–1
architecture, 12–3
make file, 12–3
navigation, 12–2
operation, 12–2
source code, 12–3
syntax, 12–1
Program files, *See also* Source files
compiling, 8–1, 8–2, 8–5
Programming
general discussion, 2–2
PROW(), 2–92
Pseudofunctions, 10–4
PUBLIC, 8–9

Q

Quit, *See* File Exit

R

READ
exiting from a, 2–95
Read Layer, 6–15, 6–23, 6–24, 6–25
@...GET, 6–8
creating, 6–17, 6–24
implementation rules, 6–16, 6–25
implementation steps, 6–17
READMODAL, 6–8
SET MESSAGE TO, 6–20
STD.H, 6–8
READEXIT(), 2–95
READVAR(), 2–96
RECALL
in a network environment, 4–11
RecLock(), 4–7, 4–10
RECNO(), 13–29
Record
processing with DBEVAL(), 2–87
scoping, 2–86, 2–87
Record pointer

and index updating, 2–89
Recursion, 2–25
REINDEX, 2–89
in a network environment, 4–13
Relational
operators, 2–30, 2–34, 2–37, 2–39, 2–40, 2–44, 2–68
precedence of operators, 2–52, 2–53
table of operators, 2–44
REPLACE, 2–47
and inline assignment, 2–48
in a network environment, 4–11
Report and label utility, *See* RL
REPORT FORM
creating, 14–1
for related files, 14–10
in a network environment, 4–8, 4–11
printing, 14–9
Reserved words, 2–8, 2–40
Restart, *See* Run Restart
RESTORE
in a network environment, 4–8
Resume, *See* File Resume
RETURN
from a user-defined function, 2–16
Return values
from a user-defined function, 2–16
RL
abandoning changes, 14–9, 14–15
creating a new label, 14–11
creating a new report, 14–2
inserting line breaks in reports, 14–4
Label Exit menu, 14–15
Label Formats menu, 14–13
Label Toggle menu, 14–11, 14–13
leaving, 14–15
make file, 11–19, 14–16
maximum number of report columns, 14–3
modifying an existing label, 14–11
modifying an existing report, 14–2
Report Columns menu, 14–3
Report Delete menu, 14–5
Report Exit menu, 14–9
Report Goto menu, 14–6
Report Groups menu, 14–7

Report Insert menu, 14–5
Report menu, 14–6
saving a label definition, 14–15
saving a report definition, 14–9
specifying a standard label format, 14–13
specifying blank label lines, 14–13
specifying fields from multiple files, 14–10
specifying label Contents, 14–13
specifying label Dimensions and Formatting, 14–12
specifying report level options, 14–6
RL.EXE, 14–1, See also RL, 14–16
   command line syntax, 14–2
   source code, 14–16
RLOCK(), 4–3, 4–4, 4–7, 4–11
RMAKE, See also Make file, 11–1
   commands, 11–8
   comments in, 11–3
   conditional make, 11–13
   configuration, 11–4
   debugging, 11–2
   dependency rule, 11–5, 11–8, 11–9
   dependency statement, 11–5
   directives, 11–13
   display warning messages, 11–3
   DOS access from, 11–16
   DOS return code, 11–3, 11–5
   DOS SETs, 11–12
   error handling, 11–3, 11–5
   examples, 11–9, 11–11, 11–12, 11–17, 11–18, 11–19
   header files, 11–15
   inference rule, 11–8, 11–11
   input files, 11–6
   internal workspace, 11–4
   macro definition, 11–17
   macro expansion, 11–17
   macros, 11–2, 11–3, 11–7, 11–8, 11–15, 11–16
   make phase, 11–5, 11–7, 11–11
   parsing phase, 11–5, 11–7, 11–13, 11–16
   PE make file, 12–3
   precedence levels, 11–17
   predefined macros, 11–11, 11–18
   RL make file, 11–19, 14–16
   searching subdirectories, 11–3, 11–6
   special macros, 11–7, 11–9, 11–13, 11–15, 11–18
   switches, 11–2
   symbol table, 11–4
   user-defined macros, 11–17
RMAKE directives
   #!, 11–16
   #else, 11–13
   #endif, 11–13
   #ifdef, 11–13
   #ifeq, 11–13
   #iffile, 11–13
   #ifndef, 11–13
   #include, 11–15
   #stderr, 11–16
   #stdout, 11–16
   #undef, 11–15
RMAKE options, See RMAKE switches
RMAKE switches
   /B, 11–2
   /D, 11–3, 11–4, 11–7, 11–13, 11–17
   /I, 11–3
   /N, 11–3
   /S, 11–3, 11–6
   /U, 11–3, 11–9
   /W, 11–3
   /XS, 11–4
   /XW, 11–4
RMAKE.EXE, See also RMAKE
   command line syntax, 11–2
Root section
   CA-Clipper modules, 9–23, 9–30, 9–33
RTLINK.CFG, 9–12
   searching for, 9–12
RTLINK.DAT, 9–3
RTLINK.EXE, See also Linker, 9–1, 9–3
   command line syntax, 9–4, 9–7, 9–37
   FREEFORMAT syntax, 9–4, 9–37
   POSITIONAL syntax, 9–7, 9–37
RTLINK.HLP, 9–3
RTLINKCMD, 9–7, 9–10, 9–11, 9–20
RTLINKST.COM, 9–3
RTLUTILS.LIB, 9–3, 9–40
   search paths, 9–18
RUN

runtime implications, 3–5
Run Animate, 10–33, 10–36, 10–37, 10–103
Run Go, 10–33, 10–35, 10–36, 10–104
Run menu, 10–31
    Animate option, 10–103
    Go option, 10–104
    Next option, 10–105
    Restart option, 10–106
    Speed option, 10–107
    Step option, 10–108
    To Cursor option, 10–109
    Trace option, 10–110
Run Next, 10–33, 10–105
Run Restart, 10–106
Run Speed, 10–33, 10–36, 10–103, 10–107
Run Step, 10–108, 10–110
Run To, 10–33, 10–109
Run Trace, 10–110
Runtime
    configuration, 3–6
    configuring expanded memory at, 3–9
    configuring number of files at, 3–9
    controlling extended cursor, 3–8
    displaying memory configuration at, 3–10
    environment settings, 3–6
    excluding available memory at, 3–11
    general discussion of environment, 3–1
    preventing idle time detection at, 3–10
    restoring EMM page frame, 3–8
    specifying dynamic overlay file handles, 3–8
    specifying swap file location at, 3–11
    specifying swap file size at, 3–11
    specifying temporary file location at, 3–11

S

SAVE
    in a network environment, 4–8
Scope
    syntax for specifying, 2–86
Script files, 10–5, 10–8, 10–25, 10–29, 10–90, 10–91
    compiling with, 8–2, 8–3, 8–5
    configuring .RTLink syntax for, 9–12, 9–22
    default extension, 9–7, 9–10
    linking with, 9–7, 9–10
    nesting, 9–7
    searching for, 9–19
Search path, 10–8
SEEK, 2–89
SELECT, 2–82
Selector
    definition, 2–79
    instance variable, 2–80
    message, 2–79
SET ALTERNATE, 2–90, 2–93
    in a network environment, 4–8
SET CENTURY, 2–34
Set Colors Window, *See* Windows
SET CONSOLE, 2–90, 2–92
SET DATE, 2–34
SET DEVICE, 2–91, 2–92
SET ESCAPE, 2–95
SET EXACT, 2–30
SET EXCLUSIVE, 4–1, 4–3, 4–5, 4–7, 4–13
SET FILTER, 13–31
SET FORMAT
    compiler implications, 8–5
SET FUNCTION, 2–96
SET INDEX, 2–88
    in a network environment, 4–7
SET KEY, 2–95, 2–96, 6–11
SET MARGIN, 2–92
SET ORDER, 2–88
SET PRINTER, 2–90, 2–92, 2–93, 4–3
    in a network environment, 4–8
SET RELATION, 13–29
    and REPORT FORMs, 14–10
SET SOFTSEEK, 2–89
SET TYPEAHEAD, 2–94
SETCANCEL(), 2–95
SETCOLOR(), 2–91
SETPRC(), 2–92
Shared mode, 4–1, 4–2, 4–3, 4–5, 4–7, 4–8, 4–11
SORT
    in a network environment, 4–8
Sorting
    arrays, 2–70
SOURCE directory, 12–3
Source files, *See also* Program files

compiling, 8–5
editing, 12–1
line numbers, 8–11, 8–12
Speed, *See* Run Speed
Standard header file
changing the, 8–13
Startup procedure, 8–11
Static overlays, 9–39
creating, 9–30
designing, 9–40
using, 9–40
STD.CH, 8–13
Step, *See* Run Step
Storage classes
defined, 2–20
STORE, 2–47
Substring
operator, 2–30, 2–44
SUM
in a network environment, 4–11
SWAPK, 3–11
SWAPPATH, 3–11

T

Target file, 11–5, 11–9
creating, 11–7
searching for, 11–7
TBColumn class
block, 5–11, 5–16
cargo, 5–10
colorBlock, 5–16, 5–19
defColor, 5–15, 5–16, 5–19
TBrowse
adding Color, 5–14
browsing search results, 5–25
browsing with Get, 5–11
Controlling highlight, 5–18
controlling scope, 5–20
Creating TBrowse Objects, 5–2
DBEVAL(), 5–25
multi-user applications, 5–7
obtaining quicker response time, 5–6
Optimization, 5–6
repositioning record pointer, 5–8

SET COLOR, 5–14
SET FILTER, 5–25
using TBColumn
cargo, 5–10
using TBrowse
cargo, 5–10
using TRANSFORM(), 5–6
viewing a key value, 5–22
TBrowse class, 5–2
autoLite, 5–18
cargo, 5–10
colorRect, 5–19
colorSpec, 5–14
deHilite, 5–18
goBottomBlock, 5–21, 5–22, 5–26
goTopBlock, 5–21, 5–22, 5–26
hilite, 5–18
refreshAll method, 5–8
refreshCurrent method, 5–7
skipBlock, 5–21, 5–22, 5–23, 5–26
stabilization, 5–4
stabilize method, 5–4, 5–6
up method, 5–8
TBrowseDB(), 5–2
TBrowseNEW(), 5–2
Temporary files
specifying a directory for, 3–4, 8–7, 8–12, 9–16
TEMPPATH, 3–11
TEXT
in a network environment, 4–8
Text files
editing, 12–1
The CA-Clipper Debugger
definition, 10–1
invoking, 10–5
quitting, 10–10, 10–57
The Callstack, 10–22, 10–42, 10–52, 10–116
TMP directory, 3–4, 8–7, 8–12, 9–16
TO FILE
in a network environment, 4–8
TO FILE clause, 2–90, 2–93
TO PRINTER clause, 2–90, 2–92
TOTAL
in a network environment, 4–8
TP, *See* Point Tracepoint

Tracepoints, 10–19, 10–33, 10–36, 10–37, 10–104
    and monitored variables, 10–74, 10–75, 10–76, 10–77, 10–79
    definition, 10–37, 10–97
    deleting, 10–38, 10–53, 10–96
    listing, 10–39, 10–67
    setting, 10–8, 10–38, 10–97
TRANSFORM(), 5–6
    in REPORT FORMs, 14–4
TSR programs, 3–8
TYPE
    in a network environment, 4–8
TYPE()
    and arrays, 2–61

U

UNLOCK, 4–3, 4–11
UPDATE
    and inline assignment, 2–48
    in a network environment, 4–8, 4–11
USE, 2–82, 2–88
    in a network environment, 4–3, 4–7, 4–13
User-defined commands, 2–72, 2–87
User-defined functions
    calling conventions, 2–4
    declaring, 2–5
    executing, 2–4
    passing arguments to, 2–11, 2–65

V

VALTYPE()
    and arrays, 2–61
    used to test for omitted parameters, 2–15
Variable names
    resolving references, 2–23
Variables
    ambiguous references to, 2–23
    changing values of, 10–19, 10–21, 10–39
    creating new, 10–40, 10–47
    creation, 2–22
    declaration, 2–5, 2–21, 2–25, 2–40
    definition of, 2–20
    displaying, 10–47
    dynamically scoped, 2–20, 2–21
    field, 2–29
    initialization, 2–25, 2–40, 2–46
    inspecting, 10–36, 10–39
    lexically scoped, 2–5, 2–20, 2–21
    lifetime of, 2–22
    local, 2–25
    monitored, 10–21, 10–36, 10–74, 10–75, 10–76, 10–77, 10–78, 10–79
    naming, 2–23
    passing as parameters, 2–12
    private, 2–27
    public, 2–28
    scoping in code blocks, 2–22, 2–73
    scoping of, 2–5, 2–21, 2–22
    static, 2–26
    undefined, 2–40
    visibility of, 2–22
View, 10–114, *See also* File Open, 13–10, 13–14, 13–19, 13–29
    definition, 13–2, 13–7
View App, 10–37, 10–115
View Callstack, 10–22, 10–42, 10–116, *See also* Callstack
View menu, 10–31
    App Screen option, 10–115
    Callstack option, 10–116
    Sets option, 10–24, 10–117
    Workareas option, 10–25, 10–118
View Sets, 10–24, 10–117
View Sets Window, *See* Windows
View Workareas, 10–25, 10–118
View Workareas Window, *See* Windows
Viewing
    header files, 10–3, 10–43, 10–58, 10–114
    preprocessed output, 10–4, 10–89
    program code, 10–43, 10–58, 10–60, 10–103, 10–114
    program output, 10–37, 10–84, 10–115
Virtual Memory Manager, 3–8, 3–10, 3–11
VMM, 3–8, 3–10, 3–11

W

Wait states
    and the keyboard buffer, 2–94
Watch Window, *See* Windows
Watchpoints, 10–19, 10–36, 10–37
    and monitored variables, 10–74, 10–75, 10–76, 10–77, 10–79
    definition, 10–37, 10–98
    deleting, 10–38, 10–53, 10–96
    listing, 10–39, 10–67
    setting, 10–8, 10–38, 10–98
WHILE condition, 2–86
Window Iconize, 10–119
Window menu, 10–31
    Iconize option, 10–119
    Move option, 10–120
    Next option, 10–121
    Prev option, 10–122
    Size option, 10–123
    Tile option, 10–124
    Zoom option, 10–125
Window Move, 10–120
Window Next, 10–121
Window Prev, 10–122
Window Size, 10–123
Window Tile, 10–124
Window Zoom, 10–125
Windows
    activating, 10–12
    closing, 10–12
    iconizing, 10–13, 10–119
    moving, 10–14, 10–120
    navigating between, 10–12, 10–121, 10–122
    opening, 10–12
    sizing, 10–13, 10–123
    the Callstack Window, 10–12, 10–21, 10–22, 10–42, 10–52, 10–116
    the Code Window, 10–4, 10–12, 10–15, 10–21, 10–43, 10–52, 10–74, 10–75, 10–76, 10–77, 10–79, 10–81, 10–84, 10–85, 10–87, 10–89, 10–93
    the Command Window, 10–12, 10–16, 10–37, 10–46
    the Help Window, 10–10, 10–23, 10–64
    the Monitor Window, 10–12, 10–21, 10–26, 10–74, 10–75, 10–76, 10–77, 10–78, 10–79
    the Set Colors Window, 10–25, 10–83
    the View Sets Window, 10–24, 10–117
    the View Workareas Window, 10–25, 10–118
    the Watch Window, 10–12, 10–19, 10–26, 10–36, 10–38, 10–39, 10–53, 10–67, 10–97, 10–98
    tiling, 10–14, 10–124
    zooming, 10–13, 10–125
Work area
    attributes in DBU, 13–7
    attributes of, 2–84
    changing, 2–82
    default aliases, 2–82
    in DBU, 13–7
    next available, 2–82
    number available, 2–82
    number available in DBU, 13–7
    table of attributes, 2–84
Workstation
    name of current, 4–4
    requirements, 4–2
WP, *See* Point Watchpoint

X

X, 3–11

Z

ZAP
    in a network environment, 4–13